Sustainable Wastewater Treatment and Pollution Control

Guest Editors

Yung-Tse Hung
Yen-Pei Fu

Basel • Beijing • Wuhan • Barcelona • Belgrade • Novi Sad • Cluj • Manchester

Guest Editors

Yung-Tse Hung
Department of Civil and
Environmental Engineering
Cleveland State University
Cleveland, OH
USA

Yen-Pei Fu
Department of Materials
Science and Engineering
National Dong Hwa
University
Shoufeng
Taiwan

Editorial Office
MDPI AG
Grosspeteranlage 5
4052 Basel, Switzerland

This is a reprint of the Special Issue, published open access by the journal *Water* (ISSN 2073-4441), freely accessible at: https://www.mdpi.com/journal/water/special_issues/wastewater_treatment.

For citation purposes, cite each article independently as indicated on the article page online and as indicated below:

Lastname, A.A.; Lastname, B.B. Article Title. *Journal Name* **Year**, *Volume Number*, Page Range.

ISBN 978-3-7258-3427-3 (Hbk)
ISBN 978-3-7258-3428-0 (PDF)
https://doi.org/10.3390/books978-3-7258-3428-0

Contents

About the Editors

Yung-Tse Hung

Prof. Dr. Yung-Tse Hung, Ph.D., P.E., DEE, Fellow-ASCE, served as a Professor of Civil Engineering at Cleveland State University from 1981 to 2024. He earned his B.Sc. and M.Sc. in Civil Engineering from Cheng Kung University, Taiwan, and his Ph.D. from the University of Texas at Austin. Prof. Hung has taught at 16 universities across 8 countries and started the public health engineering program at the University of Canterbury, New Zealand, in 1972. He has served on the faculties of numerous universities globally, including in New Zealand, the USA, Hong Kong, the UAE, Singapore, Australia, Russia, and Kyrgyzstan. Prof. Hung's research focuses on biological wastewater treatment, industrial water pollution control, and municipal wastewater treatment. He has published approximately 40 books, 242 book chapters, 198 refereed publications, and 352 other scholarly works, totaling around 811 publications and presentations. He is a Fellow of ASCE, a Diplomate of AAEE, a Fellow of the Ohio Academy of Science, a Member of AEESP, and a Life Member of WEF. He serves as Editor-in-Chief for several international journals and books and is a registered professional engineer in Ohio and North Dakota.

Yen-Pei Fu

Prof. Dr. Yen-Pei Fu is professor at the Department of Materials Science and Engineering at National Dong Hwa University, Taiwan. He holds a Ph.D. from National Tsing Hua University and specializes in areas such as solid oxide fuel cells (SOFCs), photocatalysts, and supercapacitors. His research contributions include advancements in materials science, with a focus on energy and environmental applications. Prof. Fu has numerous publications in high-impact journals. His work covers topics such as the synthesis of advanced materials, photocatalytic degradation, and the development of energy-efficient technologies. He is also actively involved in mentoring students and advancing interdisciplinary research.

Preface

Wastewater treatment has changed over the last thirty years, transforming from designing treatment technologies for suitable discharge into nature water bodies, using techniques such as conventional activated sludge and trickling filters, for solving various human health issues such as recycling wastewater, providing solutions to poor waste treatment, and preventative measures for pollution. This reprint, including 13 papers, examines challenges and innovations in wastewater treatment, environmental sustainability, and water contamination management. One paper on septic tank wastewater storage highlights *E. coli* dynamics, while another explores fecal coliform concentrations in effluent-dominated streams, underscoring public health concerns. Source-separation sanitation concepts showcase their potential for ecological and economic benefits. Studies emphasize microbial community responses to intermittent aeration, revealing shifts tied to tetracycline presence. Addressing CO_2 emissions, two papers quantify impacts from rainwater and reclaimed water use, contributing to carbon-conscious urban planning. Flocculation processes are optimized for pollutant removal, offering cost-effective advancements. Similarly, unconventional technologies in wastewater plants successfully target pharmaceuticals such as lidocaine. Innovative solutions span biosorbents such as fungal biomass for dye removal, TiO_2-modified filters enhancing treatment reactors, and high-rate anaerobic processes addressing agro-food industrial wastewater. The synthesis underscores a pressing need for advanced, sustainable, and tailored wastewater strategies amid rising contamination. Challenges such as economic viability and system efficiency persist but are gradually addressed through interdisciplinary approaches. While offering unique insights, each paper focuses on this Special Issue's ultimate goal: improving water quality and minimizing environmental impact.

Yung-Tse Hung and Yen-Pei Fu
Guest Editors

Water **2013**, *5*, 1141-1151; doi:10.3390/w5031141

OPEN ACCESS

water

ISSN 2073-4441
www.mdpi.com/journal/water

Article

Preliminary Study on the Effect of Wastewater Storage in Septic Tank on *E. coli* Concentration in Summer

Dominique Appling [1], Mussie Y. Habteselassie [1,*], David Radcliffe [2] and James K. Bradshaw [2]

[1] Department of Crop and Soil Sciences, University of Georgia Griffin Campus, 1109 Experiment Street, Griffin, GA 30223, USA; E-Mail: appling_dominique@yahoo.com
[2] Department of Crop and Soil Sciences, University of Georgia, 3111 Miller Plant Sciences Building, Athens, GA 30602, USA; E-Mails: dradclif@uga.edu (D.R.); jbradsha@uga.edu (J.B.)

* Author to whom correspondence should be addressed; E-Mail: mussieh@uga.edu; Tel.: +1-770-229-3336; Fax: +1-770-228-7271.

Received: 10 May 2013; in revised form: 5 July 2013 / Accepted: 19 July 2013 / Published: 26 July 2013

Abstract: On-site wastewater treatment systems (OWTS) work by first storing the wastewater in a septic tank before releasing it to soils for treatment that is generally effective and sustainable. However, it is not clear how the abundance of *E. coli* changes during its passage through the tank. In this study, which was conducted under the UGA young Scholar Program in summer of 2010, we examined the change in wastewater quality parameters during the passage of the wastewater through the tank and after its release into soil. We collected wastewater samples at the inlet and outlet of an experimental septic tank in addition to obtaining water samples from lysimeters below trenches where the drainpipes were buried. We report that *E. coli* concentration was higher by 100-fold in the septic tank effluent than influent wastewater samples, indicating the growth of *E. coli* inside the tank under typical Georgian summer weather. This is contrary to the assumption that *E. coli* cells do not grow outside their host and suggests that the microbial load of the wastewater is potentially enhanced during its storage in the tank. Electrical conductivity, pH and nitrogen were similar between the influent and effluent wastewater samples. *E. coli* and total coliform concentrations were mainly below detection in lysimeter samples, indicating the effectiveness of the soil in treating the wastewater.

Keywords: on-site wastewater treatment systems; *E. coli*; total coliform; growth; septic tank; piedmont

---

1. Introduction

Nationally, more than a quarter of US households employ on-site wastewater systems (OWTS) to treat and dispose wastewater [1]. In Georgia, the percent use of OWTS is higher than the national average at about 37%. These systems, also known as septic systems, are commonly designed to accumulate the waste in a two-chamber tank where solids settle while the wastewater flows to a distribution box that is connected to one or more perforated drainpipes that distribute wastewater to the soil. The drainpipes are commonly installed in trenches and surrounded by a supporting material such as gravel or polystyrene to prevent clogging. Wastewater treatment occurs in the soil via biological (predation, die-off), chemical (adsorption) and physical (filtration) mechanisms before it reaches the surrounding ground or surface waters [1,2]. OWTS must, therefore, be installed in suitable soils that can accomplish the treatment processes properly [3–5].

In general, OWTS are an effective and sustainable way of treating wastewater. OWTS can negatively impact the microbial quality of surrounding water bodies. This is mainly true if OWTS are failing, which could happen due to installation of OWTS in unsuitable soils, age of the system, excessive use of water, or poor maintenance [6,7]. Properly functioning OWTS can also contaminate surrounding water bodies at times of extreme weather [8,9]. This happens due to excessive moisture in soils that decreases the depth of the unsaturated layer where the wastewater is treated before it reaches the ground water below. In the presence of excessive soil moisture, the downward movement of the wastewater is facilitated without allowing enough time for it to interact with the soil environment for treating the contaminants.

Contaminants of concern commonly associated with OWTS are microbial pathogens and nutrients (mainly nitrogen), which are the leading causes of water quality impairments in US streams and rivers [10]. There are technologies that can be retrofitted into existing OWTS to reduce the amount of contaminants in the wastewater effluent. These technologies are commonly called advanced treatment units and work by mainly manipulating the oxygen and carbon content of the wastewater [11–13]. In pre-anoxic units, for example, the wastewater is made to pass through an aerobic unit that is retrofitted between the septic tank and the drainfield to nitrify the ammonium into nitrate. The nitrified wastewater is then recycled back to the anoxic septic tank where it is denitrified in the presence of a carbon source. In another variation, the septic system is fitted with an aerobic unit before the septic tank to facilitate the processes of nitrification and denitrification to sequentially remove nitrogen. The limited field studies carried out so far to evaluate the effectiveness of these units mainly focused on nitrogen [14,15]. The impact of these units on the microbial load of the wastewater is largely unknown. The technologies are not yet popular throughout the United States as they are expensive [16].

Previous studies that looked at the impact of OWTS on water quality had mainly focused on how well wastewater is treated in soils before it joins water bodies directly by installing monitoring wells around these systems, e.g., [9,17] or indirectly by comparing water quality in areas with varying densities of OWTS, e.g., [18]. While these approaches are sound, they do not give us any information on the kind of microbial transformation the wastewater undergoes when stored in the septic tank. This is important because it might affect the microbial contaminant load of the wastewater when it leaves the tank. In this study, which was conducted under the UGA Young Scholar Program in June of 2010, we examined the change in the quality of the wastewater before it gets to and after it leaves the septic tank but before it is released to soil. Water samples were also collected from suction lysimeters installed below the trenches of the OWTS in a typical Georgian soil (red clay soil) into which the wastewater was released for treatment. We were particularly interested in investigating whether *E. coli* cells were capable of multiplying in the septic tank.

2. Materials and Methods

2.1. Study Site and On-site Wastewater Treatment System (OWTS)

The wastewater and water samples in this study were collected from an experimental on-site wastewater treatment system (OWTS) that was installed at the Westbrook Farm of the University of Georgia, Griffin Campus in 2008 [19]. Briefly, the system consisted of an above-ground dosing tank (4170 L capacity) where residential strength wastewater obtained from Cabin Creek Wastewater Treatment Plant in Griffin, GA was stored before it was dosed to a 3875 L capacity septic tank. The retention time for the wastewater in the septic tank was 6 days. The wastewater was dosed to the drainfield at a rate of 648 L per day. The dosing schedule was three times per day every 8 h, with the total dose being divided evenly over that time period. The Cabin Creek Plant served a residential area and monthly Georgia Environmental Protection Division (EPD) reports provided by the wastewater treatment plant verified that the wastewater was residential-strength as defined by five-day biochemical oxygen demand (BOD5) and total suspended solids (TSS), respectively, 45.2 and 35 mg per L [20]. The wastewater was tested for BOD5 and TSS before dosing to make sure that they were of residential-strength. The BOD5 and TSS values showed as much as 50% variation among measurements during different sampling times in June 2010. Wastewater was collected from the inlet of the wastewater treatment plant and transported to the site twice per week. The wastewater was then released from the septic tank into three drain pipes via a distribution box. The perforated drain pipes were installed in 10 m gravel trenches in a Cecil series soil (fine kaolinitic thermic typic kanhapudult) [19]. The trenches were installed in the B-horizon, which had two layers, Bt1 and Bt2. The texture of the Bt1 and Bt2 was clay and sandy clay, with saturated hydraulic conductivity (K_s) of 5.7–65 and 22–31 cm/d, respectively. The soil porosity in the Bt1 and Bt2 layers was 39 and 30%, respectively.

The septic tank was installed in the ground with approximately 15 cm protruding from the surface to allow easy access for sampling. The septic tank was dosed every 8 hours for two years. For this particular study, however, influent (SIN) and effluent (SOUT) wastewater samples from the septic tank were collected once a week for three consecutive weeks in June 2010 (3, 10 and 17 June). Water samples were also collected from ceramic suction-cup lysimeters that were installed 15 cm below the trenches, which were approximately 70 cm below the soil surface, using a hand held vacuum pump into sterile plastic bottles. The samples were stored in a cooler with ice until they were taken to the laboratory within a few hours for testing.

2.2. Water Quality Parameters

The pH and electrical conductivity (EC) of the wastewater and water samples were measured by using a hand held probe (ORION 3 Star, Thermo Scientific, Beverly, MA, USA) in duplicates. The probes were calibrated with standard solutions before every measurement according to the instructions of the manufacturer. SIN and SOUT samples collected on 3 and 24 June 2010 were also analyzed for ammonium and nitrate. Ammonium and nitrate were determined calorimetrically using the Phenate and the Cadmium Reduction methods [21], respectively. The samples were also tested for total coliform and *E. coli* by using the IDEXX Colilert-18® kit, which has a detection limit of 1 organism per 100 ml (IDEXX Laboratories, Inc., Westbrook, ME, USA) in duplicates. Based on the number of positive wells in the 97-well tray (positive was indicated by a yellow color for total coliform and UV fluorescence for *E. coli*), the corresponding most probable number (MPN) value per 100 mL sample was obtained with manufacturer supplied MPN tables [22].

2.3 Statistical Analysis

A two-way Analysis of Variance (ANOVA) was done on the pH, EC, nitrogen, total coliform and *E. coli* data to investigate the statistical significance of the effect of time (week) and location (septic tank inlet, outlet, trench 1–3) on these parameters in SAS 9.3 (SAS Institute, Inc., City, NC, USA) at significance level of $\alpha = 0.05$. One way ANOVA was also done on individual data sets to examine the significance of the effect of one of the factors (e.g., time or position) at a time. A pair-wise t-test was also done on *E. coli* data to compare the septic tank influent and effluent samples for each time period separately. The data were either log or inverse transformed to fulfill the assumptions of the models used for analysis.

3. Results

3.1. Common Water Quality Parameters

Influent and effluent wastewater samples had significantly higher pH ($P < 0.0001$) and EC ($P < 0.0002$) values than the water samples collected from lysimeters 15 cm below the trench bottoms (Figure 1). The pH values of the influent and effluent wastewater samples were similar over the three-week time, averaging about 7.4 (Figure 1A). The EC values were also similar for the two wastewater samples, with the three-week average of 673 and 678 μs cm^{-1} for the influent and effluent samples, respectively (Figure 1B). The pH values for the trench samples ranged between 6.5 and 6.8, which were 0.6 to 0.9 units below the wastewater samples. The EC values for the trench samples ranged between 285 and 378 μs cm^{-1}, which were on average 50% lower than the wastewater samples (Figure 1). The effect of time was not significant on either pH ($P = 0.1413$) or EC ($P = 0.4577$). The average ammonium concentration for two sampling times (June 3 and 24) for SIN and SOUT samples were 34.93 and 33.57 mg NH_4^+-N per L, respectively, while nitrate concentration was below detection (0.02 mg NO_3^--N per L). There was no significant difference between the SIN and SOUT samples in regards to nitrogen. We did not see any time effect either on these nitrogen forms. The dominance of ammonium in both sample types indicates that nitrification was limited in the septic tank, which is anoxic.

Figure 1. pH (**A**) and electrical conductivity (EC); (**B**) values of wastewater samples from septic tank inlet (SIN) and outlet (SOUT), in addition to water samples from lysimeters installed 15 cm below the drainfield trenches (T1, T2 and T3). Wastewater and water samples were collected on June 3 (week 1), June 10 (week 2) and June 17 (week 3).

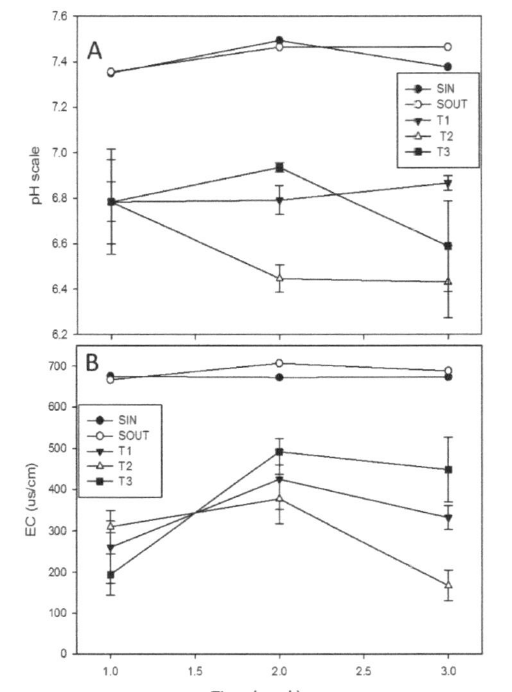

3.2. Fecal Indicator Bacteria

Based on a two-way ANOVA, the main effect of location (septic tank inlet, septic tank outlet, trench 1, 2 and 3) on *E. coli* concentration was statistically significant (P = 0.0004), while the effect of time (week 1, 2 or 3) was not (P = 0.9914). *E. coli* concentrations in the septic tank influent samples were 3.1, 3.2 and 4.3 log per 100 mL while the concentrations in the septic tank effluent were 5.3, 5.3 and 4.3 log per 100 mL for the first, second and third week of sampling, respectively, indicating a 100-fold increase in *E. coli* concentration in the effluent samples in the first and second weeks (Figure 2A). A pair-wise t-test on *E. coli* concentrations of septic tank influent and effluent samples for the individual sampling week indicated that the difference was statistically significant for weeks 1 (P = 0.0016) and 2 (p = 0.0016) but not week 3 (p = 0.6174). *E. coli* concentrations in the trenches were below detection.

Figure 2. (**A**) *E. coli*; and total coliform (**B**) counts of wastewater samples from septic tank inlet (SIN) and outlet (SOUT), in addition to water samples from lysimeters installed 15 cm below drainfield trenches (T1, T2 and T3). Wastewater and water samples were collected on June 3 (week 1), June 10 (week 2) and June 17 (week 3).

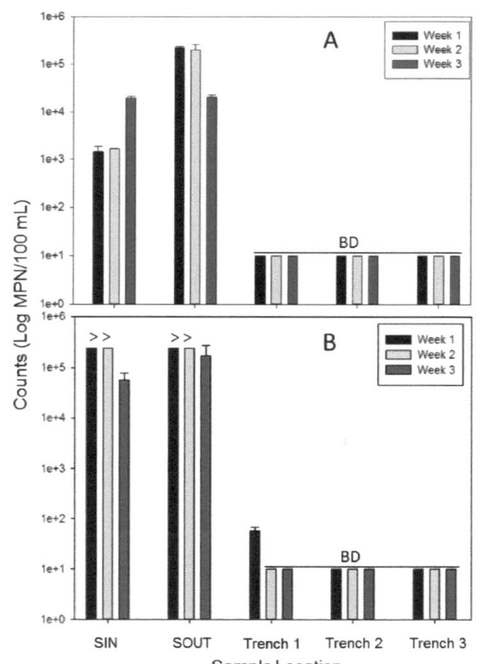

Notes: > = greater than indicated concentration; BD = below detection.

Total coliform concentrations were above the maximum level of detection (>5.38 log MPN per 100 mL) at the dilution level that was employed during testing (1:100) for the septic tank influent and effluent samples for weeks 1 and 2 (Figure 2B). For week 3, however, total coliform concentrations were higher in the septic effluent samples (5.24 log per 100 mL) than the septic influent samples (4.76 log MPN per 100 mL). Total coliform were below detection levels (BD) in trench water samples except for water samples on the first week in trench 1.

4. Discussions

As indicated by the pH, EC, ammonium and nitrate concentrations, there was not a significant change in the chemical property of the wastewater during its passage through the septic tank. These parameters are commonly used as chemical water quality indicators. There was, however, a significant change in its microbiological property. The increase in *E. coli* concentration in the effluent samples indicated that *E. coli* was able to grow inside the septic tank under typical summer weather in GA. The maximum air temperatures during the sampling days were close to the ideal growth temperature for *E. coli*, with 29.21, 31.84 and 32.86 °C for the first, second and third sampling times, respectively (Figure 3). The maximum relative humidity was also high, with 98.5%, 89.8% and 98.6% for the first, second and third sampling times, respectively (Figure 3). Temperatures in the septic tank were not measured. Since the top 15 cm of the tank was exposed, it can be expected that the temperatures in the tank were somewhat higher than the temperatures in a typical septic tank where the top is approximately 30 cm below the surface. This pattern of *E. coli* growth can also be expected to happen in other summer months, which are warmer than June in GA (e.g., July and August) [23].

Our study supports previous findings of growth of *E. coli* in tropical soils with similar type of weather [24–26]. To our knowledge, however, growth of *E. coli* in septic tanks has not been previously reported. Ottoson and Stenstrom [27] reported the growth of *Enterococci* and *Salmonella* in sterilized sediment from a settling tank of greywater in a laboratory study in Sweden but the bacteria did not grow in unsterilized sediment. This is different from our study in that greywater does not include the solid waste from the toilet and that the growth of the bacteria happened only after removal of the indigenous microorganisms through sterilization. They also did not investigate the growth of *E. coli* in the sediment. The fact that *E. coli* is capable of multiplying in environments other than the guts of warm-blooded animals undermines the original assumption under which *E. coli* was recommended to be used as indicator of microbial water quality [28,29]. The implication of *E. coli* growth in septic tanks is that the bacterium or other pathogens could be introduced into the soil environment at enhanced concentrations. This can potentially saturate the adsorption capacity of the coarse textured soils, which are widely found in coastal areas, facilitating their downward movement to groundwater sources [4]. We observed *E. coli* growth in the septic tank in the first two weeks, but not in the third week. We are not sure why as the environmental conditions such as temperature and relative humidity were not that different among the three days (3, 10 and 17 June) the samples were collected (Figure 3).

Existing technologies to enhance the performance of OWTS involve the use of advanced treatment units that manipulate the oxygen content of the wastewater to enhance the nitrification-denitrification processes to reduce the level of nitrogen (nitrate) in the wastewater before its release to the environment [11]. This has been shown to be quite effective in reducing nitrogen load of the wastewater [15]. However, its impact on microbial contaminants is not yet clear. The aeration step in this process could potentially enhance the growth of *E. coli*, which is a facultative anaerobe.

The substantial decrease in EC of trench water samples or the non-detection of *E. coli* or total coliform in the trench water samples indicated that the soil was very effective in treating the wastewater. The textures of the B horizons where the trenches were located ranged from sandy clay to clay, indicating the dominance of the clay fraction in the soil. Clay soils are highly effective in adsorbing viruses under a number of different environmental conditions, including during rainfall [30,31]. Bacteria will also be similarly affected by clay soils due to their surface charges. The capacity of the clay soils to adsorb microorganisms even under high soil moisture conditions prevents their downward migration to the ground water. Rapid movement of microbial contaminants in the soil originating from septic systems has mainly been reported in sandy soils or karst topography that is characterized by large fissures and cracks [4,32].

Figure 3. Minimum (min), average (mean) and maximum (max) daily (**A**) air temperature; and (**B**) relatively humidity for the month of June 2010 for the study site. Wastewater and water samples were collected on June 3 (week 1), June 10 (week 2) and June 17 (week 3).

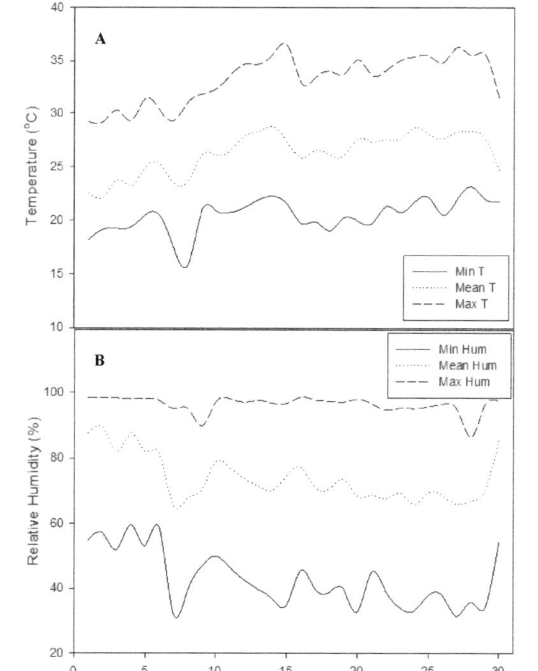

The microbial counts of the water samples from below the trenches might have been underestimated because of how they were collected. These were collected in suction samplers that have ceramic cups with an approximate pore diameter of 1.44 µm, based on the reported air-entry value of 200 kPa (Soil Moisture Equipment, Santa Barbra, CA, USA). The water samples travel through these pores into the cups because of water potential gradient between the soil and the inside of the cup. The pores are large enough to allow most bacteria to travel through, including *E. coli* whose size is about 0.55 μm in diameter [33]. However, because of the tortuous nature of the network of the pores, it is possible that the bacteria might get filtered out.

The variability of the data among the replicates for each sampling time was different for the SIN and SOUT samples. For *E. coli*, for example, the standard errors (SEs) ranged between 0.6% and 7.9% of the means for SIN samples. For the SOUT samples, the SEs ranged between 6.1% and 21.8% of the means. The variability was reasonable enough to result in significantly different means between SIN and SOUT samples for the first and second weeks. The variability increased when the three weeks *E. coli* data were pooled together. The SEs were 50% and 29% of the means for the SIN and SOUT samples, respectively, the difference mainly being between weeks 1 & 2 and week 3 (Figure 2). The large variability among the pooled data for the three-week time was probably the reason why sampling time did not have a significant effect on *E. coli* concentration (see Section 3.2). More frequent and longer sampling scheme that includes different seasons might be more appropriate to investigate the effect of time on *E. coli* growth.

5. Conclusions

The study reports the growth of *E. coli* in wastewater inside a septic tank under typical summer weather in Georgia, USA. This finding suggests that the bacterium or other pathogens could be introduced into the soil environment for treatment, as is the case for OWTS, at enhanced concentrations. This can potentially saturate the adsorption capacity of the coarse textured soils, facilitating their downward movement to groundwater sources. The growth of *E. coli* inside a septic tank is also contrary to the assumption under which *E. coli* is used as a water quality indicator. Further studies are required to identify the specific types of *E. coli* that are growing in the septic tank, in addition to the impact of retrofit technologies for enhancing OWTS performance on growth behavior of *E. coli* in septic tanks. Because of the short term and limited nature of the study, future long term studies are also needed to confirm the findings of this study over a multiple season period, preferably targeting OWTS that are in use by homeowners.

Acknowledgements

The authors would like to acknowledge Bobby Goss and Vijayalakshmi Mantripragada for their help during the field and laboratory work. The authors would also like to thank the UGA Young Scholars Program for providing summer support for Dominique Appling in 2010.

References

1. U.S. Environmental Protection Agency (USEPA). *Onsite Wastewater Treatment Systems Manual*; USEPA: Washington, DC, USA, 2002; EPA/625/R-00/008.
2. Gerba, C.; Smith, J.E., Jr. Sources of pathogenic microorganisms and their fate during land application of wastes. *J. Environ. Qual.* **2005**, *34*, 42–48.

3. Oakley, S.; Greenwood, W.P.; Lee, M. Monitoring Nitrogen and Virus Removal in the Vadose One with Suction Lysimeters. In Proceedings of 10th Northwest On-site Wastewater Treatment Short Course and Equipment Exhibition, Seattle, WA, USA, 20–21September 1999; University of Washington: Seattle, WA, USA, 1999; pp 221–232.

4. Paul, J.H.; McLaughlin, M.R.; Griffin, D.W.; Lipp, E.K.; Stokes, R.; Rose, J.B. Rapid movement of wastewater from onsite disposal systems into surface waters in the lower Florida keys. *Estuaries* **2000**, *23*, 662–668.

5. Van Cuyk, S.; Siegrist, R.L.; Lowe, K.; Harvey, R.W. Evaluating microbial purification during soil treatment of wastewater with multicomponent tracer and surrogate tests. *J. Environ. Qual.* **2004**, *33*, 316–329.

6. Ahmed, W.; Neller, R.; Katouli, M. Evidence of septic system failure determined by a bacterial biochemical fingerprinting method. *J. Appl. Microbiol.* **2005**, *98*, 910–920.

7. Keswick, B.H.; Gerba, C.P. Viruses in groundwater. *Environ. Sci. Technol.* **1980**, *14*, 1290–1297.

8. Arnade, L.J. Seasonal correlation of well contamination and septic tank distance. *Ground Water* **1999**, *37*, 920–923.

9. Habteselassie, M.Y.; Kirs, M.; Conn, K.E.; Blackwood, A.D.; Kelly, G.; Noble, R.T. Tracking microbial transport through four onsite wastewater treatment systems to receiving waters in eastern north Carolina. *J. Appl. Microbiol.* **2011**, *111*, 835–847.

10. U.S. Environmental Protection Agency. National Summary of Impaired Waters and TMDL Information. Available online: http://iaspub.epa.gov/waters10/attains_nation_cy.control?p_report_type=T (accessed on 31 May 2013).

11. Oakley, S.M.; Gold, A.J.; Oczkowski, A.J. Nitrogen control through decentralized wastewater treatment: Process performance and alternative strategies. *Ecol. Engineer.* **2010**, *36*, 1520–1531.

12. Moore, B. Innovative and Alternative System Use in Rhode Island. Personal Communication; Rhode Island Department of Environmental Management (RIDEM): Providence, RI, USA, 2008.

13. Rhode Island Department of Environmental Management (RIDEM). Rules Establishing Minimum Standards Relating to Location, Design, Construction and Maintenance of Onsite Wastewater Treatment Systems; Personal Communication; RIDEM: Providence, RI, USA, 2008.

14. Rich, B. Overview of the field test of innovative on-site wastewater treatment systems during the La Pine National Demonstration Project. *J. Hydrologic Eng.* **2008**, *13*, 752–760.

15. National Sanitation Foundation (NSF) International. *Pennsylvania ONLOT Technology Verification Program*; NSF International: Ann Arbor, MI, USA, 2009.

16. Obropta, C.C.; Berry, D. *Onsite Wastewater Treatment Systems: Alternative Technologies*; Rutgers University Cooperative Research and Extension: New Brunswick, NJ, USA, 2005.

17. Humphrey, C.P.; O'Driscoll, M.A.; Zarate, M.A. Evaluation of on-site wastewater system *Escherichia coli* contributions to shallow groundwater in costal North Carolina. *Water Sci. Technol.* **2011**, *63*, 789–795.

18. Carroll, S.; Hargreaves, M.; Goonetilleke, A. Source tracking pollution from onsite wastewater treatment systems in surface waters using antibiotic resistance analysis. *J. Appl. Microbiol.* **2005**, *99*, 471–482.

19. Bradshaw, J.K.; Radcliffe, D.E. Nitrogen Dynamics in a Piedmont Wastewater Treatment System. In Proceedings of the 2011 Georgia Water Resources Conference, Athens, GA, USA, 11–13 April 2011; Carroll, D., Ed.; Georgia Water Resources Association: Athens, GA, USA, 2011; pp 110–113.

20. Georgia Department of Human Resources—Division of Public Health (GADHR-DPH). Manual for Onsite Sewage Management Systems. Available online: http://health.state.ga.us/pdfs/environmental/LandUse/Manual/CompleteOnsiteManual.pdf (accessed on 8 March 2013).

21. American Public Health Association (APHA); American Water Works Association (AWWA); Water Pollution Control Research (WPCR). *Standard Methods for the Examination of Water and Wastewater*, 20th ed.; APHA: Washington, DC, USA, 1999.

22. Hurley, M.A.; Roscoe, M.E. Automated statistical analysis of microbial enumeration by dilution series. *J. Appl. Bacteriol.* **1983**, *55*, 159–164.

23. Georgia Weather Net Home Page. Available online: http://www.georgiaweather.net/ (accessed on 10 April 2013).

24. Byappanahalli, M.; Fujioka, R. Indigenous soil bacteria and low moisture may limit but allow faecal bacteria to multiply and become minor population in tropical soils. *Water Sci. Technol.* **2004**, *50*, 27–32.

25. Fujioka, R.S. Monitoring coastal marine waters for spore-forming bacteria of faecal and soil origin to determine point from non-point source pollution. *Water Sci. Technol.* **2001**, *44*, 181–188.

26. Ishii, S.; Ksoll, W.B.; Hicks, R.E.; Sadowsky, M.J. Presence and growth of naturalized *Escherichia coli* in temperate soils from Lake Superior watersheds. *Appl. Environ. Microbiol.* **2006**, *72*, 612–621.

27. Ottoson, J.; Stenstrom, T.A. Faecal contamination of greywater and associated microbial risks. *Water Research* **2003**, *37*, 645–655.

28. Leclerc, H.; Mossel, D.A.A.; Edberg, S.C.; Struijk, C.B. Advances in the bacteriology of the coliform group: Their suitability as markers of microbial water safety. *Annu. Rev. Microbiol.* **2001**, *55*, 201–234.

29. United States Environmental Protection Agency (USEPA). *Ambient Water Quality Criteria for Bacteria*; United States Environmental Protection Agency: Washington, DC, USA, 1986.

30. Meschke, J.S.; Sobsey, M.D. Comparative adsorption of Norwalk virus, Poliovirus 1 and F + RNA coliphage MS2 to soils suspended in treated wastewater. *Water Sci. Technol.* **1998**, *38*, 187–189.

31. Sobsey, M.D.; Dean, C.H.; Knuckles, M.E.; Wagner, R.A. Interactions and survival of enteric viruses in soil materials. *Appl. Environ. Microbiol.* **1980**, *40*, 92–101.

32. Scandura, J.E.; Sobsey, M.D. Viral and bacterial contamination of groundwater from on-site sewage treatment systems. *Water Sci. Technol.* **1997**, *35*, 141–146.

33. Trueba, F.J.; Woldringh, C.L. Changes in cell diameter during the division cycle of *Escherichia coli*. *J. Bacteriol.* **1980**, *142*, 869–878.

Water **2013**, *5*, 1006-1035; doi:10.3390/w5031006

OPEN ACCESS

water

ISSN 2073-4441

www.mdpi.com/journal/water

Article

Prospects of Source-Separation-Based Sanitation Concepts: A Model-Based Study

Taina Tervahauta [1,2,*], **Trang Hoang** [2], **Lucía Hernández** [1], **Grietje Zeeman** [2] and **Cees Buisman** [1,2]

[1] Wetsus, Centre of Excellence for Sustainable Water Technology, P.O. Box 1113, 8900CC Leeuwarden, The Netherlands; E-Mails: lucia.hernandez@wetsus.nl (L.H.); cees.buisman@wetsus.nl (C.B.)

[2] Sub-Department Environmental Technology, Wageningen University, P.O. Box 17, 6700AA Wageningen, The Netherlands; E-Mails: hoangminhtrang0203@yahoo.com (T.H.); grietje.zeeman@wur.nl (G.Z.)

* Author to whom correspondence should be addressed; E-Mail: taina.tervahauta@wetsus.nl; Tel.: +31-0-58-284-3000; Fax: +31-0-58-284-3001.

Received: 27 April 2013; in revised form: 13 June 2013 / Accepted: 14 June 2013 / Published: 8 July 2013

Abstract: Separation of different domestic wastewater streams and targeted on-site treatment for resource recovery has been recognized as one of the most promising sanitation concepts to re-establish the balance in carbon, nutrient and water cycles. In this study a model was developed based on literature data to compare energy and water balance, nutrient recovery, chemical use, effluent quality and land area requirement in four different sanitation concepts: (1) centralized; (2) centralized with source-separation of urine; (3) source-separation of black water, kitchen refuse and grey water; and (4) source-separation of urine, feces, kitchen refuse and grey water. The highest primary energy consumption of 914 MJ/capita(cap)/year was attained within the centralized sanitation concept, and the lowest primary energy consumption of 437 MJ/cap/year was attained within source-separation of urine, feces, kitchen refuse and grey water. Grey water bio-flocculation and subsequent grey water sludge co-digestion decreased the primary energy consumption, but was not energetically favorable to couple with grey water effluent reuse. Source-separation of urine improved the energy balance, nutrient recovery and effluent quality, but required larger land area and higher chemical use in the centralized concept.

Keywords: centralized sanitation; source-separation-based sanitation; energy balance; water balance; nutrient recovery; chemical use; effluent quality; land area requirement

1. Introduction

Separation of different domestic wastewater streams and targeted on-site treatment of these streams for resource recovery has been recognized as one of the most promising concepts to re-establish the balance in carbon, nutrient and water cycles [1–4]. Domestic wastewater can be divided into two major streams: concentrated stream of black water (feces and urine) and kitchen refuse, and less concentrated stream of grey water from washing activities, such as laundry, shower and bath. Black water can be further divided into urine and feces using urine diverting toilets or urinals. Energy and nutrients can be recovered primarily from the concentrated streams, while the less concentrated stream serves as an alternative water source.

Key technology for energy recovery from source-separated streams is anaerobic treatment of black water or feces and kitchen refuse in an up-flow anaerobic sludge blanket (UASB) reactor [4,5]. Nutrient recovery and pollutant removal from the UASB reactor effluent can be established by struvite precipitation, autotrophic nitrogen removal using oxygen limited anaerobic nitrification denitrification (OLAND) reactor and a post-treatment, such as a trickling filter (TF), to remove remaining organic material [4]. Due to operational conditions, such as a lower buffer capacity of the OLAND reactor effluent compared to the UASB reactor effluent, the struvite precipitation is preferred after the nitrogen removal [6].

Urine separation can be employed in two different approaches: in the source-separation-based sanitation and coupled with the existing centralized sanitation. Separation and direct reuse of urine on agricultural land can be used to increase nutrient recovery, improve wastewater effluent quality and to decrease operational energy consumption, due to lower nutrient concentrations in wastewater [7]. However, collection and reuse of source-separated waste streams, urine in particular, also involves social and cultural issues requiring attention when implementing new technology [8].

Commonly used treatment systems to remove organic material and nutrients from grey water include sequencing batch reactor (SBR) [9] and constructed wetlands (CW) [10]. Due to the considerably high land area requirement, the use of CW is not suitable for densely populated areas, such as the Netherlands [11]. One option could be, however, to implement CW as a green roof [10]. To utilize the organic material present in grey water, excess sludge from the grey water treatment system can be potentially co-digested in the UASB reactor instead of using energy-intensive sludge transport and disposal [12]. However, the possible inhibitory effect of surfactants present in grey water sludge on anaerobic digestion should be investigated [13]. To avoid extensive mineralization of grey water sludge, a bio-flocculation unit, such as a high loaded membrane bioreactor (MBR) or A-trap from the AB-process [14], can be used to concentrate grey water at short hydraulic and sludge retention times (HRT and SRT). A post-treatment system (such as TF) can be applied to remove the remaining organic material from grey water effluent prior to reuse.

Quantitative tools, such as Material Intensity per Service unit (MIPS), exergy analysis and Life Cycle Assessment (LCA) have been used to draw energy and material balances of different centralized and source-separation-based sanitation concepts [15–17]. These studies present data on energy consumption and production, material intensity, and emissions of source-separated feces, urine and grey water treatment and centralized wastewater treatment with and without urine separation. For more in depth insight into the urban water cycle, Makropoulos *et al.* [18] developed an Excel/Matlab-based decision support tool for sustainable integrated urban water management, including domestic wastewater streams and rain water. Extensive information was provided on different household components for water use and options for water treatment and reuse, producing a complete water balance. A study on economic viability and critical influencing factors of different implementation scales of black water and grey water source-separation compared to the centralized sanitation was conducted by Thibodeau *et al.* [19]. Van Beuzekom *et al.* [20] conducted a social cost-benefit analysis on different sanitation concepts in Geerpark Heusden, a neighborhood in the Netherlands. This study compared centralized sanitation with different levels of source-separation of wastewater and different scales for the treatment of source-separated wastewater in terms of livability, safety, health, biodiversity and affordability. No studies, however, have investigated the influence of urine separation combined with different grey water treatment configurations and grey water sludge co-digestion on the energy and material balances of the sanitation concepts. The objective of this study was to present energy and water balances, nutrient recovery, chemical use, effluent quality and land area requirement of the centralized and source-separation-based sanitation concepts with and without urine separation, and with different configurations of grey water treatment.

2. Materials and Methods

2.1. Construction of the Model

An Excel-based model was developed based on literature data for the comparison of four sanitation concepts: (1) centralized sanitation; (2) centralized sanitation with source-separation of urine; (3) source-separation of black water, kitchen refuse and grey water; and (4) source-separation of urine, feces, kitchen refuse and grey water (Figure 1), from which Concept 1 has been applied on a full scale, and Concepts 2, 3 and 4 have been demonstrated on a pilot or lab scale. These concepts were compared in terms of energy consumption and production, water saving and reuse, nutrient recovery, chemical use, effluent quality and land area requirement. The energy and material balances were based on collection, transport and treatment of wastewater, leaving out the energy and materials used in the construction and maintenance of the required infrastructure. The model was tailored for European circumstances with a specific focus on the Netherlands. However, with small modifications on data input, the model is applicable also in other circumstances.

The model was constructed from location-specific data on environmental temperature, tap water temperature and distances to a sewage sludge incineration plant and agricultural land, general data on water consumption of different appliances and wastewater characteristics, and treatment system-specific data on operational conditions, reactor performance, sludge production, energy consumption and energy production. The energy and water balance, recovered nutrients, chemicals used, effluent quality and land area requirement for each treatment system was then calculated using energy and mass balances based on the selected data.

Figure 1. Sanitation Concepts (**1–4**) included in the model with wastewater streams and corresponding treatment systems (AS = activated sludge process; SBR = sequencing batch reactor, MBR = membrane bioreactor; A-trap = A-stage of AB-process; TF = trickling filter; UASB = up-flow anaerobic sludge blanket reactor; OLAND = oxygen limited anaerobic nitrification denitrification).

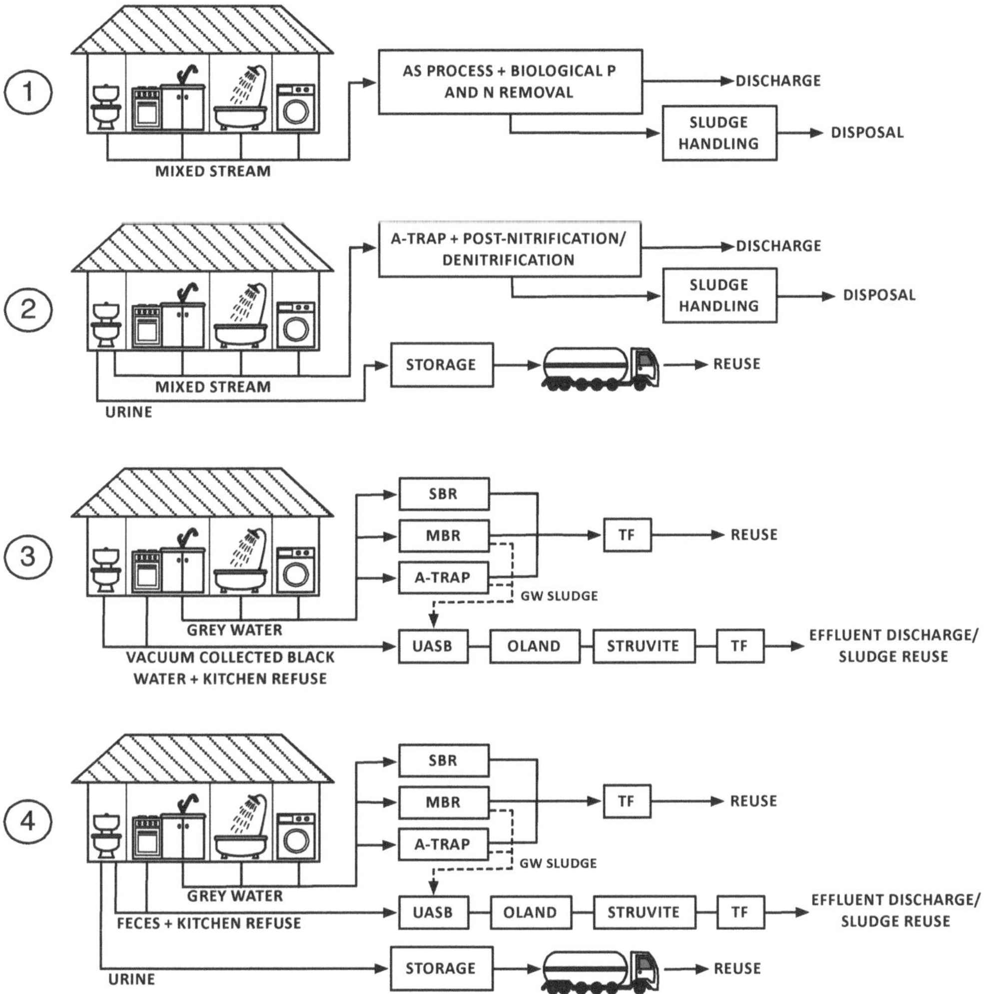

2.2. Data Inventory: Location Specific Data

Wastewater in Concepts 1 and 2 were considered to be treated centralized (10,000 or more people), and the urine collection (Concepts 2 and 4) and the treatment of black water or feces, kitchen refuse and grey water were considered to be community-on-site (100–10,000 people). Average environmental temperature of 10 °C [21] and tap water temperature of 12 °C [22] of the Netherlands were used. The distance from the centralized wastewater treatment plant to the sewage sludge incineration plant was set to 10 km [22], and the distance from the on-site collection to agricultural land was assumed to be 50 km, as a typical distance in the Netherlands. The influence of the transport distance on feasibility of the sanitation concepts was further discussed in the sensitivity analysis.

2.3. Data Inventory: General Data

The toilet type selected for Concept 1 was a normal flush toilet, for Concept 2, a urine diverting toilet (gravity), for Concept 3, a vacuum toilet, and for Concept 4, a urine diverting toilet (gravity/vacuum). The water consumption of different toilets and kitchen grinders is presented in Table 1.

Table 1. Water consumption of different toilets and kitchen grinders.

Parameter	Unit	Water use
Normal flush toilet (Concept 1)	L/cap/d	34 [1]
Vacuum toilet (Concept 3)	L/cap/d	6 [2],*
Urine diverting toilet (gravity) (Concepts 2 and 4)	L/cap/d	5 [3],*
Urine diverting toilet (vacuum) (Concept 4)	L/cap/d	2 **
Kitchen grinder (Concepts 1, 2, 3 and 4)	L/cap/d	0.6 [2]

Notes: [1] [23]; [2] [24]; [3] [25] (0.2 L for urine and 4 L (assumed) for feces per flush); * based on production of one time feces and five times urine per day; ** based on 0.2 L for urine [25] and 1 L for feces per flush [24].

As a common practice in the centralized approach, the wastewater influent in Concepts 1 and 2 was considered to consist of domestic wastewater, rain water runoffs and some industrial effluents, ending up with a daily flow of 300 L/cap [22]. For better comparison between centralized and source-separation-based sanitation concepts, the pollutant loading in the wastewater influent was considered to originate only from the domestic wastewater streams of urine, feces, kitchen refuse and grey water and sludge rejection water from sludge dewatering, forming a daily loading of 176 gCOD(Chemical Oxygen Demand)/cap, 21 gTN(Total Nitrogen)/cap and 3.6 gTP(Total Phosphorus)/cap (Concept 1), similar to the study of Wilsenach and van Loosdrecht [22]. Although kitchen refuse was not included in the study of Wilsenach and van Loosdrecht [22], the pollutant loading from kitchen refuse was considered to replace the pollutant loading from industrial effluents in this study.

Table 2 presents the characteristics of different domestic wastewater streams. In every sanitation concept, the pollutant loading in the wastewater influent was calculated as a sum of the according sub-streams, and in Concepts 3 and 4, the daily flow was calculated as a sum of the pollutant loading and the water consumption of the toilet and the kitchen grinder.

Table 2. Domestic wastewater characteristics. (COD = Chemical Oxygen Demand; BOD_5 = Biochemical Oxygen Demand; TSS = Total Suspended Solids; TN = Total Nitrogen; NH_4^+-N = Ammonium Nitrogen; TP = Total Phosphorus; PO_4^{3-}-P = Phosphate Phosphorus; K = Potassium).

Parameter	Unit	Feces	Urine	Kitchen refuse	Grey water
Temperature	°C	37 *	37 *	20 *	32 [2]
Volume	L/cap/d	0.1 [1]	1.4 [1]	0.2 [1]	79 [4]
COD	g/cap/d	50 [1]	11 [1]	59 [1]	52 [1]
BOD_5	g/cap/d	24 [1]	5.5 [1]	37 **	27 [1]
TSS	g/cap/d	30 [1]	40 [1]	79 [1]	55 [1]
TN	g/cap/d	1.8 [1]	9 [1]	1.7 [1]	1.2 [1]
NH_4^+-N	g/cap/d	1.2 [3]	9 [5]	-	0.1 ***
TP	g/cap/d	0.5 [1]	0.8 [1]	0.2 [1]	0.4 [1]
PO_4^{3-}-P	g/cap/d	0.2 [3]	0.3 [3]	-	0.1 ***
K	g/cap/d	0.9 [1]	2.8 [1]	0.2 [1]	0.8 [1]

Notes: [1] [24]; [2] [26]; [3] [27] (NH_4^+-N/TN ratio of 0.7); [4] [23]; [5] [22] (TN = NH_4^+-N in urine); * based on body temperature (feces and urine) and average room temperature; ** based on COD/BOD ratio of 1.6 [28]; *** based on NH_4^+-N/TN ratio of 0.1 and PO_4^{3-}-P/TP ratio of 0.35 [26].

2.4. Data Inventory: Treatment System Specific Data

The wastewater treatment system in Concept 1 was based on an activated sludge process (AS process) with biological phosphate and nitrogen removal and, in Concept 2 on an A-trap (A-stage of AB-process [14]) with a post-nitrification/denitrification step according to the study of Wilsenach and van Loosdrecht [22]. As the wastewater in Concept 2 was without the input of urine, a high loaded process with a short SRT and a post-treatment step was assumed to be sufficient for pollutant removal. Urine in Concepts 2 and 4 was considered to be collected on-site with a collection degree of 75% [8], first stored for six months on-site and, then, transported to agricultural land to be used as a fertilizer by spreading. As a result of the breakdown of urea during storage, the high ammonium content and the increased pH ensures the hygienization of urine [29] and is recommended by the World Health Organization (WHO) for safe use of urine in agriculture [30]. The risk of ammonia emissions is prevented by using non-ventilated storage and handling. The treatment systems applied for black water or feces, kitchen refuse and grey water in Concepts 3 and 4 are presented in Figure 1. Table 3 presents the pollutant removal efficiencies of the different treatment systems. The removal efficiencies in the AS process in Concept 1 were according to existing wastewater treatment plants in the Netherlands.

Incineration was selected for excess sludge treatment in Concepts 1 and 2, as it is the most common practice in the Netherlands [31]. Complete sludge treatment consisted of anaerobic digestion to produce methane, sludge dewatering, transport of dewatered sludge to an incineration plant and sludge incineration. Sludge rejection water from sludge dewatering was recycled back to the influent. Excess sludge from the UASB reactor and the SBR (Concepts 3 and 4) was considered to be transported to agricultural land for spreading without dewatering.

Table 3. Pollutant removal efficiencies of biological reactors in Concept 1 and 2, and of up-flow anaerobic sludge blanket reactor (UASB), oxygen limited anaerobic nitrification denitrification (OLAND), struvite precipitator, trickling filter (TF), sequencing batch reactor (SBR), A-stage of AB-process (A-trap) and membrane bioreactor (MBR) in Concepts 3 and 4.

Parameter	Unit	Concept 1	Concept 2	Concepts 3 and 4 Black water/feces and kitchen refuse					Grey water		
				UASB	OLAND	Struvite	TF	Total	SBR	A-trap	MBR
COD	%	92 [1]	92 [7]	83 [2]	53 [2]	-	85 [3]	99 **	90 [4]	42 [5]	75 [6]
BOD_5	%	98 [1]	92 *	83 [2]	53 *	-	85 [3]	99 **	90 *	42 *	75 [6]
TSS	%	95 [1]	92 *	83 [2]	-	-	85 [3]	97 **	76 [4]	42 *	≥95 [6]
TN	%	80 [1]	72 [7]	1 [2]	73 [2]	9 [8]	-	76 **	35 [4]	36 [5]	81 [6]
TP	%	82 [1]	79 [7]	33 [2]	-	96 [8]	-	98 **	28 [4]	40 [5]	65 [6]

Notes: [1] [32]; [2] [33]; [3] [28] (based on standard rate filter with hydraulic loading of 1–4 m^3/m^2*d); [4] [9]; [5] [34]; [6] [12]; [7] [22]; [8] [35]; * assumed based on COD removal; ** calculated as total removal efficiency of UASB, OLAND, Struvite and TF.

2.5. Calculations for Energy Balance

The total primary energy consumption in the sanitation concepts was calculated according to Equation (1):

$$E_{total} = E_{collection} + E_{treatment} + E_{urine/sludge\ transport} - E_{methane} \tag{1}$$

where $E_{collection}$ was the energy requirement for the collection and transport of wastewater, $E_{treatment}$ was the energy requirement for all the biological, chemical and physical treatment units for mixed wastewater stream, excess sludge and source-separated urine, black water/feces, kitchen refuse and grey water, $E_{urine/sludge\ transport}$ was the energy requirement for urine and excess sludge transport and $E_{methane}$ was the energy production as methane. The detailed description of the energy parameters is presented in the Appendix. All the energy parameters were calculated as primary energy by converting the electrical energy (collection, aeration, mixing and pumping) using efficiency of 0.31 based on the European electricity mix [36].

2.6. Calculations for Chemical Use

In Concepts 1 and 2, polymers were used for sludge dewatering, and calcium oxide (CaO) was used for flu gas treatment after sludge incineration. The dose of CaO was 30 kg/t Dry Matter (DM) and the dose of polymers was 7.1 kg/t DM [37]. Methanol (CH_3OH) was consumed 1.48 kg/cap/year in the post-denitrification step in Concept 2 [22]. In Concepts 3 and 4, sodium hydroxide (NaOH) and magnesium chloride ($MgCl_2$) were used in struvite precipitation to increase the pH and the supersaturation state. Consumption of NaOH was calculated from stoichiometry to increase the pH of influent to the operational pH (see Appendix). Consumption of $MgCl_2$ was calculated from the influent phosphate concentration using a Mg/PO_4-P ratio of 1.5 [35].

2.7. Calculations for Reactor Dimensions and Land Area Requirement

Total land area requirement for Concepts 1 and 2 consisted of the volume of the biological reactors, secondary settling tank, digester, biogas storage tank and the urine storage tank (Concept 2). The land use of the incineration process was not taken into account, due to lack of data. Total land area requirement for Concepts 3 and 4 consisted of the volume of the buffer tank (for UASB, SBR, A-trap and MBR), reactors (UASB, OLAND, Struvite, black water TF, SBR/A-trap/MBR and grey water TF), biogas storage tank and the urine storage tank (Concept 4). The detailed calculations for reactor dimensions and land area requirement are described in the Appendix.

3. Results and Discussion

3.1. Energy Balance

Figure 2 presents the total primary energy consumption in the sanitation concepts. The highest primary energy consumption of 914 MJ/cap/year is attained in the centralized sanitation concept (Concept 1), and by applying urine separation within the centralized concept, the primary energy consumption is decreased to 687 MJ/cap/year, creating a yearly energy saving of 227 MJ/cap. The lowest primary energy consumption of 437 MJ/cap/year is attained in the source-separation of urine, feces, kitchen refuse and grey water (Concept 4 vacuum) using the A-trap for grey water treatment. Urine separation in the source-separation-based sanitation concept creates a yearly energy saving of 200 MJ/cap using the SBR, 180 MJ/cap using the A-trap and 203 MJ/cap using the MBR in Concept 4 with gravity separation, 212 MJ/cap using the SBR, 187 MJ/cap using the A-trap and 200 MJ/cap using the MBR in Concept 4 with vacuum separation. Bio-flocculation of grey water in the A-trap and sub-sequent grey water sludge co-digestion in the UASB reactor creates a yearly energy saving of 143 MJ/cap in Concept 3, 123 MJ/cap in Concept 4 (gravity) and 118 MJ/cap in Concept 4 (vacuum) compared to the use of the SBR for grey water treatment. The high primary energy consumption of Concept 1 originates mainly from the high energy input to mineralize organic matter in the AS process and the resulting low energy recovery as methane. The low primary energy consumption of Concept 4 originates from the low water consumption of the urine diverting toilets, resulting in low energy demand of collection and treatment of feces and kitchen refuse. In addition, by grey water sludge co-digestion in the UASB reactor, high energy consumption for sludge transport can be avoided, while simultaneously increasing energy recovery as methane.

The energy parameters, together with the sludge production and urine collection, are presented in the Appendix in Table A1 (Concepts 1 and 2) and in Tables A2 and A3 with UASB influent characteristics (Concepts 3 and 4). The most prominent parameters in the energy balance in Concepts 1 and 2 are energy consumption for the collection of wastewater and aeration of the biological reactors. The collection contributes 27% in Concept 1 and 30% in Concept 2 to the total primary energy consumption, and the aeration contributes 40% in Concept 1 and 23% in Concept 2. Furthermore, transporting of urine in Concept 2 contributes 18% to the total primary energy consumption. Due to the shorter SRT in the A-trap (0.8 d) compared to the AS process (12 d), the energy consumption for aeration is significantly lower in Concept 2 compared to Concept 1. However, short SRT increases the excess sludge production, leading to an increase in the energy requirement for heating of the digester. Nevertheless, the higher excess sludge production together with the low mineralization of organic matter creates almost twice as high methane production in Concept 2 compared to Concept 1. Compared to the study of Wilsenach and van Loosrecht [22], both concepts have higher total primary energy consumption, mainly due to the energy consumption for the collection that is included in this study and the higher energy consumption for the transporting of collected urine compared to the treatment of urine and sludge rejection water in struvite precipitation and the Single reactor system for High activity Ammonium Removal Over Nitrite (SHARON) processes, used in the study of Wilsenach and van Loosdrecht [22]. However, direct reuse of urine provides a clean route for nutrient recovery, while the mixing of sludge rejection water with urine might deteriorate the quality of the produced struvite with heavy metals from sewage.

Figure 2. Total primary energy consumption in sanitation concepts with different grey water treatment configurations.

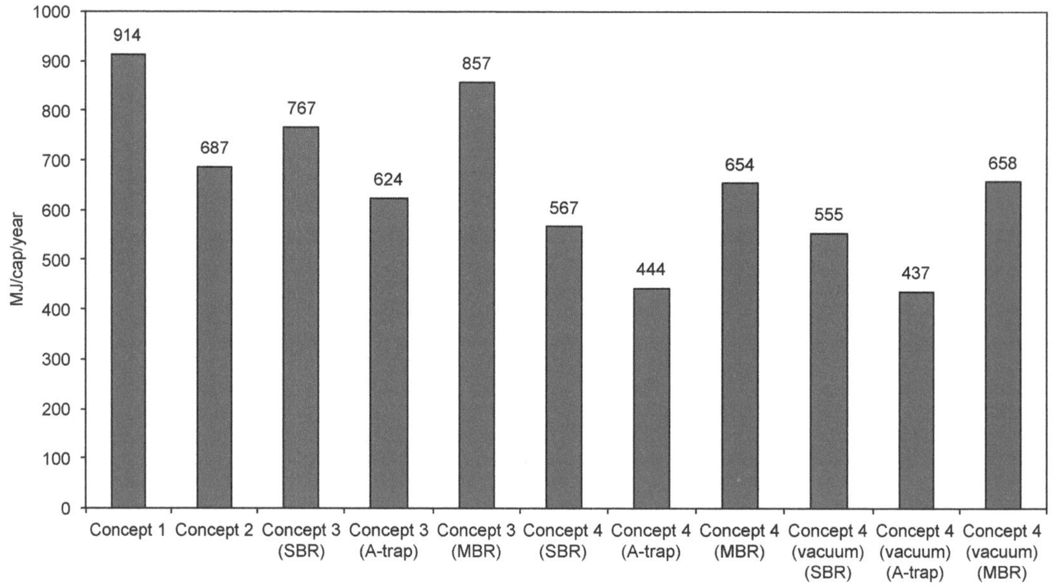

The most prominent parameters in the energy balance in Concepts 3 and 4 are energy consumption for the vacuum collection and transport of black water and kitchen refuse and heating of the UASB reactor. The vacuum collection and transport contributes 27%–35% in Concept 3 and 15%–20% in Concept 4 (vacuum) to the total primary energy consumption, and heating of the UASB reactor contributes 33%–46% in Concept 3, 36%–53% in Concept 4 (gravity) and 24%–43% in Concept 4 (vacuum). Furthermore, transporting of collected urine in Concept 4 contributes 17%–23% to the total primary energy consumption.

Urine separation in the source-separation-based sanitation concept (Concept 4) has the potential to decrease the total energy consumption, due to a lower energy demand of the feces collection and the post-treatment of UASB reactor effluent in the OLAND reactor, struvite precipitator and TF compared to Concept 3. In addition, separation of urine from feces and kitchen refuse and the low water consumption of the urine diverting toilets decreases the UASB reactor influent volume and, thus, the energy used for heating of the reactor. However, urine separation has an extra energy consumption for transporting of collected urine. Although vacuum collection of feces and kitchen refuse increases the energy demand of collection compared to gravity collection, vacuum separation of urine presents the energetically most favorable option, due to the smallest UASB reactor influent volume.

A significant fraction of the energy consumption for the SBR originates from the high aeration demand at the long SRT (15 d [9]). By decreasing the SRT to 0.6 d using the A-trap [34] or to 1 d using the MBR [12]), the energy consumption for the grey water treatment system can be decreased. The energy consumption for the MBR, however, is four times higher than for the A-trap, due to the higher energy requirement of membrane technology. When grey water sludge is co-digested in the UASB reactor, the total energy consumption can be decreased, as no transporting of grey water sludge is required. Furthermore, methane production in the UASB reactor can be increased, due to the higher loading of the reactor and the higher methanization level of grey water sludge compared to black water, feces and kitchen refuse. However, co-digestion of grey water sludge increases the heating energy required for the reactor as a result of a higher influent volume and a lower influent temperature, originating from the lower grey water sludge temperature that was assumed to be the environmental temperature. Consequently, bio-flocculation of grey water in the MBR and sub-sequent grey water sludge co-digestion in the UASB reactor is not energetically favorable compared to grey water treatment in the SBR, due to the high sludge production in the MBR and the resulting high heating energy requirement for the UASB reactor. However, to decrease the volume of the MBR sludge, a settler can be implemented to increase the concentration of the sludge.

3.2. Water Reuse

Table 4 presents the calculated effluent quality of the different grey water treatment systems and the standards for non-potable grey water reuse suggested by Li *et al.* [38]. The reuse standards were divided into recreational impoundments, such as ornamental fountains and lakes, and urban reuse, such as toilet flushing, laundry and irrigation. Unrestricted reuse is considered in close contact with people and restricted reuse in areas without public access. Due to high nutrient concentrations in the effluent, none of the treatment systems fulfilled the reuse standards for recreational impoundments. The SBR and the MBR with TF as a post-treatment step fulfilled the standards for urban reuse, but only the effluent from the SBR-TF was according to the unrestricted reuse. The better effluent quality from the SBR-TF in terms of BOD_5 can be explained by the longer SRT and, thus, more extensive degradation of organic material. However, membrane technology has the potential to produce grey water effluent free of solids and, therefore, benefit from the use of advanced
post-treatment systems, such as UV and ozonation, for removing micro-pollutants and pathogens. Nevertheless, the costs of advanced post-treatment systems have to be related to the actual need for high quality water, rather than striving to fulfill the most stringent standards.

Table 4. Calculated effluent quality of grey water treatment systems and suggested standards for water reuse.

| Parameter | Unit | Grey water effluent quality (This study) | | | Suggested reuse standards [38] | | | |
| | | | | | Recreational impoundments | | Urban reuse | |
		SBR-TF	A-Trap-TF	MBR-TF	Restricted	Unrestricted	Restricted	Unrestricted
BOD_5	mg/L	5	30	14	30	10	30	10
TSS	mg/L	25	60	6	30	-	30	-
TN	mg/L	10	10	3	1	1	-	-
TP	mg/L	4	3	2	0.05	0.05	-	-

3.3. Nutrient Recovery

Nutrients, such as nitrogen, phosphorus and potassium, can be recovered using urine separation in the centralized concept and in the source-separation-based sanitation concepts. Nutrients were considered to be recovered through urine spreading on agricultural land in Concepts 2 and 4, and thus, all the nutrients present in the collected urine (collection degree of 75%) were considered to be recovered. Struvite ($MgNH_4PO_4$ $6H_2O$) precipitation is used to recover nutrients from the effluent of the OLAND reactor in Concepts 3 and 4. Struvite is produced 2.13 kg/cap/year from which 0.27 kg is phosphorus and 0.12 kg is nitrogen in Concept 3 and 1.0 kg/cap/year from which 0.13 kg is phosphorus and 0.06 kg is nitrogen in Concept 4. In Concepts 3 and 4, nutrients were also considered to be recovered from the excess sludge of the UASB reactor and the SBR through sludge reuse on agricultural land. Nitrogen and phosphorus removed in the UASB reactor and the SBR were considered to be trapped in the sludge and, in this way, recovered. Figure 3 presents the nutrients recovered in Concepts 2–4 with different grey water treatment configurations. As most of the nutrients are present in urine, source-separation and direct reuse of urine brings forth a major contribution to the total nutrient recovery. The choice between the different grey water treatment configurations (SBR/A-trap/MBR) has only a slight effect on the total amount of nutrients recovered. The maximum nutrient recovery can be achieved with Concept 4, where nutrient recovery from sludge increases the recovery of nitrogen and phosphorus compared to Concept 2.

Compared to artificial fertilizers, direct reuse of urine in agriculture, as suggested here, has an advantage of acting as a multicomponent fertilizer. However, direct reuse of urine also has disadvantages, such as transporting of urine to agricultural land and the possible adverse effect of high salt content of urine on soil, especially in low rainfall areas. Several technologies have been presented to overcome these issues by indirectly recovering the resources from urine. Nutrients can be recovered from urine by struvite precipitation [7,8] or using algae for nutrient up-take from urine and subsequent reuse of algae biomass [39]. In the study of Kuntke *et al.* [40], a microbial fuel cell was used to simultaneously produce energy (3.46 kJ/gN) and recover ammonium (3.29 gN/d/m^2) from urine. By replacing the urine transport with a microbial fuel cell, the total primary energy consumption can be decreased by 19% in Concept 2 and 17%–23% in Concept 4, indicating a promising new direction for urine treatment.

According to the current Dutch guidelines for sewage sludge reuse in agriculture (BOOM), reuse of black water sludge is prohibited, due to elevated concentrations of copper and zinc [41]. However, as black water is predominantly human originated (urine, feces and tap water), the applicability of sewage sludge reuse guidelines on the reuse of black water sludge can be argued. Furthermore, the amount of heavy metals related to the phosphorus content of sludge is significantly higher in cow manure [42] and in artificial phosphorus fertilizers in the case of cadmium, chromium and nickel [43]. The heavy metal content of grey water sludge and the effect of grey water sludge co-digestion on the excess sludge quality of the UASB reactor needs to be further investigated to decide whether or not to mix these streams.

Figure 3. Nutrient recovery in Concepts 2–4 with different grey water treatment configurations.

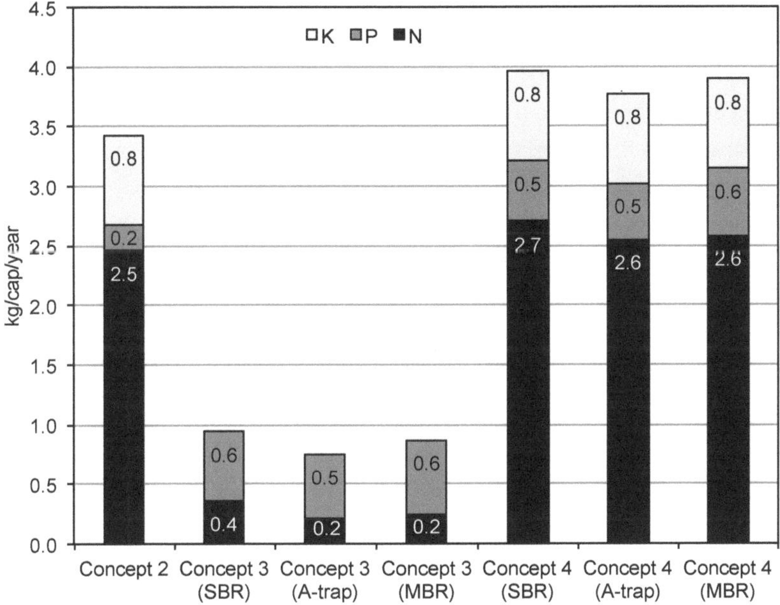

3.4. Energy Balance Including Water Saving and Reuse and Nutrient Recovery

Compared to the normal flush toilet in Concept 1, the use of a urine diverting toilet or a vacuum toilet saves water of a drinking quality. The vacuum toilet saves 28 L/cap/day, the urine diverting toilet (gravity) saves 29 L/cap/day and the urine diverting toilet (vacuum) saves 32 L/cap/day. Considering a primary energy consumption of 5.4 MJ/m^3 for drinking water production and distribution [31] (using efficiency of 0.31 [36]), 57 MJ/cap/year can be indirectly gained in Concepts 2 and 4 (gravity), 55 MJ/cap/year in Concept 3 and 63 MJ/cap/year in Concept 4 (vacuum). Furthermore, by reusing grey water effluent for toilet flushing, laundry and irrigation, drinking water can be saved and energy can be indirectly gained in Concepts 3 and 4. By assuming full reuse of grey water effluent (29 m^3/cap/year), energy can be indirectly gained as 157 MJ/cap/year by using either the SBR-TF for unrestricted or the MBR-TF for restricted urban reuse. As the water use for toilet flushing and laundry is only 8 m^3/cap/year [23], 73% of the SBR-TF effluent is left for irrigation. Grey water effluent from the MBR-TF can only be used for urban reuse applications without public access, such as irrigation of restricted areas.

Through the recovery of nutrients, energy can be indirectly gained in the production of artificial fertilizers. Considering a primary energy requirement of 45 MJ/kgN, 29 MJ/kgP and 11 MJ/kgK for fertilizer production [7], energy can be indirectly gained 129 MJ/cap/year in Concept 2, in Concept 3, 33 MJ/cap/year with SBR, 25 MJ/cap/year with A-trap and 29 MJ/cap/year with MBR and in concept 4, 145 MJ/cap/year with SBR, 137 MJ/cap/year with A-trap and 141 MJ/cap/year with MBR.

Figure 4 presents the total primary energy consumption with and without the indirect energy gain from water saving and reuse, and nutrient recovery. The most prominent energy gain can be achieved with the recovery of nutrients through urine separation (Concepts 2 and 4) and the reuse of grey water effluent using either the SBR or the MBR (Concepts 3 and 4). Due to the significant energy gain from the grey water effluent reuse, grey water treatment in the SBR becomes energetically more favorable than bio-flocculation of grey water in the A-trap and subsequent grey water sludge co-digestion in the UASB reactor. Besides water and nutrient recovery, there is an increasing interest to recover the heat content of wastewater [44]. Heat recovery on-site from source-separated grey water using a heat exchanger would be an energy-efficient option to preheat the incoming tap water, as no electricity is needed.

When the indirect energy gain is taken into account, urine separation applied in the centralized sanitation creates even higher yearly energy saving of 413 MJ/cap compared to Concept 1. The lowest energy consumption in Concept 3 (522 MJ/cap/year) and Concept 4 (208 MJ/cap/year (gravity) and 190 MJ/cap/year (vacuum)) is attained when the SBR is used. By applying urine separation in the source-separation-based sanitation, 294–331 MJ/cap/year can be saved with indirect energy gain.

Figure 4. Total primary energy consumption in sanitation concepts with and without indirect energy gain from water saving and reuse, and nutrient recovery.

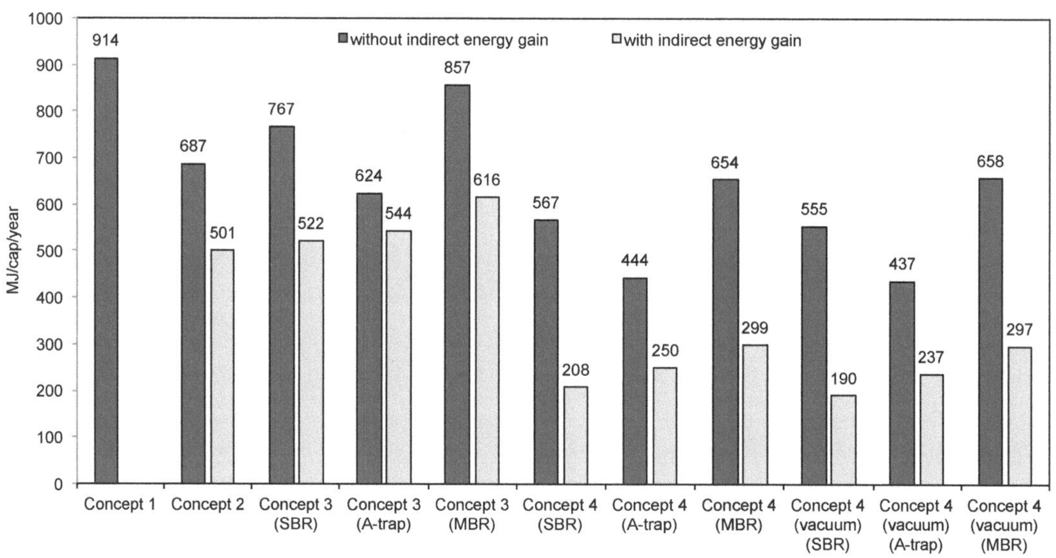

3.5. Chemical Use

Figure 5 presents the chemical use in Concepts 1–4 with different grey water treatment configurations. The chemical use in Concepts 1 and 2 is considerably higher than in Concepts 3 and 4, due to the high sludge production in aerobic processes and the resulting consumption of polymers for sludge dewatering and CaO for flu gas treatment after sludge incineration. As the sludge production in Concept 2 is higher than in Concept 1 (due to the shorter SRT in the aerobic process), the chemical use is accordingly higher. Furthermore, additional chemical use in Concept 2 originates from the consumption of methanol in the post-denitrification step. As the amount of NaOH is calculated to be negligible, the only chemical taken into account in the struvite precipitation in Concepts 3 and 4 is $MgCl_2$. The use of $MgCl_2$ is the highest in Concept 3, due to the highest phosphate concentration in the OLAND reactor effluent. Grey water treatment in the MBR and the sub-sequent grey water sludge co-digestion in the UASB reactor slightly increases the $MgCl_2$ consumption, due to the increased phosphate loading. The use of either a gravity or vacuum urine diverting toilet does not influence the chemical use in Concept 4. Contrary to the centralized concept, urine separation in the source-separation-based concept decreases the chemical use.

Figure 5. Chemical use in Concepts 1–4 with different grey water treatment configurations.

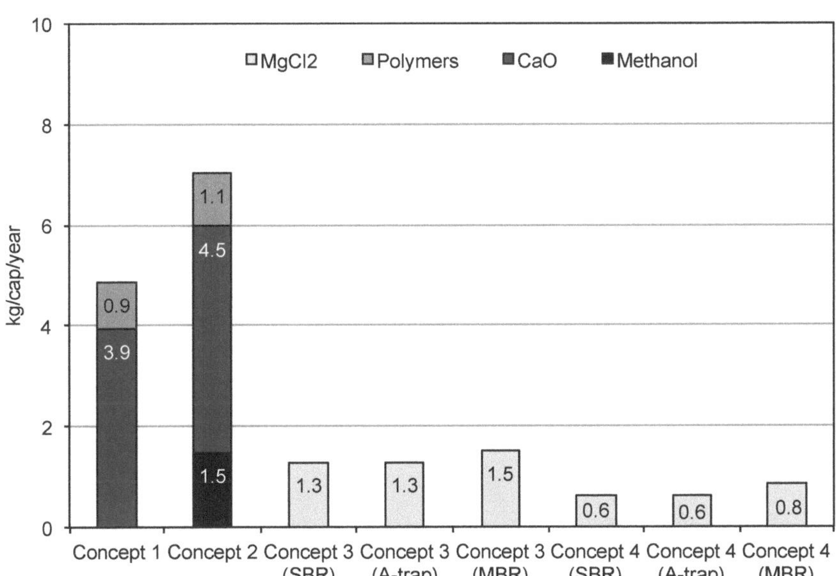

3.6. Effluent Quality

Within the European Union, the discharge of wastewater effluent is controlled by the pollutant removal efficiencies of the treatment systems and the final effluent concentrations per connected person, according to the EU Water Framework Directive 91/271/EEC [45]. Table 5 presents the calculated effluent quality of the different sanitation concepts and the discharge standards. In Concepts 3 and 4, only the effluent discharge of the source-separated concentrated stream is taken into account, leaving out the grey water effluent that is considered to be reused. For simplicity, the effluent quality presented in

Concepts 3 and 4 is the average of the different grey water treatment configurations (without co-digestion using the SBR or with co-digestion using the A-trap/MBR). The pollutant concentrations in the effluent of the concentrated stream are higher and the pollutant loadings are lower without grey water sludge co-digestion, due to the lower UASB reactor influent volume compared to co-digestion.

As the total pollutant removal efficiencies in Concepts 3 and 4 are mostly higher than in Concepts 1 and 2 (Table 3), the higher pollutant concentrations in the effluent in Concepts 3 and 4 originate from the higher concentrations in the source-separated streams. Consequently, according to the current discharge standards that are based on pollutant concentrations rather than pollutant loadings, the discharge of effluent in Concepts 3 and 4 is prohibited. However, as the pollutant loadings in the effluent in Concepts 3 and 4 decrease by up to 90% compared to Concepts 1 and 2, the future discharge standards ought to consider also the total pollutant load discharged from wastewater treatment. With urine separation (Concepts 2 and 4), both nutrient (N and P) concentrations and loadings are decreased.

Table 5. Calculated effluent quality in sanitation concepts and discharge standards.

Parameter	Unit	Concept 1	Concept 2	Concept 3	Concept 4	Discharge standards [45]
COD	mg/L	46	44	155	187	125
BOD$_5$	mg/L	6	24	83	100	25
TSS	mg/L	34	47	393	385	35
TN	mg/L	9	6	350	70	15
TP	mg/L	4	1	27	17	2
COD	g/cap/y	5037	4802	599	551	-
BOD$_5$	g/cap/y	657	2619	321	297	-
TSS	g/cap/y	3723	5129	1520	1148	-
TN	g/cap/y	986	655	1392	221	-
TP	g/cap/y	438	109	104	51	-

The COD/BOD$_5$ ratio of the effluent loading in Concept 1 is higher than in other concepts, originating from the high BOD$_5$ removal efficiencies in the existing wastewater treatment plants in the Netherlands, applied in Concept 1. More data on the actual BOD$_5$ removal efficiencies in the A-trap in Concept 2 and in the OLAND reactor in Concepts 3 and 4 is required to confirm the actual COD/BOD$_5$ ratio of the effluent loading. To deal with the current discharge standards, further treatment of effluent in Concepts 3 and 4 need to be considered. However, according to the COD:N:P ratio of 100:20:1 necessary for biological treatment [28], the effluent is short in organic matter with a ratio of 100:226:18 (Concept 3) and 100:38:9 (Concept 4) and requires an alternative treatment method or a source of organic matter.

3.7. Land Area Requirement

The total volume of the treatment systems in Concept 1 is 0.32 m^3/cap and in Concept 2 is 0.53 m^3/cap, of which 0.38 m^3/cap originates from the urine storage tank. The total volume of the treatment systems for black water and kitchen refuse (Concept 3) is 0.15–0.22 m^3/cap and for feces and kitchen refuse (Concept 4) is 0.13–0.17 m^3/cap, the lowest value being without grey water sludge co-digestion and the highest with grey water sludge co-digestion using the MBR for bio-flocculation of grey water. Grey water treatment in the SBR-TF requires a total volume of 0.16 m³/cap, the A-trap-TF requires 0.29 m³/cap and the MBR-TF requires 0.14 m³/cap. The total volume of the treatment systems in Concept 3 is 0.31–0.47 m³/cap and in Concept 4 is 0.67–0.81 m³/cap, reaching the highest volumes with the A-trap and the lowest with the SBR. Urine separation in both centralized and source-separation-based sanitation concepts increases the land area requirement, due to the large volume of the urine storage tank. In addition, the land use of the incineration process (Concept 1 and 2) will further increase the land area requirement. The lowest land area requirement is achieved with source-separation of black water and kitchen refuse, and by using the SBR for grey water treatment.

3.8. Sensitivity Analysis

The SRT applied in the high loaded biological reactors, such as the A-trap, can have significant influence on the pollutant removal efficiencies and resulting effluent quality. For example, the removal efficiencies of the A-trap used for sewage treatment in Concept 2 are significantly higher than of the A-trap used for grey water treatment in Concepts 3 and 4 (Table 3). The A-trap used for sewage treatment is according to the study of Wilsenach and van Loosdrecht [22] in which an SRT of 0.8 d was assumed to attain the highest effluent quality, while the SRT of the A-trap used for grey water treatment is according to the actual SRT of 0.6 d applied at the demonstration site of [34], resulting in lower removal efficiencies similar to the ones reported by Böhnke [14]. Consequently, if the SRT of the A-trap for grey water treatment is increased to 0.8 d, the pollutant removal efficiencies could be increased, resulting in higher effluent quality. Furthermore, effluent from the A-trap with higher quality could be reused according to the urban reuse standards, resulting in a significant indirect energy gain from water reuse, and turning the use of the A-trap and subsequent grey water sludge co-digestion into an energetically more favorable option than the use of the SBR. However, due to limited experimental data and the different composition of grey water and sewage, more research is required to confirm the relation between the SRT of the A-trap and the pollutant removal efficiencies.

A significant part of the total energy consumption in the sanitation concepts originates from the energy used for heating the digester and the UASB reactor. Location-specific data on the environmental temperature and the tap water temperature have a major effect on the energy demand of heating, as the tap water temperature defines the amount of energy used for heating up the influent, and the environmental temperature defines the amount of energy used to compensate heat loss through reactor walls. For example, if the tap water and environmental temperature is increased to 15 °C (as an average annual temperature in the south of Europe), the primary energy consumption for heating decreases by 13%–20% in all sanitation concepts. In contrast, if the tap water and environmental temperature is decreased to 6 °C (as an average annual temperature in the north of Europe), the primary energy consumption for heating increases by 15%–21% in all sanitation concepts. The location and the according temperatures may therefore affect the feasibility of grey water sludge co-digestion in the UASB reactor, especially when grey water is concentrated in the MBR with high sludge production.

The transport distance of urine and excess sludge is another location-specific parameter significantly influencing the energy balance of the sanitation concepts. Accessibility and the demand for fertilizers on agricultural land in the vicinity determines the transport distance of urine and excess sludge. In the case of centralized sanitation, the critical distance to agricultural land at which urine transport (Concept 2) becomes unfavorable compared to Concept 1 is 410 km, including the indirect energy gain from water saving and nutrient recovery. This distance covers transport of urine from the Netherlands to France and is higher than any actual distance to accessible agricultural land. However, to avoid high energy consumption of transporting, collected urine should be concentrated at long distances. When considering the use of a vapor compression distillation process with an average primary energy consumption of 337 MJ/m^3 [46], the critical distance at which evaporation of urine becomes more favorable than transporting of urine is 90 km. In the case of source-separation-based sanitation, the critical distance to agricultural land at which urine and excess sludge transport (from the UASB reactor and the SBR) becomes unfavorable compared to Concept 1 is 140 km in Concept 3 and 150 km in Concept 4, including the indirect energy gain from water saving and reuse, and nutrient recovery. Furthermore, by using the A-trap for bio-flocculation of grey water and subsequent grey water sludge co-digestion in the UASB reactor, the critical distance is increased to 300 km in Concept 3 and to 180 km in Concept 4, covering the transport within the Netherlands. Although the transport of urine and excess sludge over long distances is never the optimal solution for nutrient recovery, the long critical distances presented above realizes the possibilities of implementing nutrient recovery technologies in locations surrounded by agricultural lands with a surplus of nutrients.

According to the study of Thibodeau *et al.* [19], one of the most critical factors influencing the economic viability of source-separation of black and grey water is the water consumption for vacuum toilet. Reduction in the vacuum toilet flow has a major effect, not only on the heating energy used for the UASB reactor, but also on the energy consumption for the vacuum collection and transport of wastewater. For example, if the water consumption for the vacuum toilet used for black water is decreased to 1.5 L/cap/d (0.25 L per flush) and the energy consumption for the vacuum collection is assumed to decrease by 75% ($\frac{1.5}{6}$ L), the energy consumption in Concept 3 can be decreased by 35%–55%, attaining the lowest primary energy consumption (156 MJ/cap/year using the SBR) of all the sanitation concepts, including the indirect energy gain from water saving and reuse, and nutrient recovery.

3.9. Outlook

This study provides insight into the influence of urine separation and different grey water treatment configurations [with (A-trap/MBR) and without (SBR) grey water sludge co-digestion] on the energy and material balances of centralized and source-separation-based sanitation concepts. The energy and material balances are based on collection, transport and treatment of wastewater, leaving out the energy and materials used in the construction and maintenance of the required infrastructure. However, according to Tidåker *et al.* [47], the energy use for the source-separation infrastructure is significant, and further research is therefore needed to complete the total lifecycle of the sanitation concepts.

This study emphasizes the direct reuse of source-separated urine as a multicomponent fertilizer in agriculture. Besides the downside of urine transport, direct reuse also involves concerns about the contamination of soil and plants by pharmaceutical residues present in urine [48]. Further research on technologies for indirect resource recovery from urine would help to address both of these issues.

Nevertheless, micro-pollutants are widely measured also from wastewater effluents and receiving water bodies, posing an actual contamination risk on the surrounding agriculture and drinking water production [49]. Clearly, micro-pollutants are of concern, not only in the reuse of source-separated waste streams, but in the whole urban water cycle.

To guarantee the optimal energy recovery from domestic wastewater streams, the influence of grey water sludge co-digestion on the UASB reactor performance, in particular, the effect of surfactants on the digestion process, needs to be further investigated. In addition, the effect of grey water sludge co-digestion on the excess sludge quality in terms of heavy metals and micro-pollutants should be determined.

Beside struvite recovery, further research should focus on alternative phosphorus recovery technologies to minimize the chemical use and to produce other phosphorus products, such as calcium phosphate, more suited for the needs of current fertilizer industries. Furthermore, to promote the full closing of carbon and nutrient cycles, a better understanding on the origin of heavy metals in the excess sludge of the UASB reactor is required. By targeted and functional standards for the sludge reuse in agriculture, resources from the source-separated waste streams can be recovered in such a way that the soil quality is improved.

4. Conclusions

The highest primary energy consumption of 914 MJ/cap/year is attained within the centralized sanitation concept. By coupling the centralized concept with source-separation of urine, the energy consumption is decreased to 687 MJ/cap/year and, further, to 501 MJ/cap/year with an indirect energy gain from water saving and nutrient recovery.

Source-separation of black water, kitchen refuse and grey water results in a primary energy consumption of 767 MJ/cap/year, and in a consumption of 522 MJ/cap/year with indirect energy gain from water saving and reuse, and nutrient recovery. Urine separation within the source-separation-based sanitation concept decreases the energy consumption to 567 MJ/cap/year with a gravity urine diverting toilet and to 555 MJ/cap/year with a vacuum urine diverting toilet. With the indirect energy gain from water saving and reuse, and nutrient recovery, the energy consumptions are further decreased, reaching the lowest energy consumptions of 208 MJ/cap/year (gravity) and 190 MJ/cap/year (vacuum) of all the sanitation concepts.

Source-separation of urine not only improves the energy balance and nutrient recovery, but also increases the effluent quality in terms of nutrient concentrations and the overall pollutant loading in both centralized and source-separation-based sanitation concepts. However, larger land area and higher chemical use in the centralized concept is required.

Grey water bio-flocculation in the A-trap and subsequent grey water sludge co-digestion in the UASB reactor decreases the primary energy consumption by 19% in the source-separation of black water and 22% (gravity) and 21% (vacuum) in the source-separation of urine and feces, compared to grey water treatment in the SBR without grey water sludge co-digestion. However, as grey water effluent from the A-trap does not comply with the water reuse standards, in contrast to effluent from the SBR, the use of the SBR for grey water treatment becomes energetically more favorable than the A-trap when indirect energy gain from water reuse is taken into account. Although grey water effluent from the MBR is applicable for water reuse, the high sludge production and the resulting high energy consumption makes the use of the MBR energetically unfavorable.

Acknowledgments

This work was performed in the TTIW cooperation framework of Wetsus, center of excellence for Sustainable Water Technology. Wetsus is funded by the Dutch Ministry of Economic Affairs, the European Union Regional Development Fund, the Province of Fryslân, the City of Leeuwarden and the EZ/Kompas program of the 'Samenwerkingsverband Noord-Nederland. The authors would like to thank the participants of the research theme "Separation at Source" for the fruitful discussions and their financial support.

A. Appendix

A.1. Calculations for Energy Balance

$E_{collection}$ was the energy requirement for the gravity sewers with lifting stations (20 kWh/cap/y) [50] in Concepts 1 and 2, for the vacuum collection and transport of black water and kitchen refuse (25 kWh/cap/y) [4] in Concept 3 and for the vacuum collection and transport of feces and kitchen refuse (8 kWh/cap/y) in Concept 4 (assumed to be $\frac{1}{3}$ of the energy requirement for the black water vacuum collection according to the water consumption ratio of $\frac{2}{6}$ L). Urine separation in Concept 2 was assumed not to have a significant effect on the total wastewater flow and, thus, on the energy requirement for the collection. Due to short wastewater transport distances in semi-centralized sanitation, the energy requirement for the gravity urine diverting toilet was assumed to be insignificant. The collection also included the energy consumption for the kitchen grinder (5 kWh/cap/y) [4] in all of the sanitation concepts.

In Concepts 1 and 2, $E_{treatment}$ consisted of the following energy parameters. $E_{aeration}$ was the aeration energy required to oxidize organic matter and nitrogen in the AS process, A-trap and post-nitrification step and was calculated based on an energy requirement of 2.2 MJ/kgCOD$_{converted}$ and 14 MJ/kgN$_{converted}$ [7]. The aeration energy was calculated based on the fraction of oxidized COD of the total COD removed (43% in Concept 1 and 22% in Concept 2) and the fraction of nitrified N of the total N removed (94% in Concept 1 and 76% in Concept 2) [22]. E_{mixing} was the energy requirement for mixing of the biological reactors and the anaerobic digester, and $E_{pumping}$ was the energy requirement for pumping of the internal flows, return activated sludge and excess sludge to the anaerobic digester [22]. In Concept 1, additional mixing energy of 5 MJ/kg P$_{removed}$ originated from the biological phosphorus removal [7]. $E_{heating(digester)}$ was the energy required to heat up the influent (excess sludge) to the operational temperature of the digester and to compensate heat loss through the digester walls. The primary energy required to heat up the influent was calculated according to Equation (A1):

$$\Delta Q = m * C * \Delta T \tag{A1}$$

where ΔQ is the required energy (J), m is the mass of liquid (g), C is the specific heat capacity of water (4.2 J/g °C) and ΔT is the temperature difference between the influent temperature and the operational temperature of the reactor. The influent temperature of the digester (Concept 1 and 2) was considered to be the tap water temperature (12 °C). The primary energy required to compensate heat loss was calculated according to Fourier's law presented in Equation (A2):

$$E_{heat} = \Phi = -\lambda * A * \frac{dT}{dx} \tag{A2}$$

where Φ is the heat transfer (W), λ is the thermal conductivity of the isolation material (W/m*k), A is the heat transfer area, dT is the temperature difference across the isolation material (K) and dx is the thickness of the isolation material (m). Mineral wool with thermal conductivity of 0.04 W/m*k and thickness of 0.05 m was considered to be used as isolation material [5]. The area of heat transfer was considered to be the surface area of the reactor (calculated from the volume and dimensions of the reactor presented under the sub-chapter *Calculations for reactor dimensions and land area requirement*), and the temperature difference was considered to be the difference between the environmental temperature (10 °C) and the operational temperature of the reactor (35 °C). $E_{dewatering}$ and $E_{incineration}$ were the primary energy requirements for dewatering of the digested sludge and for incinerating the dewatered sludge according to the study of Wilsenach and van Loosdrecht [22], from which they were recalculated to primary energy using an efficiency of 0.31. The heat production in the incineration of sludge was taken into account in the energy requirement.

$E_{sludge\ transport}$ was the energy requirement for transporting of dewatered sludge to the incineration plant and was calculated based on a primary energy requirement of 4.8 MJ/t/km (including empty return trip) [22]. $E_{urine\ transport}$ was the energy requirement for transporting of urine from the on-site collection to agricultural land and was calculated based on the energy requirement of transporting described above.

$E_{methane}$ was the energy produced as methane in the digestion of excess sludge and was calculated by taking into account the different excess sludge compositions in Concepts 1 and 2, originating from the different SRTs (12 d and 0.8 d, respectively). As presented in the study of Wilsenach and van Loosdrecht [22], excess sludge from the A-trap was considered to consist of 25% adsorbed substrate and 75% biomass. The methanization level of the adsorbed substrate was assumed to be 73% [51]. No adsorbed substrate was considered in Concept 1, due to the high SRT. The fraction of biodegradable biomass in Concept 1 was assumed to be 45% and in Concept 2 65%, and the methanization level of this fraction was considered to be 90% [22]. The volume of the produced methane was calculated using a theoretical methane production of 0.35 L/gCOD, and the primary energy production from methane was calculated using the volume of methane and the calorific value of methane (35.8 MJ/m^3) [28].

The sludge production in the AS process (Concept 1) and the A-trap (Concept 2) was calculated according to Tchobanoglous *et al.* [28] [Equation (A3)]:

$$P = Y * Q * (S_0 - S) \tag{A3}$$

where P is the sludge production (kgVSS/d), Y is the sludge yield (kgVSS/kg BOD$_{removed}$), Q is the influent flow (m^3/d), S_0 is the influent BOD concentration (mg/L) and S is the effluent BOD concentration (mg/L). A sludge yield of 0.58 kgVSS/kg BOD$_{removed}$ was used for the AS process (SRT 12 d) and 0.85 kgVSS/kg BOD$_{removed}$ for the A-trap (SRT 0.8 d) at 12 °C. The sludge production as total solids was calculated using a VSS/TSS ratio of 0.85 [28]. The total wet sludge production was calculated using a dry solid content of 2.5%, and the total dry sludge production (after dewatering) was calculated using a dry solid content of 20% [22]. In Concept 1, additional sludge production of 3.3 kgTSS/kg $P_{removed}$ was assumed to originate from the biological phosphorus removal [7].

The composition of the sludge rejection water (COD, TN and TP) was defined as the difference between the digester influent (excess sludge from the AS process and A-trap) and the COD converted into methane and nitrogen and phosphorus incorporated into the anaerobic biomass. The amount of biomass produced in the digester was calculated using a biomass yield of 0.08 gVSS/g $COD_{converted}$, and the amount of nitrogen and phosphorus incorporated into the biomass was calculated using fractions of 0.12 gN/g VSS and 0.03 gP/g VSS, respectively [28]. All of the nitrogen and phosphorus in the sludge rejection water was considered to be in the inorganic form of NH_4^+ and PO_4^{3-}.

In Concepts 3 and 4, $E_{treatment}$ consisted of the following energy parameters. $E_{heating(UASB)}$ was the energy required to heat up the influent to the operational temperature of the reactor and to compensate heat loss through the reactor walls, calculated as described above with the digester in Concepts 1 and 2. The influent temperature of the UASB reactor was calculated from the mass proportions of the according wastewater sub-streams (Table 2). In the case of grey water sludge co-digestion in the UASB reactor, the influent temperature was adjusted with the temperature of grey water sludge that was assumed to be the environmental temperature (10 °C). No heating energy for other treatment steps were taken into account. E_{OLAND} was the energy requirement for the OLAND reactor and was derived from the rotating power requirement of the rotating biological contactor according to Fujie *et al.* [52] [Equation (A4)].

$$P(w) = \lambda_1 * N^2 * D^2 * A \tag{A4}$$

where A is the surface area of the discs (m²), λ_1 is the frictional constant ($8.6*10^{-6}$ kWmin²/min⁴), N is the rotational speed of a disc (min⁻¹) and D is the disc diameter (m). The surface area of the discs was calculated from the total nitrogen load and the biofilm load (6300 mgN/m²/d [53]). The disc rotational speed of 3 min⁻¹ [6] and the disc diameter of 1 m [52] were selected. $E_{Struvite}$ was the energy requirement for the struvite precipitation and was calculated based on an electricity consumption of 3.8 kWh/kgN$_{influent}$ [53]. E_{TF} was the energy requirement for the trickling filter as a post-treatment step in both black water and grey water treatment lines and was calculated based on an average electricity consumption of 3 kW/1000 m³$_{influent}$ [28]. E_{MBR} was the energy requirement for the MBR and was calculated based on an average electricity consumption of 0.3 kWh/m³$_{greywater}$ [54]. The electricity consumption for the OLAND reactor, struvite precipitator, TF and the MBR was converted to primary energy using an efficiency of 0.31 [36]. E_{SBR} and E_{A-trap} were the energy requirements for grey water treatment in the SBR and the A-trap, respectively, consisting of energy consumption for pumping and aeration. Energy consumption for pumping was calculated with Equation (A5) according to Karassik *et al.* [55]:

$$E_{pump}(kW) = \frac{Q(m^3/d) * H(m) * specific\ gravity\ of\ fluid}{367.7 * \eta} \tag{A5}$$

where Q is the flow rate, H is the pump head and η is the pump efficiency. For the SBR, the pump head was considered to be the height of wastewater in the reactor. For the A-trap, the pump head was considered to be the height of the buffer tank for influent pump and the height of the aerated grit chamber and settling tank for the two intermediate pumps (calculations for pump head are presented in the sub-chapter *Calculations for reactor dimensions and land area requirement*). The specific gravity of fluid was considered to be one and η was set to 0.68, according to the study of Wilsenach and van Loosdrecht [22]. The total energy consumption for pumping in the SBR was calculated from the energy consumption for two pumps: influent and effluent pump, feeding and discharge time of 15 min each and a total cycle time of 360 min [9]. The total energy consumption for pumping in the A-trap was calculated by assuming the pumping to be continuous. The energy requirement for pumping of the UASB influent was calculated to be insignificant and was not included in the energy balance. The energy consumption for aeration in the SBR and the A-trap was calculated according to the energy requirement of 2.2 MJ/kgCOD$_{converted}$ [7]. The amount of oxidized COD in the SBR was calculated by defining the total amount of biodegradable COD removed in the reactor using a COD$_{biodegradable}$/BOD$_5$ ratio of 1.6 g/g [28] and excluding the amount of COD removed in the sludge using a sludge yield of 0.12 kgVSS/kgCOD [9] and a COD/VSS ratio of 1.4. The amount of oxidized COD in the A-trap was assumed to be 11% of the incoming COD [34]. Nitrogen removal in the SBR and A-trap was assumed to take place only through the excess sludge removal.

$E_{sludge\ transport}$ and $E_{urine\ transport}$ were the energy requirements for transporting of excess sludge from the UASB reactor and the SBR and urine, respectively, from the on-site collection to agricultural land, and was calculated based on the primary energy requirement of 4.8 MJ/t/km (including empty return trip) [22].

$E_{methane}$ was the energy produced as methane in the UASB reactor. The volume of produced methane was calculated from the COD load of the reactor, the methanization level of the influent and the theoretical methane production of 0.35 L/gCOD. The methanization level of the influent was calculated as a mass proportion of the methanization levels of the sub-streams (70% for black water with kitchen refuse, 78% for feces with kitchen refuse [5] and 88% for grey water sludge [12]). The primary energy production from methane was calculated using the volume of methane and the calorific value of methane (35.8 MJ/m^3) [28].

The sludge production in the UASB reactor was calculated according to Zeeman and Lettinga [56] [Equation (A6)]:

$$X_p = O * SS * R * (1 - H) \tag{A6}$$

where X_p is the sludge production (kgCOD/m^3/d), O is the organic loading rate (2.98 kgCOD/m^3/d [33]), SS is the fraction of suspended solids in the influent (COD$_{ss}$/COD$_{total}$) (0.76 with a mixture of black water and kitchen refuse, and 0.88 with a mixture of feces and kitchen refuse [5]), R is the fraction of COD$_{ss\ removed}$ (0.96 [33]) and H is the level of hydrolysis of the removed solids (0.7 [5]). The total wet sludge production was calculated using the volume of the UASB reactor (calculations for the reactor volume are presented in the sub-chapter *Calculations for reactor dimensions and land area requirement*) and the sludge concentration (34 gCOD/L [27]). The sludge production in the SBR was calculated using a sludge yield of 0.12 kgVSS/kgCOD$_{removed}$ and a sludge concentration of 5.5 gVSS/L [9]. The sludge production in the A-trap was calculated using a sludge yield of 0.73 kgVSS/kgCOD$_{removed}$ and a sludge concentration of 6.3 gVSS/L [34]. The sludge production in the MBR was calculated from the flow mass balance of the system using a SRT of 1 d and HRT of 1.9 h [12].

A.2. Calculations for Chemical Use

Consumption of NaOH in struvite precipitation was calculated using Equation (A7):

$$m_{NaOH} = M_{NaOH} * 10^{-14}(10^{pH_b} - 10^{pH_a})$$ (A7)

where m_{NaOH} is the mass of NaOH (g/L), M is the molecular mass (g/mol), pH_a is the influent pH of 7.7 [33] and pH_b is the operational pH of 9 [35]. Consumption of 33% NaOH was further determined from the mass of NaOH.

A.3. Calculations for Reactor Dimensions and Land Area Requirement

The volume of the biological reactors and secondary settling tanks were according to Wilsenach and van Loosdrecht [22], and the volume of the buffer tanks, urine storage tank and reactors (digester/UASB, struvite, MBR and A-trap) were determined using the influent flow rate and the storage time or the HRT. The volume of the A-trap consisted of three parts: aerated grit chamber, A-trap reactor and settling tank. The storage time was 1 d for the UASB buffer tank (assumed), 0.3 d for the SBR, A-trap and MBR buffer tanks (assumed) and six months for the urine collection tank [7]. The HRT was 15 d for the digester [22], 0.08 d for the struvite reactor [35], 1.9 h for the MBR [12], 4 min and 54 min for the aerated grit chamber and settling tank, respectively [57], and 1.9 h for the A-trap reactor [34]. The HRT of the UASB reactor was calculated according to Zeeman and Lettinga [56] [Equation (A8)]:

$$HRT = C * \frac{SS}{X} * R * (1 - H) * SRT$$ (A8)

where C is the influent, COD_{total} concentration (gCOD/L), X is the sludge concentration in the reactor (34 gCOD/L [27]), SS is the fraction of suspended solids in the influent (COD_{ss}/COD_{total}) (0.76 with a mixture of black water and kitchen refuse and 0.88 with a mixture of feces and kitchen refuse [5]), R is the fraction of $COD_{ss\ removed}$ (0.96 [33]), H is the level of hydrolysis of the removed solids (0.7 [5]), and SRT is the sludge retention time (d) calculated from the sludge production (kgCOD/m^3/d) and the sludge concentration in the reactor.

The volume of the biogas storage tank was calculated using the volume of produced methane, the fraction of methane in biogas (65% [28]) and storage time of 1 d [5]. The volume of the SBR was calculated using the volume of wastewater per cycle (360 min) and a volume$_{wastewater}$/volume$_{total}$ ratio of 0.3 m^3/m^3 [28]. The volume of a single-stage TF was determined according to Tchobanoglous *et al.* [28] [Equation (A9)]:

$$V = \frac{W}{\left(\frac{100}{e*(1+0.4432)}\right)^2}$$ (A9)

where W is the BOD$_5$ loading and e is the BOD$_5$ removal efficiency. The depth of the filter was set to 2.1 m as the average depth in standard rate filters.

The volume of the OLAND reactor was determined from the length, width and height of the reactor. The length of the reactor was determined by the length of the shaft and the width and height by the disc diameter. To calculate the length of the shaft, the total number of discs was defined from the total surface area of discs and the disc diameter (determined previously with the energy requirement of OLAND). The length of the shaft was calculated using a disc thickness of 0.5 cm and a disc interspace of 1 cm [6]. The length, width and height of the reactor was then determined using the length of the shaft and the disc diameter, respectively, with 15% of the disc diameter as extra space.

Height of the buffer tanks, digester, UASB reactor and SBR was calculated using Equation (A10), which was derived from the equation for cylinder volume using f as a height/diameter ratio.

$$H = \sqrt[3]{\frac{4 * V_{cylinder} * f^2}{\pi}}, \; f = \frac{H}{d} \tag{A10}$$

where $V_{cylinder}$ is the volume of the reactor and f is the height/diameter ratio that was assumed to be three with the exception of the SBR with a ratio of 1. The diameter was calculated using an assumed maximum height of 5 m as a boundary condition.

The height of the aerated grit chamber and settling tank of the A-trap was calculated using Equation (A11), which was derived from the sum of cube volume and pyramid volume using f as the height$_{pyramid}$/height$_{vessel}$ ratio:

$$H = \frac{V_{vessel}}{A * (1 - \frac{2}{3} * f)}, \; f = \frac{H_{pyramid}}{H_{vessel}} \tag{A11}$$

V_{vessel} is the volume of the aerated grit chamber and settling tank, A is the surface area and f is the height$_{pyramid}$/height$_{vessel}$ ratio of 0.1 for the aerated grit chamber and 0.5 for the settling tank. The surface area of the aerated grit chamber was calculated using a maximum surface loading of 30 m^3/(m^{2}h), and the surface area of the settling tank was calculated using a maximum surface loading of 1.5 m^3/(m^{2}h) [57]. The height of the A-trap reactor was considered to be the difference between the height of the vessel and the height of the pyramid.

A.4. Energy Balance

Table A1 presents the sludge production, urine collection, and the energy consumption and production (methane) in Concepts 1 and 2.

Table A2 presents the UASB reactor influent characteristics, sludge production in the UASB reactor, SBR, A-trap and MBR, and the urine collection in Concepts 3 and 4.

Table A1. Sludge production, urine collection, and energy consumption and production (methane) in Concepts 1 and 2 (primary energy presented as bolded figures).

Parameter	Unit	Concept 1	Concept 2
Urine collection	kg/cap/y	-	743
Sludge production	kgWS/cap/y	1048	1201
	kgDS/cap/y	131	150
$E_{collection}$	kWh/cap/y	25	25
	MJ/cap/y	**288**	**288**
$E_{aeration}$	MJ/cap/y	135	68
	MJ/cap/y	**432**	**218**
E_{mixing}	MJ/cap/y	37	17
	MJ/cap/y	**118**	**54**
$E_{pumping}$	MJ/cap/y	20	15
	MJ/cap/y	**64**	**48**
$E_{heating(digester)}$	**MJ/cap/y**	**104**	**114**
$E_{dewatering}$	**MJ/cap/y**	**5**	**5**
$E_{sludge\ transport}$	**MJ/cap/y**	**6**	**7**
$E_{incineration}$	**MJ/cap/y**	**54**	**52**
$E_{urine\ transport}$	**MJ/cap/y**	**-**	**178**
$E_{methane}$	**MJ/cap/y**	**157**	**277**
E_{total}	**MJ/cap/y**	**914**	**687**

Notes: WS = Wet Sludge; DS = Dry Sludge.

Table A2. UASB influent characteristics, sludge production and urine collection in Concepts 3 and 4 with different grey water treatment configurations (without co-digestion using the SBR or with co-digestion using the A-trap/MBR) (UASB = up-flow anaerobic sludge blanket reactor, OLAND = oxygen limited anaerobic nitrification denitrification, struvite precipitator, TF = trickling filter, SBR = sequencing batch reactor, A-trap = A-stage of AB-process and MBR = membrane bioreactor).

Parameter	Unit	Concept 3			Concept 4					
					Gravity toilet			Vacuum toilet		
		SBR	A-trap	MBR	SBR	A-trap	MBR	SBR	A-trap	MBR
UASB influent										
Volume	m³/cap/y	3	4	5	2	3	4	1	2	3
Temperature	°C	16	15	13	12	11	11	11	11	10
Methanization level	%	70	79	80	78	79	80	78	79	80
Sludge production										
UASB reactor	kg/cap/y	277	321	365	299	343	394	299	343	394
SBR/A-trap/MBR	kg/cap/y	373	682	2128	373	682	2128	373	682	2128
Urine collection	kg/cap/y	-	-	-	743	743	743	743	743	743

Table A3 presents the energy consumption and production (methane) in Concepts 3 and 4 with different grey water treatment configurations (without co-digestion using the SBR or with co-digestion using the A-trap/MBR).

Table A3. Energy consumption and production (methane) in Concepts 3 and 4 (primary energy presented as bolded figures) (UASB = up-flow anaerobic sludge blanket reactor, OLAND = oxygen limited anaerobic nitrification denitrification, struvite precipitator, TF = trickling filter, SBR = sequencing batch reactor (SBR), A-trap = A-stage of AB-process and MBR = membrane bioreactor).

| Parameter | Unit | Concept 3 | | | Concept 4 | | | | | |
| | | | | | Gravity toilet | | | Vacuum toilet | | |
		SBR	A-trap	MBR	SBR	A-trap	MBR	SBR	A-trap	MBR
$E_{collection}$	kWh/cap/y	30	30	30	5	5	5	13	13	13
	MJ/cap/y	**346**	**346**	**346**	**58**	**58**	**58**	**150**	**150**	**150**
$E_{heating(UASB)}$	**MJ/cap/y**	**341**	**422**	**584**	**305**	**385**	**547**	**199**	**280**	**441**
E_{OLAND}	kWh/cap/y	1.3	1.6	2.2	0.2	0.3	0.4	0.3	0.4	0.8
	MJ/cap/y	**15**	**18**	**25**	**2**	**3**	**5**	**3**	**5**	**9**
$E_{Struvite}$	kWh/cap/y	4.4	5.4	7.5	0.8	1.0	1.5	0.9	1.5	2.7
	MJ/cap/y	**51**	**62**	**86**	**9**	**12**	**17**	**10**	**17**	**31**
$E_{TF\ (BW)}$	kWh/cap/y	0.2	0.3	0.4	0.2	0.2	0.3	0.1	0.1	0.2
	MJ/cap/y	**2**	**3**	**5**	**2**	**2**	**3**	**1**	**1**	**2**
$E_{sludge\ transport}$	**MJ/cap/y**	**156**	**77**	**88**	**161**	**83**	**95**	**161**	**83**	**95**
E_{SBR}	MJ/cap/y	33	-	-	33	-	-	33	-	-
	MJ/cap/y	**106**	**-**	**-**	**106**	**-**	**-**	**106**	**-**	**-**
E_{A-trap}	MJ/cap/y	-	7.2	-	-	7.2	-	-	7.2	-
	MJ/cap/y	**-**	**23**	**-**	**-**	**23**	**-**	**-**	**23**	**-**
E_{MBR}	kWh/cap/y	-	-	8.7	-	-	8.7	-	-	8.7
	MJ/cap/y	**-**	**-**	**100**	**-**	**-**	**100**	**-**	**-**	**100**
$E_{TF\ (GW)}$	kWh/cap/y	2.1	2.1	2.1	2.1	2.1	2.1	2.1	2.1	2.1
	MJ/cap/y	**24**	**24**	**24**	**24**	**24**	**24**	**24**	**24**	**24**
$E_{urine\ transport}$	**MJ/cap/y**	**-**	**-**	**-**	**178**	**178**	**178**	**178**	**178**	**178**
$E_{methane}$	**MJ/cap/y**	**274**	**352**	**401**	**278**	**324**	**373**	**278**	**324**	**373**
E_{total}	**MJ/cap/y**	**767**	**624**	**857**	**567**	**444**	**654**	**555**	**437**	**658**

References

1. Larsen, T.A.; Alder, A.C.; Eggen, R.I.; Maurer, M.; Lienert, J. Source separation: Will we see a paradigm shift in wastewater handling? *Environ. Sci. Technol.* **2009**, *43*, 6121–6125.

2. Otterpohl, R.; Braun, U.; Oldenburg, M. Innovative technologies for decentralized water-, wastewater and biowaste management in urban and peri-urban areas. *Water Sci. Technol.* **2003**, *48*, 23–32.

3. Wilsenach, J.A.; Maurer, M.; Larsen, T.A.; van Loosdrecht, M.C.M. From waste treatment to integrated resource management. *Water Sci. Technol.* **2003**, *48*, 1–9.

4. Zeeman, G.; Kujawa, K.; de Mes, T.; Hernandez, L.; de Graaff, M.; Abu-Ghunmi, L.; Mels, A.; Meulman, B.; Temmink, H.; Buisman, C.; *et al.* Anaerobic treatment as a core technology for energy, nutrients and water recovery from source-separated domestic waste(water). *Water Sci. Technol.* **2008**, *57*, 1207–1212.

5. Kujawa, K. *Anaerobic Treatment of Concentrated Wastewater in DESAR Concepts*; STOWA: Utrecht, the Netherlands, 2005.

6. Vlaeminck, S.E.; Terada, A.; Smets, B.F.; Linden, D.V.D.; Boon, N.; Verstraete, W.; Carballa, M. Nitrogen removal from digested black water by one-stage partial nitritation and anammox. *Environ. Sci. Technol.* **2009**, *43*, 5035–5041.

7. Maurer, M.; Schwegler, P.; Larsen, T. Nutrients in urine: Energetic aspects of removal and recovery. *Nutr. Remov. Recover.* **2003**, *48*, 37–46.

8. Larsen, T.A.; Lienert, J. *NoMix–A New Approach to Urban Water Management*; Novaquatis Final Report; Eawag: Dübendorf, Switzerland, 2007.

9. Hernández Leal, L.; Temmink, H.; Zeeman, G.; Buisman, C.J. Comparison of three systems for biological greywater treatment. *Water* **2010**, *2*, 155–169.

10. Avery, L.M.; Frazer-Williams, R.A.D.; Winward, G.; Shirley-Smith, C.; Liu, S.; Memon, F.A.; Jefferson, B. Constructed wetlands for grey water treatment. *Ecohydrol. Hydrobiol.* **2007**, *7*, 191–200.

11. Brix, H.; Arias, C.A. The use of vertical flow constructed wetlands for on-site treatment of domestic wastewater: New Danish guidelines. *Ecol. Eng.* **2005**, *25*, 491–500.

12. Hernández Leal, L.; Temmink, H.; Zeeman, G.; Buisman, C.J.N. Bioflocculation of grey water for improved energy recovery within decentralized sanitation concepts. *Bioresour. Technol.* **2010**, *101*, 9065–9070.

13. García-Morales, J.L.; Nebot, E.; Romero, L.I.; Sales, D. Comparison between acidogenic and methanogenic inhibition caused by linear alkylbenzene-sulfonate (LAS). *Chem. Biochem. Eng. Q.* **2001**, *15*, 13–20.

14. Böhnke, B. *Energieminimierung durch das Adsorptions-Belebungsverfahren*; Gewässerschutz Wasser Abwasser No. 49; Instituts für Siedlungswasserwirtschaft der RWTH: Aachen, Germany, 1981.

15. Otterpohl, R.; Grottker, M.; Lange, J. Sustainable water and waste management in urban areas. *Water Sci. Technol.* **1997**, *35*, 121–133.

16. Hellström, D. Exergy analysis: A comparison of source separation systems and conventional treatment systems. *Water Environ. Res.* **1999**, *71*, 1354–1363.

17. Benetto, E.; Nguyen, D.; Lohmann, T.; Schmitt, B.; Schosseler, P. Life cycle assessment of ecological sanitation system for small-scale wastewater treatment. *Sci. Total Environ.* **2009**, *407*, 1506–1516.

18. Makropoulos, C.; Natsis, K.; Liu, S.; Mittas, K.; Butler, D. Decision support for sustainable option selection in integrated urban water management. *Environ. Model. Softw.* **2008**, *23*, 1448–1460.

19. Thibodeau, C.; Monette, F.; Glaus, M.; Laflamme, C.B. Economic viability and critical influencing factors assessment of black water and grey water source-separation sanitation system. *Water Sci. Technol.* **2011**, *64*, 2417–2424.

20. Van Beuzekom, I.; Dekker, S.; Teunissen, M. *Duurzame afvalwaterbehandeling Geerpark Heusden*; Grontmij: De Bilt, the Netherlands, 2010.

21. KNMI (Koninklijk Nederlands Meteorologisch Instituut). *Annual Report 2010: KNMI round the Clock*; KNMI: De Bilt, the Netherlands, 2010.

22. Wilsenach, J.A.; van Loosdrecht, M.C.M. Integration of processes to treat wastewater and source-separated urine. *J. Environ. Eng.* **2006**, *132*, 331–341.

23. Statistics Netherlands. *Urban Wastewater Treatment per Province and River Basin District*; Statistics Netherlands: Den Haag, the Netherlands, 2011.

24. Kujawa-Roeleveld, K.; Zeeman, G. Anaerobic treatment in decentralised and source-separation-based sanitation concepts. *Rev. Environ. Sci. Biotechnol.* **2006**, *5*, 115–139.

25. Larsen, T.A.; Peters, I.; Alder, A.; Eggen, R.; Maurer, M.; Muncke, J. Re-engineering the toilet for sustainable waste water management. *Environ. Sci. Technol.* **2001**, *35*, 192–197.

26. Hernández Leal, L.; Temmink, H.; Zeeman, G.; Buisman, C. Characterization and anaerobic biodegradability of grey water. *Desalination* **2011**, *270*, 111–115.

27. De Graaff, M.S.; Temmink, H.; Zeeman, G.; Buisman, C.J.N. Anaerobic treatment of concentrated black water in a UASB reactor at a short HRT. *Water* **2010**, *2*, 101–119.

28. Tchobanoglous, G.; Burton, F.L.; Stensel, H.D. *Wastewater Engineering: Treatment and Reuse*, 4th ed.; McGraw-Hill Series in Civil and Environmental Engineering; McGraw-Hill: New York, NY, USA, 2004.

29. Vinnerås, B.; Nordin, A.; Niwagaba, C.; Nyberg, K. Inactivation of bacteria and viruses in human urine depending on temperature and dilution rate. *Water Res.* **2008**, *42*, 4067–4074.

30. World Health Organization (WHO). *Guidelines for the Safe Use of Wastewater, Excreta and Greywater, Volume 3: Wastewater and Excreta Use in Aquaculture*; World Health Organization: Geneva, Switzerland, 2006.

31. Frijns, J.; Mulder, M.; Roorda, J. *Op weg Naar een Klimaatneutrale Waterketen*; STOWA: Utrecht, the Netherlands, 2008.

32. RIONED Foundation. *Urban Drainage Statistics 2009–2010*; RIONED Foundation: Ede, the Netherlands, 2009.

33. Kujawa-Roeleveld, K.; Weijma, J.; Nanninga, T. *Nieuwe Sanitatie op Wijkniveau: Resultaten en Ervaringen Demo-site Sneek en Perspectieven voor Opschaling*; LeAF (Lettinga Associates Foundation): Wageningen, the Netherlands, 2012.

34. *Reactor Performance of AB-Process at the DeSaH Demonstration Site*; DeSaH: Sneek, the Netherlands, 2010.

35. De Graaff, M.S.; Temmink, H.; Zeeman, G.; Buisman, C.J.N. Energy and phosphorus recovery from black water. *Water Sci. Technol.* **2011**, *63*, 2759–2765.

36. UCPTE (Union pour la coordination de la production et du transport de l'électricité). *Yearly Report 1993*; UCPTE: Vienna, Austria, 1994.

37. Houillon, G.; Jolliet, O. Life cycle assessment of processes for the treatment of wastewater urban sludge: Energy and global warming analysis. *J. Clean. Prod.* **2005**, *13*, 287–299.

38. Li, F.; Wichmann, K.; Otterpohl, R. Review of the technological approaches for grey water treatment and reuses. *Sci. Total Environ.* **2009**, *407*, 3439–3449.

39. Adamsson, M. Potential use of human urine by greenhouse culturing of microalgae (*Scenedesmus acuminatus*), zooplankton (*Daphnia magna*) and tomatoes (Lycopersicon). *Ecol. Eng.* **2000**, *16*, 243–254.

40. Kuntke, P.; Śmiech, K.; Bruning, H.; Zeeman, G.; Saakes, M.; Sleutels, T.; Hamelers, H.; Buisman, C. Ammonium recovery and energy production from urine by a microbial fuel cell. *Water Res.* **2012**, *46*, 2627–2636.

41. De Graaff, M. Resource Recovery from Black Water. Ph.D. Thesis, Wageningen University, Wageningen, the Netherlands, 2010.

42. Van Dooren, H.J.C.; Hanegraaf, M.C.; Blanken, K. *Emission and Compost Quality of on Farm Composted Dairy Slurry. PraktijkRapport Rundvee 68*; Animal Sciences Group, Wageningen University, Wageningen, the Netherlands, 2005.

43. Remy, D.I.C.; Ruhland, I.A. Ecological Assessment of Alternative Sanitation Concepts with Life Cycle Assessment. Technical University Berlin, Berlin, Germany, 2006; Volume 55.

44. Verstraete, W.; Vlaeminck, S.E. ZeroWasteWater: Short-cycling of wastewater resources for sustainable cities of the future. *Int. J. Sustain. Dev. World Ecol.* **2011**, *18*, 253–264.

45. European Commission. *EU Water Framework Directive 91/271/EEC*; European Commission: Brussels, Belgium, 2013.

46. Maurer, M.; Pronk, W.; Larsen, T. Treatment processes for source-separated urine. *Water Res.* **2006**, *40*, 3151–3166.

47. Tidåker, P.; Mattsson, B.; Jönsson, H. Environmental impact of wheat production using human urine and mineral fertilisers—A scenario study. *J. Clean. Prod.* **2007**, *15*, 52–62.

48. Winker, M.; Clemens, J.; Reich, M.; Gulyas, H.; Otterpohl, R. Ryegrass uptake of carbamazepine and ibuprofen applied by urine fertilization. *Sci. Total Environ.* **2010**, *408*, 1902–1908.

49. Ternes, T.; Joss, A. *Human Pharmaceuticals, Hormones and Fragrances: The Challenge of Micropollutants in Urban Water Management*; International Water Association: London, UK, 2006.

50. Van Buuren, J.C.L. Sanitation Choice Involving Stakeholders. Ph.D. Thesis, University of Wageningen, Wageningen, The Netherlands, 2010.

51. Elmitwalli, T.A.; Soellner, J.; de Keizer, A.; Bruning, H.; Zeeman, G.; Lettinga, G. Biodegradability and change of physical characteristics of particles during anaerobic digestion of domestic sewage. *Water Res.* **2001**, *35*, 1311–1317.

52. Fujie, K.; Bravo, H.E.; Kubota, H. Operational design and power economy of a rotating biological contactor. *Water Res.* **1983**, *17*, 1153–1162.

53. Meulman, B.; Zeeman, G.; Buisman, C.J.N. Treatment of Concentrated Black Water on Pilot Scale: Options and Challenges. In Proceedings of Sanitation Challenge, Wageningen, the Netherlands, 19–21 May 2008.

54. Melin, T.; Jefferson, B.; Bixio, D.; Thoeye, C.; de Wilde, W.; de Koning, J.; van der Graaf, J.; Wintgens, T. Membrane bioreactor technology for wastewater treatment and reuse. *Desalination* **2006**, *187*, 271–282.

55. Karassik, I.J.; Messina, J.P.; Cooper, P.; Heald, C.C. *Pump Handbook*, 3rd ed.; McGraw-Hill: New York, NY, USA, 2001.

56. Zeeman, G.; Lettinga, G. The role of anaerobic digestion of domestic sewage in closing the water and nutrient cycle at community level. *Water Sci. Technol.* **1999**, *39*, 187–194.

57. Dorussen, H.L. *Ontwerp awzi Dokhaven*; DBW/RIZA: Lelystad; Stora: Den Haag, the Netherlands, 1980.

Water **2013**, *5*, 893-916; doi:10.3390/w5030893

OPEN ACCESS

water

ISSN 2073-4441

www.mdpi.com/journal/water

Article

A Preliminary Investigation of Wastewater Treatment Efficiency and Economic Cost of Subsurface Flow Oyster-Shell-Bedded Constructed Wetland Systems

Rita S.W. Yam [1,2]**, Chia-Chuan Hsu** [1]**, Tsang-Jung Chang** [1,2,]*** and Wen-Lian Chang** [1,2]

[1] Department of Bioenvironmental Systems Engineering, National Taiwan University, Taipei 106, Taiwan; E-Mails: ritayam@ntu.edu.tw (R.S.W.Y.); enockhsu@ntu.edu.tw (C.-C.H); wenlian@ntu.edu.tw (W.-L.C.)

[2] Ecological Engineering Research Center, National Taiwan University, Taipei 106, Taiwan

* Author to whom correspondence should be addressed; E-Mail: tjchang@ntu.edu.tw; Tel.: +886-2-23635854; Fax: +886-2-23635854.

Received: 25 April 2013; in revised form: 3 June 2013 / Accepted: 19 June 2013 / Published: 28 June 2013

Abstract: We conducted a preliminary investigation of wastewater treatment efficiency and economic cost of the oyster-shell-bedded constructed wetlands (CWs) compared to the conventional gravel-bedded CW based on field monitoring data of water quality and numerical modeling. Four study subsurface (SSF) CWs were built to receive wastewater from Taipei, Taiwan. Among these sites, two are vertical wetlands, filled with bagged- (VA) and scattered- (VB) oyster shells, and the other two horizontal wetlands were filled with scattered-oyster shells (HA) and gravels (HB). The BOD, NO_3^-, DO and SS treatment efficiency of VA and VB were higher than HA and HB. However, VA was determined as the best option of CW design due to its highest cost-effectiveness in term of BOD removal (only 6.56 US$/kg) as compared to VB, HA and HB (10.88–25.01 US$/kg). The results confirmed that oyster shells were an effective adsorption medium in CWs. Hydraulic design and arrangement of oyster shells could be important in determining their treatment efficiency and cost-effectiveness. A dynamic model was developed to simulate substance transmissions in different treatment processes in the CWS using AQUASIM 2.1 based on the water quality data. Feasible ranges of biomedical parameters involved were determined for characterizing the importance of different biochemical treatment processes in SSF CWs. Future work will involve extending the experimental period to confirm the treatment efficiency of the oyster-shell-bedded CW systems in long-term operation and provide more field data for the simulated model instead of the literature values.

Keywords: subsurface flow (SSF) constructed wetland; ecological adsorbent medium; natural wastewater treatment systems; water quality simulation; AQUASIM

1. Introduction

Constructed wetlands (CWs) are recognized as a low-cost, eco-technology system [1–5], commonly suggested for small towns that cannot afford expensive conventional treatment systems. Recently, more and more studies have reported that CWs, as engineered systems integrating wetland vegetation, soil and their microbial assemblages to facilitate wastewater treatment, could serve as the natural practical alternatives for wastewater treatment through various physical, chemical and biological processes including adsorption, nitrification-denitrification, plant and microbial assimilation [2,3,6]. In general, there are two major types of CWs including subsurface flow (SSF) constructed wetland and free water surface (FWS) constructed wetland. As suggested by USEPA [6,7], SSF CWs have the advantages of occupying less land area and isolating the wastewater from vectors to animals and humans. On the other hand, FWS CWs allow the provision of wildlife habitats for supporting high biodiversity and recreational areas for public uses. In Taiwan, as in many island countries, land area is an important resource as there is high population density living on a limited land area. In Taiwan, 23 million people live in 36,000 km^2 of land area resulting in the second highest population density in the world. As the cost of land is expensive, SSF CWs could be a better approach for the low-cost wastewater treatment in Taiwan and probably other island countries [4].

In Taiwan, shellfish farming activities occupy 129.5 km^2 of coastal ocean and result in approximately 28,200 tons of oyster shells every year, this has caused serious environmental problems of oyster shell disposal and health hazards in Taiwan [8]. The main chemical components of oyster shells include calcium and protein, *i.e.*, aspartic acid and glycine. Previous studies on the physical and chemical properties of oyster shells suggested that oyster shells could be suitable adsorbent medium in CWs. Moreover, as the cost of imported gravels often represented 50% of the building cost of CWs [7,9], replacing expensive gravels with oyster shells as the adsorption medium in CWs could reduce the capital cost of CWs. Therefore, oyster shells can serves as environmental-friendly waste adsorption medium in the biofilter systems of CWs that enables local sustainability of CWs through reducing disposal cost of oyster shells and avoiding the purchase of expensive adsorption materials [10].

Previous studies investigating the wastewater treatment efficiency of oyster-shell-bedded CWs were primarily based on laboratory experiments. Seo *et al.* [11] used oyster shells as the filter medium (internal diameter: 21 mm and height: 365 mm) and examined the phosphorus capacity of those filtering columns. Results showed that oyster shells enabled extending the phosphorus saturation in CWs. Park and Polprasert [12] built an integrated constructed wetland system, which consisted of a polyethylene tank with a volume of 0.187 m^3 and a post-filter unit filling with oyster shells as the adsorption medium for wastewater treatment. Their results suggested that such a system could help to minimize eutrophication. Also, Lin and Jing [9] confirmed the water purification ability of small-scale oyster-shell system (~0.2 m^3) on wastewater and sludge. However, few field studies on the "real" oyster-shell-bedded CWs of practical size have been done to investigate their wastewater treatment efficiency and economic cost. Moreover, numerical modeling of water quality in the oyster-shell-bedded CWs is generally lacking.

To fill this gap, we aimed to conduct a preliminary investigation of wastewater treatment efficiency and economic cost of the oyster-shell-bedded CWs compared to the conventional gravel-bedded system based on field monitoring data of water quality and numerical modeling. Numerical modeling is usually regarded as a valuable tool for scientific investigation. In this study, we aimed to use numerical modeling based on field monitoring data for providing further information which cannot be easily obtained from direct experimental observation to help investigate the reasons accounting for the waste removal quantity of different biochemical processes in these natural wastewater treatment systems. Consequently, if we can enhance these essential biochemical processes by wetland settings, the efficiency of decontamination will be increased. Furthermore, numerical modeling can estimate outcomes before carrying out many complicated, time consuming, and high-cost experiments. This can thus provide decision makers different potential directions for cost-effective design and management. In this study, four unvegetated study SSF CWs including two vertical (VA and VB filled with bagged and scattered oyster shells respectively) and two horizontal subsurface wetlands (HA and HB filled with scattered oyster shells and gravels respectively) were built to receive municipal wastewater in Taipei, Taiwan. The treatment efficiency and the cost-effectiveness of these four types of study wetland were compared. Since this investigation was the first attempt to study the waste removal efficiency of oyster shells in SFF CWs, it was important to reduce the possible confounding factors in the systems for better understanding of the performance of oyster shells in the wastewater treatment process, no macrophyte was planted in these CWs [2,13]. A dynamic model was then developed within AQUASIM 2.1 platform [14]. The model contained seven variables and five submodels, which could be used to estimate water quality change and biochemical reactions in CWs. Based on the experimental results, parameter regression and sensitivity analysis were performed to determine the feasible range of each parameter, and sensitivity of each biochemical process in CWs.

2. Materials and Methods

2.1. Field Experiment

2.1.1. Site Description

Four SSF CWs (latitude 25°4'17" N, longitude 121°27'31" E, absolute altitude = 3.6 m a.s.l.) were established in the floodplain of Ta-Han Stream in Taipei, Taiwan as our study sites (Figure 1a). These wetlands were built to receive municipal wastewater from Taipei City. The wastewater flowing to Ta-Han Stream mainly comes from domestic discharge (>90%) only with minor contribution from industrial and agricultural sewage [15]. Thus, the contamination sources are dominated by organic pollutants coming from black water (fecal sewage) and gray water (wastewater from dishwashers, washing machines, sinks, and baths).

Figure 1. (**a**) Configuration; and (**b**) arrangement plan of the constructed wetlands (CWs) [16]. VA, VB, HA and HB were built with bagged oyster shells, scattered oyster shells, scattered oyster shells and gravels as adsorption media respectively.

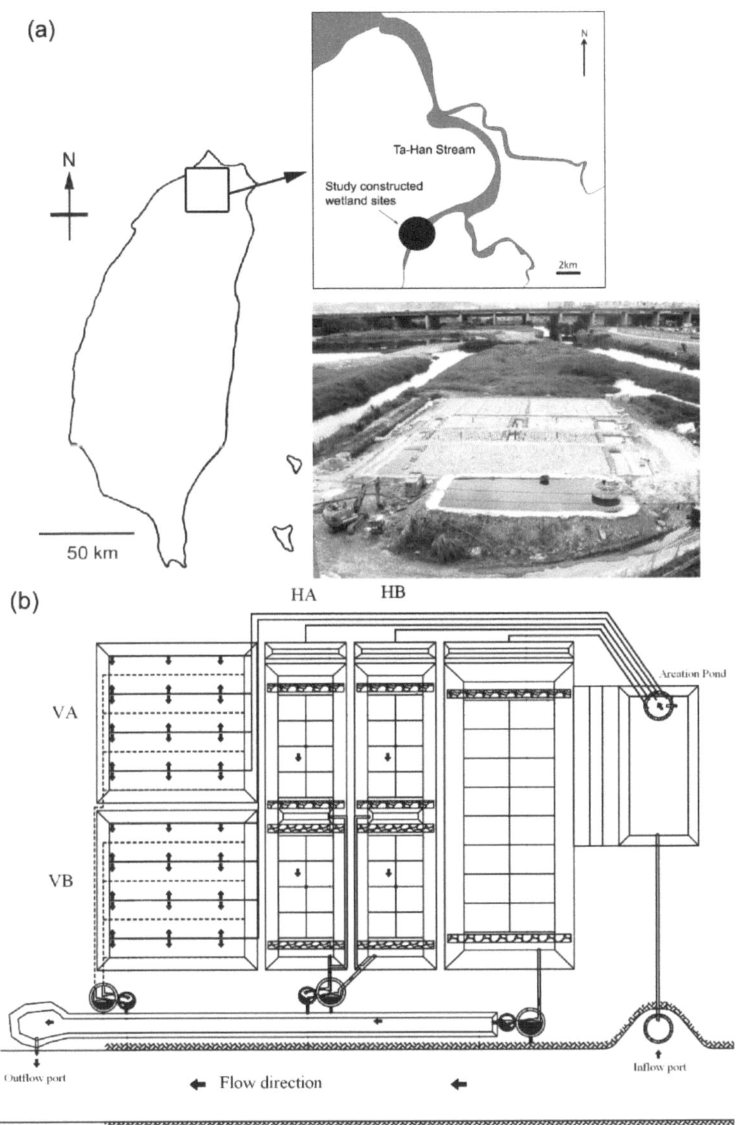

2.1.2. Configuration of the Four Study SSF CWs

In order to investigate the wastewater treatment efficiency of the oyster-shell-bedded CW systems as compared to the conventional gravel-bedded CW, we established four study SSF CWs built into two different types of water-flowing systems, including two vertical SSF CWs and two horizontal SSF CWs, packed with different arrangements of oyster shells and gravels. The two vertical SSF CWs were measured 8.4 m × 8.4 m × 1 m (L × W × D) in size and filled with oyster shells with average dimensions of 6.76 cm long, 4.23 cm wide, and 0.27 cm thick. Bagged and scattered stacking methods of oyster shells were applied to these two vertical SSF CWs named as VA and VB respectively. The original purpose of bagged oyster-shell arrangement included fixing the void ratio in the unit and stabilizing the system of filtering medium under high wastewater discharge. The two horizontal SSF CWs were named as HA and HB (Figure 1b). The dimension of each horizontal SSF constructed wetland was 12 m × 3.4 m × 1 m (L × W × D). HA wetland was filled with oyster shells as the adsorbent medium while HB wetland was a conventional gravel-bedded constructed wetland. Due to the difference in physical properties between oyster shells and gravels (Table 1), the resulting difference in the waste removal quantity between HA and HB wetlands could therefore be indicative of the waste treatment performance between these two filtering materials.

Table 1. Summary of physical properties of oyster shells and gravels [9].

Item	Oyster shells	Gravels
True density (kg/m^3)	1273	2283
Bulk density (kg/m^3)	289	1365
Porosity (%)	77	40
Special surface area (m^2/kg)	0.96	0.23
Special surface area (m^2/m^3)	1217	527

The inflow discharge of wastewater was maintained consistently for the four study SSF constructed wetlands, *i.e.*, between 101 and 225 m^3/day, to simulate the natural condition of Ta-Han Stream floodplain. The water outlet of each study SSF constructed wetland was designed as a gravitational jet form to increase the aeration effect. Same source of the inflow wastewater was directed to the four SSF CWs. The inflow water quality and the operation procedures of all four sites were maintained identical (Table 2). During the study period, inflow wastewater was first pumped through the aeration tank for oxygenation and allowed for precipitation as pre-treatment, it was then flowed into each study constructed wetland separately so that we could monitor the water quality of the outflow to determine the treatment efficiency and waste removal quantity of the four study SSF CWs. The purposes for the two-stage pre-treatment included removing the suspended solids through precipitation and oxidizing most of the ammonium into nitrates through aeration to enable the denitrification of nitrates into nitrogen in the anaerobic environment of the four SSF CWs.

Table 2. Water quality parameters of the wetland influent in this study.

Descriptive statistics	BOD	DO	TP	SS	NH$_4^+$	NO$_3^-$	pH	Temp
Average	14.7	2.34	0.99	32.8	9.09	0.75	7.05	28.5
SD	4.53	0.56	0.37	6.21	3.79	0.61	0.24	0.83
Maximum	27.7	3.90	1.93	65.0	24.6	2.83	7.68	32.4
Minimum	5.48	0.20	0.49	12.0	1.88	0.04	6.74	26.9

Notes: Unit of BOD, DO, NH$_4^+$, NO$_3^-$, TP and SS = mg/L; unit of Temp = °C.

The present work was the pioneer study of application of oyster shells as the adsorbent medium in the "real" SSF CWs, our experiment was carried out for 55 days during June 25–August 18, 2008 to provide preliminary data of the waste treatment performance and cost effectiveness of the oyster-shell-bedded constructed wetland systems. Operational data of the four study CWs were collected twice per day by measuring eight water quality parameters including temperature ($^{\circ}$C), pH, concentrations of biochemical oxygen demand (BOD, mg/L), dissolved oxygen (DO, mg/L), total phosphorous (TP, mg/L), suspended solids (SS, mg/L), ammonium (NH_4^+, mg/L), and nitrate (NO_3^-, mg/L). Measurement of these water quality parameters were based on the standard methods [17].

As the inflow wastewater was dominated by organic pollutants, BOD was selected as the key parameter for assessing waste removal quantity and treatment efficiency for the organic wastewater by the four SSF CWs. The waste removal quantity and treatment efficiency were evaluated at 35-day and 55-day periods during the wetland operation. The mean hydraulic retention time (HRT) of the four SSF CWs were 0.2 day [range = 0.09 (HB)–0.28 (VB)] and 0.12 day [range = 0.07 (HB)–0.19 (VA)] at 35-day and 55-day operation periods respectively (Table 3).

Table 3. Average wastewater removal quantity ($g/m^3/day$) and average treatment efficiency (%) and hydraulic properties of the four study wetlands calculated at 35-day and 55-day operation period in the present study.

Removal quantity ($g/m^3/day$)	BOD	DO	TP	SS	NH_4^+	NO_3^-	Q (CMD)	HRT (day)
Average wastewater removal quantity in 35 days								
HA(oyster shells)	13.52	6.27	0.55	60.18	−0.91	1.50	114.98	0.16
HB(gravels)	9.81	5.59	0.89	49.65	2.19	0.61	101.21	0.09
VA(bagged oyster shells)	9.24	3.50	0.46	39.76	−0.97	0.60	122.99	0.26
VB(scattered oyster shells)	8.03	3.35	0.20	37.85	0.61	0.41	111.67	0.28
Average wastewater removal quantity in 55 days								
HA	23.17	8.89	0.35	114.65	2.00	2.42	178.79	0.09
HB	12.61	5.69	0.34	64.60	−0.10	0.60	110.09	0.07
VA	21.50	6.92	0.34	100.62	−4.45	1.51	224.91	0.13
VB	17.59	4.32	0.10	57.76	−1.55	0.54	137.36	0.19

Rate (%)	BOD	DO	TP	SS	NH_4^+	NO_3^-
Average treatment efficiency in 35 days						
HA	24.23	46.97	4.23	38.83	−7.53	24.32
HB	21.97	49.01	18.20	34.23	4.93	8.87
VA	24.97	44.29	5.96	39.18	−7.15	19.42
VB	19.13	49.66	−0.90	44.10	−2.67	4.82
Average treatment efficiency in 55 days						
HA	22.14	47.46	1.89	41.28	−3.43	26.86
HB	19.68	49.70	5.48	38.73	−5.03	7.42
VA	28.61	49.30	3.15	43.68	−11.00	19.98
VB	22.22	51.42	−0.87	48.11	−10.60	10.90

2.1.3. Cost-Effectiveness Analysis

The total cost of wastewater treatment consists of two aspects including the capital cost (*i.e.*, construction cost) and operation and maintenance (O&M) cost. The capital costs of our four study SSF CWs included the construction materials and building services was determined from the actual expenses involved in establishing these CWs. However, the expense of land was neglected in the present study as our field experiment was conducted on the land owned by the local Government and it was impossible to estimate the cost of land rental. As BOD is commonly regarded as an important index of wastewater treatment in Taiwan and many other countries [6,15], so the cost per mass BOD removed was selected as the measure of wastewater treatment performance in the present cost-effectiveness analysis. In our study, the BOD treatment performance was estimated during the operation time of our experimental period of 35 and 55 days. The cost-effectiveness values of the four study SSF CWs in 55-day period were calculated. However, it would be important to consider the cost-effectiveness of wastewater treatment of CWs in long-term operation. We therefore broke the capital costs into 20-year annuity (w) by the following equation:

$$w = \frac{20P(1+r)r}{20(1+r)-1} \tag{1}$$

where P is the capital cost and r is the interest rate which was assumed to be 0.05 [4]. The total costs per annuity of the four study wetlands were obtained by the summation of their capital costs per annuity and O&M costs.

2.1.4. Statistical Analysis

Data of inflow and outflow water quality of the four study SSF CWs were compared and used for determination of their wastewater treatment efficiency. Water quality data were first checked for normality and homogeneity of variance test, one-way Analysis of Variance (ANOVA) was then used to test the difference in each water chemistry parameter among the four SSF CW. Student-Newman-Keuls *Post-hoc* test (S-N-K test) was applied when significant among-site difference in water chemistry parameter was detected by the 1-way ANOVA. All statistical analysis was carried out using SPSS Statistics 17.0.

2.2. Simulation Model

In the CWs, biochemical reactions, such as mineralization, nitrification, respiration, biofilm adsorption, biomass decay, and sediments consumption are important wastewater treatment mechanisms [2]. Many computer programs such as CW2D [18] and WASP/EUTRO5 [19] were developed for describing complex reactions in CWs. In this study, we used the program AQUASIM 2.1, which was originally designed for identification and simulation of aquatic systems under varied situations [14]. The major reason for choosing AQUASIM was due to its flexible operational platform for easily simulation of the above biochemical processes in CWs, especially biofilm adsorption. In water quality modeling, we assumed water was well mixed in SSF CWs, so that the mixed reactor compartment, a water quality simulation tool in AQUASIM, was applied to describe well-mixed domains. Then, temporal variations of BOD, DO, TP, SS, NH_4^+, and NO_3^- concentrations during the wastewater treatment processes could be then simulated. Details of the water quality operational equations, biochemical processes, five water chemistry submodels of C-cycle, O-cycle, N-cycle, P-cycle, and suspended solids from our water quality model were reported in the following sections [Tables 4 and 5; see also the definition of each process rate in Equations (2) to (7)].

Table 4. Parameters and reaction kinetics of different nutrient cycle processes involved in the CW waste treatment [20].

Process rate	Definition	Rate
C-cycle		
r_{BOD}	Biochemical degradation	$k_{BOD} \cdot C_{BOD} \cdot \theta_{BOD}^{(T-20)}$
r_{R_BOD}	Microorganism respiration	$R_{BOD} \cdot \theta_R^{(T-20)} \cdot \mu_{BOD} \cdot \dfrac{C_{BOD}}{(K_{BOD}+C_{BOD})} \cdot \dfrac{C_{DO}}{(K_{DO}+C_{DO})} \cdot X_H$
r_{Decay_BOD}	Biomass decay	$k_{Decay\_BOD} \cdot \theta_{Decay}^{(T-20)} \cdot X_H$
r_{B_BOD}	Biofilm adsorption	$k_{sob\_BOD} \cdot LF_{model} \cdot C_T \cdot A_S \cdot C_{BOD}$
O-cycle		
r_{DO}	Biochemical degradation	$(k_{d\_C} \cdot C_{BOD_d} + k_{s\_C} \cdot C_{BOD_s}) \cdot \theta_{BOD}^{(T-20)} \cdot \dfrac{C_{DO}}{(K_{DO}+C_{DO})}$
r_{N_DO}	Nitrification	$k_N \cdot C_{NH_4^+} \cdot \theta_N^{(T-20)}$
r_{R_DO}	Microorganism respiration	$R_{DO} \cdot \theta_R^{(T-20)} \cdot \mu_{DO} \cdot \dfrac{C_{DO}}{(K_{DO}+C_{DO})} \cdot X_H$
r_{SOD}	Sediment consumption	$k_{sed} \cdot \dfrac{SOD}{H} \cdot \dfrac{C_{DO}}{(HS_{DO}+C_{DO})}$
r_{B_DO}	Biofilm adsorption	$k_{sob\_DO} \cdot LF_{model} \cdot C_T \cdot A_S \cdot C_{DO}$
P-cycle		
r_P	Phosphorous utilization by microorganisms	$k_P \cdot \theta_R^{(T-20)} \cdot \dfrac{C_{TP}}{(K_P+C_{TP})} \cdot \dfrac{C_{PO_4^{3-}}}{(K_P+C_{PO_4^{3-}})} \cdot X_H$
$r_{Settling\_P}$	Phosphorous settling	$k_{Settling\_P} \cdot \dfrac{C_{TP}}{H}$
r_{Decay_P}	Biomass decay	$k_{Decay\_P} \cdot \theta_{Decay}^{(T-20)} \cdot X_H \cdot i_{P,BM}$
r_{B_P}	Biofilm adsorption	$k_{sob\_P} \cdot LF_{model} \cdot C_T \cdot A_S \cdot C_{TP}$
Suspended solids		
$r_{Fitration}$	Filtration	$k_F \cdot \dfrac{Q_{in}}{A} \cdot \left(\dfrac{C_{SS}}{(1-p) \cdot d_C} \right)$
$r_{Settling\_SS}$	Settling	$k_{Settling\_SS} \cdot d_{SS}^2 \cdot \dfrac{v_W}{H} \cdot \dfrac{\rho_S - \rho_w}{\rho_w} \cdot C_{SS}$
r_{Decay_SS}	Biomass decay	$k_{Decay\_SS} \cdot \theta_{Decay}^{(T-20)} \cdot X_H$
r_{B_SS}	Biofilm adsorption	$k_{sob\_S} \cdot LF_{model} \cdot C_T \cdot A_S \cdot C_{SS}$
N-cycle		
r_{N_N}	Nitrification	$\dfrac{k_{N\_N}}{Y_n} \cdot \dfrac{C_{NH_4^+}}{\left(K_{NH4}+C_{NH_4^+}\right)} \cdot \dfrac{C_{DO}}{(K_{nDO}+C_{DO})} \cdot \theta_N^{(T-20)} \cdot C_{pH} \cdot C_{NH_4^+}$
r_{G_NH4}	Ammonia utilization by microorganisms	$k_{G\_NH4} \cdot \mu_{max,20} \cdot \theta_{growth}^{(T-20)} \cdot \dfrac{C_{NH_4^+}}{(K_{NH4}+C_{NH_4^+})} \cdot X_H$
r_{G_NO3}	Nitrate utilization by microorganisms	$k_{G\_NO3} \cdot \mu_{max,20} \cdot \theta_{growth}^{(T-20)} \cdot \dfrac{C_{NO_3^-}}{(K_{NO3}+C_{NO_3^-})} \cdot X_H$
r_{Reg}	Ammonia regeneration	$k_{reg} \cdot S_{Naggr}$
r_{Min}	Mineralization	$k_{Min} \cdot S_{ON} \cdot \dfrac{C_{DO}}{(K_{nDO}+C_{DO})}$
r_{DN}	Denitrification	$k_{DN} \cdot \theta_{DN}^{(T-20)} \cdot C_{NO_3^-}$
r_{Decay_N}	Biomass decay	$k_{Decay\_N} \cdot \theta_{Decay}^{(T-20)} \cdot X_H \cdot i_{N,BM}$
r_{B_N}	Biofilm adsorption	$k_{sob\_N} \cdot LF_{model} \cdot C_T \cdot A_S \cdot (C_{NH_4^+}+C_{NO_3^-})$
C_T^*	Temp. dependent factor	$e^{\varphi(T-20)}$
C_{pH}^*	pH growth-limiting factor	*If pH<7.2 then (1-0.833•(7.2-pH)) else 1*

Also, as accumulated studies on CWs have suggested that biofilm was an important factor associated with the water quality of CWs, e.g., [21,22], the biofilm reactor compartment was established in our model for estimating the biofilm population dynamics. Detailed descriptions of the model coefficients, parameters and constants of the biofilm reactor compartment are given in Table 5.

Table 5. Summary of parameters and constants of different biochemical processes of CW waste treatment involved in the simulated model (experimental data, C-cycle, O-cycle, P-cycle and SS removal, N-cycle, biofilms, temperature coefficients and half-saturation constants).

Parameter	Description	Literature range	Unit	Source
Experimental data				
A	Cross-sectional area	-	m^2	Field monitoring data
A_S	Special surface area of media	-	m^2/m^3	[9]
BOD_d	Dissolve BOD	-	mg/L	Field monitoring data
BOD_s	Suspended BOD	-	mg/L	Field monitoring data
d_c	Diameter of collector	-	m	Field monitoring data
H	Depth	-	m	Field monitoring data
p	Porosity	-	%	Field monitoring data
Q_{in}	Inflow	-	m^3/day	Field monitoring data
S_{Nagger}	Nitrogen in aggregates	-	mg/L	Field monitoring data
S_{ON}	Organic nitrogen	-	mg/L	Field monitoring data
C-cycle				
k_{BOD}	Biochemical degradation rate of BOD	0.3	day^{-1}	[23]
$k_{Decay\ BOD}$	Biomass decay rate	0.15	day^{-1}	[24]
$k_{sob\ BOD}$	Biofilm adsorption coefficient of BOD	-	m^{-3}day^{-1}	-
R_{BOD}	Microorganisms respiration coefficient	-	-	-
μ_{BOD}	Max growth rate of hetero. at 20 °C	0.8–6	day^{-1}	[25]
φ_{BOD}	Empirical constant of BOD	0.098	°C^{-1}	[20]
O-cycle				
HS_{DO}	Sediment oxygen demand constant	2.5	mg/L	[24]
$k_{d\ C}$	Degradation rate for BOD_d	0.3	day^{-1}	[23]
k_N	Nitrification rate at 20 °C	0.05	day^{-1}	[23]
$k_{s\ C}$	Degradation rate for BOD_s	0.3	day^{-1}	[23]
k_{sed}	Sedimentation coefficient	0.1	-	[23]
$k_{sob\ DO}$	Biofilm adsorption coefficient of DO	-	m^{-3}day^{-1}	-
R_{DO}	Heterotrophic respiration coefficient	0.1	-	-
SOD	Sediment oxygen demand	0.1	gO$_2$/m^{2}day	[23]
μ_{DO}	Max growth rate of hetero. at 20 °C	0.015–0.2	day^{-1}	[26,27]
φ_{DO}	Empirical constant of DO	0.098	°C^{-1}	[20]
P-cycle				
$i_{P,BM}$	Phosphorus content of biomass	0.02	mg$_P$/mg$_{BM}$	[28]
$k_{Decay\ P}$	Biomass decay rate	0.15	day^{-1}	[24]
k_P	Biochemical degradation rate	-	day^{-1}	-
$k_{Settling\ P}$	Phosphorous settling coefficient	0.03	m^{-1}day^{-1}	-
$k_{sob\ P}$	Biofilm adsorption coefficient of TP	-	m^{-3}day^{-1}	-
φ_P	Empirical constant of TP	0.098	°C^{-1}	[20]

Table 5. *Cont.*

Parameter	Description	Literature range	Unit	Source
Suspended solids				
d_{SS}	Diameter of settling particle	0.1–4	mm	[29]
$k_{Decay\ SS}$	Biomass decay rate	0.15	day^{-1}	[24]
k_F	Filtration coefficient	-	-	-
$k_{Settling\ SS}$	Settling coefficient	-	m^{-3}	-
$k_{sob\ SS}$	Biofilm adsorption coefficient of SS	-	m^{-3}day^{-1}	-
α	Sticking coefficient	0.0008–0.012	-	[30]
ρ_s	Density of settling particle	1050–1500	kg/m^3	[31]
ρ_W	Density of water	995.69	kg/m^3	[31]
v_W	Kinematic viscosity of water	0.0867	m^2/day	[32]
φ_{SS}	Empirical constant of SS	0.098	°C^{-1}	[20]
N-cycle				
$i_{N,BM}$	Nitrogen content of biomass	0.07	mg$_N$/mg$_{BM}$	[28]
$k_{Decay\ N}$	Biomass decay rate	0.15	day^{-1}	[24]
k_{DN}	Denitrification rate at 20 °C	0–1	day^{-1}	[33]
$k_{G\ NH4}$	NH$_4^+$ uptake preference factor	-	-	-
$k_{G\ NO3}$	NO$_3^-$ uptake preference factor	-	-	-
k_{Min}	Mineralization rate	0.0005–0.143	day^{-1}	[34]
k_{N_N}	Growth rate of nitrosomonas by nitrification	0.33–2.21	day^{-1}	[35]
k_{Reg}	NH$_4^+$ regeneration rate	0.085	day^{-1}	[36]
k_{sob_N}	Biofilm adsorption coefficient of NH$_4^+$ and NO$_3^-$	-	m^{-3}day^{-1}	-
Y_n	Nitrosomonas yield coefficient	0.03–0.13	mg$_{VSS}$/mg$_N$	[37]
$\mu_{max,20}$	Max. growth rate of bacteria at 20 °C	0.18	day^{-1}	[38]
φ_N	Empirical constant	0.098	°C^{-1}	[20]
Biofilms				
b_{X1}	Microorganism heterotroph decay rate	0.3	day^{-1}	-
b_{X2}	Microorganism nitrosomonas decay rate	0.3	day^{-1}	-
D_{NH4}	Diffusion coefficient of NH$_4^+$	1.71×10^{-4}	m^2/day	[39]
D_{NO3}	Diffusion coefficient of NO$_3^-$	$(4.5–27.9) \times 10^{-6}$	m^2/day	[40]
D_{TOC}	TOC diffusion coefficient	1.56×10^{-5}	m^2/day	[41]
D_X	Microorganism diffusion coefficient	-	m^2/day	-
LF_{model}	Biofilms thickness	-	m	Biofilm model result
X_H	Heterotrophic organisms	-	mg/L	Biofilm model result
Y_1	Yield constant of heterotroph	0.6	-	[27]
Y_2	NH$_4^+$ yield constant of nitrosomonas	0.13	-	[27]
Y_3	NO$_3^-$ yield constant of nitrosomonas	0.03	-	[27]
μ_{X1}	Max growth rate of heterotroph	3–6	day^{-1}	[28,42]
μ_{X2}	Max growth rate of nitrosomonas	0.33–2.21	day^{-1}	[35]
Temperature coefficient				
θ_{BOD}	Temp. coefficient of degradation	1.09	-	[23]
θ_{Decay}	Temp. coefficient of biomass decay	-	-	-
θ_R	Temp. coefficient of respiration	-	-	-
θ_{DN}	Temp. coefficient of denitrification	1.15	-	[43]
θ_{growth}	Temp. coefficient of microorganisms growth	1.08–1.12	-	[31]
θ_N	Temp. coefficient of nitrification	1.1	-	[23]

Table 5. *Cont.*

Parameter	Description	Literature range	Unit	Source
Half- saturation (Half-sat.) constant				
K_{BOD}	Half-sat. constant of BOD	2	mg/L	[23]
K_{DO}	Half-sat. constant of DO	2	g_{O2}/m^3	[23]
K_P	Half-sat. constant of TP	0.02	mg/L	[44]
K_n	Half-sat. constant of NH_4^+ nitrosomonas	0.05	mg/L	[23]
K_{nDO}	Half-sat. constant of DO nitrosomonas	0.13–1.3	mg/L	[35]
K_{NH4}	Half-sat. constant of NH_4^+	2	g_{COD}/m^3	[27]
K_{NO3}	Half-sat. constant of NO_3^-	0.15–0.5	g_N/m^3	[26,45]

2.2.1. Carbon Cycle

Organic matters usually exist in five different types in CWs, e.g., dissolved phase, suspended phase, bottom phase, biomass, and inertia carbon [45]. Microorganisms play the principal roles of organic matter removal in CWs through their utilization and respiration. The temporal and spatial variability of BOD in CWs are controlled by the following equation (Tables 4 and 5):

$$\frac{d(C_{BOD})}{dt} = \frac{I_{in,BOD}}{V_R} - \frac{Q_{out}}{V_R}C_{BOD} - r_{BOD} - r_{R\_BOD} + r_{Decay\_BOD} - r_{B\_BOD} \tag{2}$$

where $I_{in,BOD}$ is loading of BOD into the reactor (mass per unit per time), V_R is the reactor volume, Q_{out} is the volumetric outflow, and C_{BOD} is the concentration of BOD. Other process rates are shown in Table 4.

2.2.2. Oxygen Cycle

DO is one of the most important water quality indicators as many biochemical processes require the participation of oxygen. As the flow velocity is relatively low and water surface area for gaseous exchange is small in SSF CWs, oxygen cannot enter its water bodies by diffusion. Moreover, there is no other aeration mechanism such as photosynthesis, root-zone effect and artificial aeration in these wetlands. Therefore, DO is further diminished by the processes associated with sediment oxygen demand, bacteria respiration, nitrification, and oxidation of BOD as described by the following equation (Tables 4 and 5):

$$\frac{d(C_{DO})}{dt} = \frac{I_{in,DO}}{V_R} - \frac{Q_{out}}{V_R}C_{DO} - r_{DO} - r_{N\_DO} - r_{R\_DO} - r_{SOD} - r_{B\_DO} \tag{3}$$

where $I_{in,DO}$ is loading of DO into the reactor (mass per unit per time), and C_{DO} is the concentration of DO.

2.2.3. Phosphorus Cycle

Removal rates of TP in CWs are dominated by plant uptake [28]. In addition, phosphorus can combined with heavy metal, adsorbed by suspended solids and utilized by microorganisms in wetlands. The mass balance equation for TP is given in the following (Tables 4 and 5):

$$\frac{d(C_{TP})}{dt} = \frac{I_{in,TP}}{V_R} - \frac{Q_{out}}{V_R} C_{TP} - r_P - r_{Settling\_P} - r_{B\_P} + r_{Decay\_P} \tag{4}$$

where $I_{in,TP}$ is loading of TP into the reactor (mass per unit per time) and C_{TP} is the concentration of TP.

2.2.4. Suspended Solids

Multiple physical processes relating to filtration and precipitation control the temporal variability of SS in CWs. In SSF CWs, SS can be blocked, trapped and intercepted when they pass through stems/roots of plants, sandstones, and other media. In our simulated model, we also considered the adsorption of biofilm as a momentous process for SS removal. The mass balance for SS in wetlands is given as follows (Tables 4 and 5):

$$\frac{d(C_{SS})}{dt} = \frac{I_{in,SS}}{V_R} - \frac{Q_{out}}{V_R} C_{SS} - r_{Filtration} - r_{Settling\_SS} + r_{Decay\_SS} - r_{B\_SS} \tag{5}$$

where $I_{in,SS}$ is loading of suspended solids into the reactor (mass per unit per time), and C_{SS} is the concentration of suspended solids.

2.2.5. Nitrogen Cycle

In natural environment, nitrogen involves in many biochemical processes and it exists in many different forms from the most oxidized form nitrates (NO_3^-) to the most reduced form ammonium (NH_4^+). Organic nitrogen in wetlands is first transformed into NH_4^+ through mineralization, and then converted into NO_3^- via the two stages of nitrification [2]. During the removal process of NH_4^+, part of the NH_4^+ is converted into NO_3^- and remains in wetlands. In this study, we therefore considered dissolved nitrogen (NH_4^+ and NO_3^-) as the major forms of nitrogen in the study CWs. The mass balance for NH_4^+ and NO_3^- are given as follows (Tables 4 and 5):

$$\frac{d(C_{NH_4^+})}{dt} = \frac{I_{in,NH_4^+}}{V_R} - \frac{Q_{out}}{V_R} C_{NH_4^+} - r_{N\_N} - r_{G\_NH4} + r_{Reg} + r_{Min} - r_{B\_N} + r_{Decay\_N} \tag{6}$$

$$\frac{d(C_{NO_3^+})}{dt} = \frac{I_{in,NO_3^-}}{V_R} - \frac{Q_{out}}{V_R} C_{NO_3^-} + r_{N\_N} - r_{DN} - r_{G\_NO3} - r_{B\_N} + r_{Decay\_N} \tag{7}$$

where I_{in,NH_4^+} is loading of NH_4^+ into the reactor (mass per unit per time), I_{in,NO_3^-} is loading of NO_3^- into the reactor (mass per unit per time), $C_{NH_4^+}$ is the concentration of NH_4^+, and $C_{NO_3^-}$ is the concentration of NO_3^-.

2.2.6. Biofilm Reactor Compartment

The biofilm model is developed based on the one-dimensional mixed culture biofilm model [14,46] (Table 5). The one-dimensional conservation laws are formulated by AQUASIM 2.1 to describe the transmission processes of dissolved substances and suspended solids in biofilms (solid matrix and pore water). The growth or decay of organisms was expressed by the expansion or contraction of biofilms.

2.2.7. Sensitivity Analysis

The wastewater treatment efficiency in each wetland was obtained based on the monitoring data of inflow and outflow water quality in the study SSF CWs. The influence of different biochemical processes on the wastewater treatment efficiency were quantified by inputting the field monitoring data of the four wetlands into the water quality model for sensitivity analysis. The sensitivity analysis was used to determine the sensitivity and relative importance of each biochemical process [35,47]. We first applied the absolute-relative sensitivity function [Equation (8)] provided by AQUASIM to measure the sensitive value (*SensAR*) of each parameter:

$$SensAR = p\frac{dy}{dp} \tag{8}$$

where *SensAR* is sensitive value; p is a model parameter and y is a state variable.

Since BOD removal is one of the main functions of CWs and BOD loading are commonly considered as an important factor for assessing wetland operation [1,6,7], BOD was taken as the basis for the evaluation of the sensitivity of parameters in this study. Also, as identification of parameters is necessary for improving the accuracy of water quality simulations, the feasible range of all parameters in oyster-shell-bedded CWs were determined in AQUASIM.

3. Results and Discussion

3.1. Field Experiment

3.1.1. Cost-Effectiveness Analysis

The capital costs, O&M costs, total costs and cost-effectiveness in 55-day- and 20-year annuity period for each study SSF CW were shown in Table 6. The original capital costs were 19 times of the capital cost in 20-year annuity. Therefore, the capital costs made up of the majority (~97%) of the total costs of all four CWs when the operation period was 55 days. However, the capital cost per annuity was only 20% of HA, VA and VB, and 24% of HB of the total costs when these wetlands were assumed to be operated for 20 years. Moreover, the total cost for all the CWs operated for in 55-day-period [range = US$10711 (HA)–13586 (HB)] were 2.9–3.7 times higher as compared to the total cost for 20-year annuity [range = US$2737 (VB)–2869 (HB)]. Also, the cost per mass BOD removed was 25–30 times higher in all wetlands for the 55-day than 20-year annuity period. Our results highlighted that the economic returns of CWs would be higher for long-term operation.

Among the four study SSF CWs, the capital cost and total cost of gravel bedded site HB were 16%–28% higher than the other three filled with oyster shells. However, the total BOD removal quantity of HB was only one forth to half of HA, VA and VB. Our results showed that, upon long term operation (20-year annuity), the treatment cost of 1kg BOD was US$25.01 in HB but only US$6.56 was required for VA wetland. VA also demonstrated the highest cost-effectiveness among the three oyster-shell filled CW systems (HA = US$13.04; VB = US$10.88). This confirmed that oyster shells were the cost-effective adsorption medium in SSF CW as compared to the conventional gravel-bedded SSF CW.

Table 6. Results of cost-effectiveness analysis of the four study SSF CWs.

Cost	HA	HB	VA	VB
Capital cost				
Suppose engineering	916	987	1046	1046
Civil engineering	570	614	651	651
Pumping well	254	273	290	290
Aeration pond	851	916	971	971
Diversion cut	1740	1880	0	0
Reverse-flushing system	1260	1167	0	0
Water distribution pipe	0	0	560	560
Sludge pipe	0	0	1700	1700
Antiseep engineering	1406	1514	1606	1606
Collection drains	1960	2111	2239	2239
Media paving	282	303	322	322
Water quality monitoring pipe	133	133	100	100
Gravels	0	3360	0	0
Oyster shell transport	1007	0	1007	1007
Bagged	0	0	984	0
Original capital cost (US$)	10379	13258	11475	10491
Capital cost—20-year annuity (US$/yr)	545	696	602	551
O&M cost				
55-day-operation-period (US$)	332	328	335	329
Per year (US$/yr)	2205	2173	2226	2186
Total cost				
55-day-operation-period (US$)	10711	13586	11810	10820
20-year annuity (US$/yr)	2749	2869	2828	2737
Total waste removal quantity of BOD during the operation time				
55-day-operation-period (kg)	31.77	17.29	64.97	37.92
Per year (kg/yr)	210.83	114.74	431.17	251.63
Cost-effectiveness value (Cost per mass BOD removed)				
55-day-operation-period (US$/kg)	337.15	785.79	181.78	285.38
20-year annuity (US$/kg)	13.04	25.01	6.56	10.88

3.1.2. Treatment Efficiency Analysis

As no macrophyte was planted in the four study SSF CWs, the wastewater treatment mechanisms were dominated by physical deposition and biochemical decomposition, including settling, filtration, regeneration, nitrification, denitrification, mineralization, sediment consumption, biomass decay, microorganism respiration, biochemical degradation, biofilm adsorption, and microorganism utilization. The waste removal quantity of BOD, DO, TP, SS, NH_4^+, and NO_3^- showed inconsistent trend in HA, HB, VA, and VB wetlands during our study period (Figure 2a–f; Table 7). Removal quantity of BOD, SS and NO_3^- were higher in HA and VA wetlands. But, the removal of TP was not significant in these wetlands. In wastewater purification processes, DO was consumed continuously by the aerobic biochemical reactions, resulting in the low DO concentration in outflow from all four wetlands (Figure 2b).

The average waste removal quantity of most wastewater parameters increased slightly from 35-day to 55-day-periods but the average treatment efficiency of all wastewater parameters remained fairly constant between 35-day and 55-day-periods. Our preliminary findings suggested that increasing the operation time could enhance the success of CWs in terms of wastewater treatment efficiency. However, further confirmation would be needed for the four types of study CWs in Taiwan by extending the length of study period.

Figure 2. Waste removal quantity ($g/m^3/day$) of (**a**) oxygen demand (BOD); (**b**) dissolved oxygen (DO); (**c**) total phosphorous (TP); (**d**) suspended solids (SS); (**e**) NH_4^+; and (**f**) NO_3^- in the four study SSF wetlands (HA = black circles; HB = grey circles; VA = inverted grey triangle; VB = white triangle).

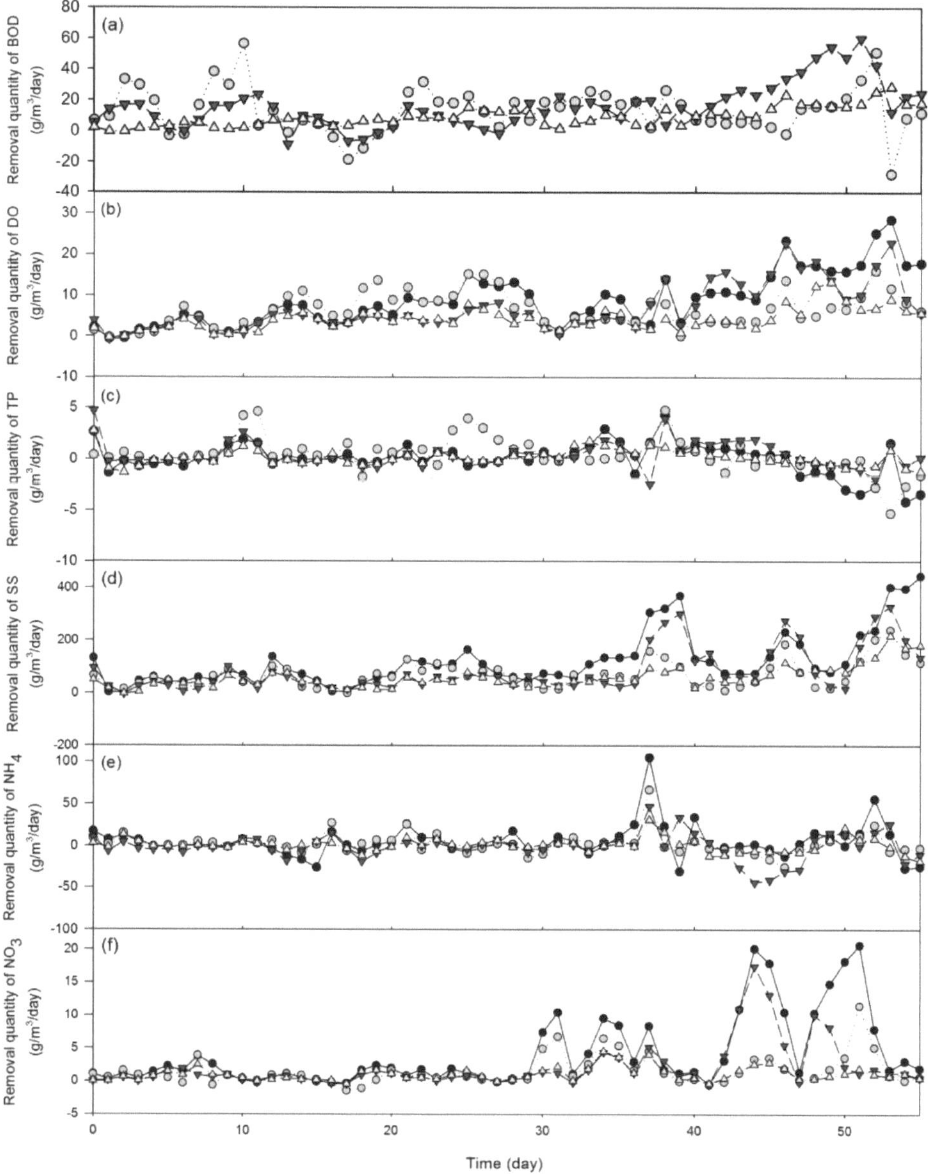

Table 7. Results of one-way Analysis of Variance (ANOVA) assessing the waste removal quantity of each wastewater parameter among the four study SSF CWs. * $p < 0.05$.

Wastewater parameter	F-value	P-value
BOD	2.655	0.049*
DO	8.498	0.000*
TP	0.380	0.767
SS	6.727	0.000*
NH_4^+	1.388	0.247
NO_3^-	4.233	0.006*

Our results highlighted that there were significant differences in the waste removal quantity of BOD, DO, NO_3^-, SS among the four wetlands (BOD: $F_{3,220} = 2.655$, p = 0.049; DO: $F_{3,220} = 8.498$, $p < 0.001$; NO_3^-: $F_{3,220} = 4.233$, p = 0.006; SS: $F_{3,200} = 6.727$, $p < 0.001$) (Table 7). *Post-hoc* S-N-K comparisons between HA and HB wetlands showed that waste removal quantity in HA (23.17 g/m^3/day for BOD, 2.42 g/m^3/day for NO_3^-, and 114.65 g/m^3/day for SS) was significantly higher than HB (12.61 g/m^3/day for BOD, 0.6 g/m^3/day for NO_3^-, and 64.6 g/m^3/day for SS) (Table 3). Thus, the treatment efficiency of HA was higher than HB in BOD, NH_4^+, NO_3^-, and SS. However, the BOD, NO_3^-, DO and SS treatment efficiency of both HA and HB were lower than VA and VB primarily due to the difference in the site infrastructure (Figure 1b) and the size of biofilm reactor compartment (VA and VB > HA and HB).

Comparing HA and VB wetlands, HA had 1.92 g/m^3/day NO_3^- and 56.89 g/m^3/day SS of waste removal quantity which were significantly higher than VB. However, VB showed slightly higher treatment efficiency than HA because the reactor volume of VB was larger than HA. On the other hand, despite the SS removal quantity in VA wetland was significantly higher than VB (100.62 g/m^3/day and 57.76 g/m^3/day respectively) (Tables 3 and 8), treatment efficiency of SS remained relatively similar among the four study CWs.

Table 8. Results of sensitivity analysis of BOD removal quantity from all biochemical processes in the simulated model.

Parameters	SensAR		SensAR		SensAR		SensAR	
	RMS	Mean	RMS	Mean	RMS	Mean	RMS	Mean
	HA		HB		VA		VB	
k_{sob_BOD}	3.535	−3.086	0.920	−0.794	3.341	−3.053	3.338	−3.179
k_{Decay_BOD}	1.578	1.392	0.571	0.538	1.753	1.663	0.958	0.930
k_{BOD}	0.556	−0.492	1.856	−1.713	0.098	0.083	0.280	−0.257
R_{BOD}	0.274	−0.231	0.009	0.004	0.146	−0.241	0.204	−0.187

Comparison between bagged (VA) and scattered (VB) arrangement oyster-shell-bedded CW indicated that the waste removal quantity and treatment efficiency between these two wetlands were generally similar. However, VA wetland demonstrated significantly highest BOD treatment efficiency among all study CWs. Our results indicated that oyster shells were an effective adsorption medium in SSF CW because of its lower cost and better wastewater treatment performance as compared to the conventional gravel-bedded SSF CW. But, the site infrastructure, hydraulic patterns and arrangement of oyster shells could be important in determining the waste removal efficiency and cost-effectiveness of CWs. The observed effectiveness of oyster shells as biofilter substrates in CWs due to their higher porosity and surface area/volume ratio as compared to gravels, and thus providing larger contact area for efficient nutrient treatment (Table 1) [9].

In addition, as denitrification usually occurs in low dissolved oxygen or anaerobic conditions because denitrifying bacteria are usually anaerobic and heterotrophic, the efficiency of denitrification is also limited by the source of carbon in the environment [48]. Among the four study wetlands, HA showed the highest NO_3^- removal efficiency probably due to its low DO environment (Figure 3). In general, horizontal SSF CWs were predominantly anaerobic, but the oxygen supply is usually higher in vertical SSF CWs which show higher rates of bio-decomposition of organic carbon [49]. This could therefore explain the higher BOD removal efficiencies in VA and VB wetlands.

Figure 3. Simulated BOD, DO, TP, SS, NH_4^+ and NO_3^- outflow results and measured data in (**a**) HA; (**b**) HB; (**c**) VA; and (**d**) VB wetlands.

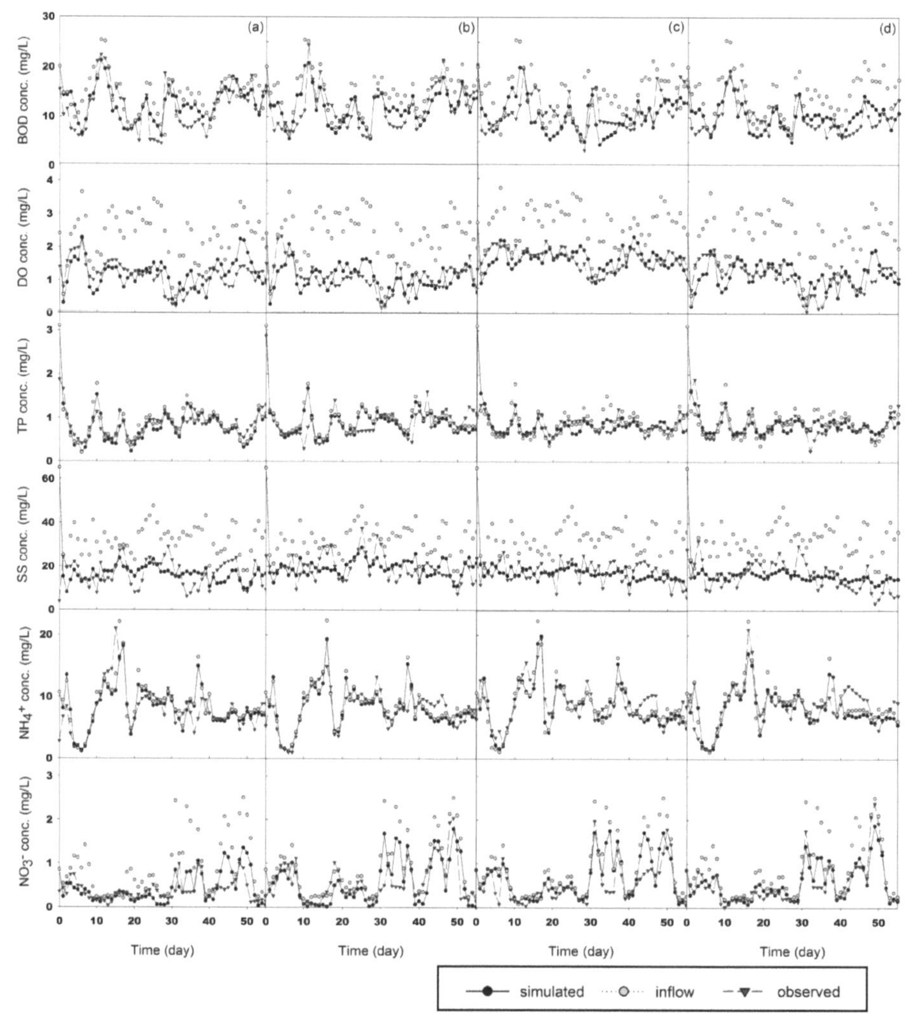

3.2. Simulation

3.2.1. Sensitivity Analysis

The biofilm adsorption coefficient of BOD (k_{sob_BOD}) had the highest *SensAR* in most of the treatment units except for the gravel-bedded constructed wetland (HB) among the four key parameters (k_{sob_BOD}, k_{Decay_BOD}, k_{BOD} and R_{BOD}) (Table 8). This suggested that the biofilm adsorption was the most effective process for BOD removal quantity in all three oyster-shell-bedded wetlands (*i.e.*, HA, VA and VB wetlands) as *SensAR* represented the waste removal quantity of each biochemical processes (Table 8). Hence, biofilm adsorption was the major mechanism for wastewater treatment in the three oyster-shell-bedded CWs as previous studies confirmed that oyster shells provided more area for microbial propagation than gravels [50,51]. Therefore, the treatment efficiency in CWs can be enhanced by using oyster shells as an adsorption medium.

3.2.2. Feasible Range of Parameters in Oyster-Shell-Bedded CWs

Field monitoring data of water quality from the three oyster-shell-bedded CWs were input to the water quality model to determine the feasible range of each parameter. In this part, we avoided changing values of constants such as half-saturation constant, max growth rate of bacteria, and others, which were obtained from microorganism experiments. Thus, three sets of parameters were obtained by model fitting of the experimental data of the three oyster-shell-bedded wetlands. The model fitting results of HA, VA, and VB wetlands are given in Figure 3.

For the three sets of parameters, we took the maximum value as the upper bound and the minimum value as the lower bound of each parameter. The upper and lower bounds were integrated to set the feasible range of each parameter in Table 9. The feasible range could provide a reference for simulation and prediction in further studies.

Table 9. Feasible range of parameters in the oyster-shell bedded CWs.

Submodel	Parameter	HA	VA	VB	Feasible range
C-cycle	k_{BOD}	0.680	0.894	0.752	0.680–0.894
	k_{Decay_BOD}	5.977	8.764	9.335	5.977–9.335
	k_{sob_BOD}	12.65	34.13	32.69	12.65–34.13
	R_{BOD}	6.704	17.97	13.08	6.704–17.97
	θ_{BOD}	0.931	0.898	0.826	0.826–0.931
	θ_{Decay}	0.794	0.764	0.706	0.706–0.794
	θ_R	0.729	0.771	0.745	0.729–0.771
	μ_{BOD}	3.000	3.000	3.000	3.000
	φ_{BOD}	0.081	0.104	0.097	0.081–0.104
O-cycle	HS_{DO}	3.000	3.000	3.000	3.000
	k_{d_C}	0.076	0.268	0.010	0.010–0.268
	k_N	0.001	0.001	0.001	0.001
	k_{s_C}	0.578	1.001	0.572	0.572–1.001
	k_{sed}	0.921	0.742	0.897	0.742–0.897
	k_{sob_DO}	23.36	26.14	45.01	23.36–45.01
	R_{DO}	10.55	23.02	8.224	8.224–10.56

Table 9. *Cont.*

Submodel	Parameter	HA	VA	VB	Feasible range
O-cycle	SOD	0.100	0.100	0.100	0.100
	θ_{BOD}	0.905	0.759	0.854	0.759–0.905
	θ_N	0.688	0.600	0.600	0.600–0.688
	θ_R	0.881	0.836	0.799	0.799–0.881
	μ_{DO}	0.163	0.163	0.163	0.163
	φ_{DO}	0.010	0.010	0.010	0.010
P-cycle	$i_{P,BM}$	0.020	0.020	0.020	0.020
	k_{Decay_P}	5.662	11.68	10.83	5.662–11.68
	k_P	0.472	2.127	0.921	0.472–2.127
	$k_{Settling\_P}$	0.020	0.025	0.024	0.020–0.025
	k_{sob_P}	0.058	0.593	0.309	0.058–0.593
	θ_{Decay}	0.725	0.903	0.848	0.725–0.903
	θ_R	0.757	0.718	0.742	0.718–0.757
	φ_P	0.130	0.127	0.092	0.092–0.130
SS	d_{SS}	0.001	0.001	0.001	0.001
	k_{Decay_SS}	1.514	3.937	2.413	1.514–3.937
	k_F	0.007	0.017	0.007	0.007–0.017
	$k_{Settling\_SS}$	0.100	0.300	0.315	0.100–0.315
	k_{sob_SS}	8.390	16.74	28.24	8.390–28.24
	S_g	1500	1500	1500	1500
	α	0.007	0.007	0.007	0.007
	ρ_S	1300	1300	1300	1300
	ρ_W	995.7	995.7	995.7	995.7
	v_W	0.087	0.087	0.087	0.087
	θ_{Decay}	1.043	0.996	1.028	0.996–1.043
	φ_{SS}	0.018	0.134	0.010	0.010–0.134
N-cycle	$i_{N,BM}$	0.070	0.070	0.070	0.070
	k_{Decay_N}	0.035	0.086	0.178	0.035–0.178
	k_{DN}	1.582	0.051	0.771	0.051–1.582
	k_{G_NH4}	0.354	0.032	0.032	0.032–0.354
	k_{G_NO3}	0.727	2.934	2.477	0.727–2.934
	k_{Min}	0.100	0.569	0.228	0.100–0.569
	k_{N_N}	0.873	0.808	0.915	0.808–0.915
	k_{Reg}	0.100	0.731	0.291	0.100–0.731
	k_{sob_N}	0.934	1.498	1.347	0.934–1.498
	Y_n	0.130	0.130	0.130	0.130
	φ_P	0.130	0.127	0.092	0.092–0.130
	θ_{Decay}	0.894	0.852	0.882	0.852–0.884
	θ_{DN}	1.181	1.198	0.958	0.958–1.198
	θ_{Growth}	0.871	0.903	0.900	0.871–0.903
	θ_N	0.939	0.768	0.77	0.768–0.939
	$\mu_{max,20}$	0.180	0.180	0.180	0.180
	φ_N	0.098	0.098	0.098	0.098

3.2.3. Applications of Our Model

Many wetland models presented previously often used diffusion coefficient in sublayer (m^2/day), diffusivity of substrate in biofilm (m^2/day), and sublayer thickness (m), along with the experiment results of the biofilm thickness (from 1.46×10^{-3} to 1.62×10^{-3} m), to estimate reaction of biofilm adsorption [36,50]. In contrast, our model utilized the biofilm compartment in AQUASIM to perform an initial dynamic modeling of the biofilm time variation, as one of the referencing conditions for water quality modeling. After that, a sensitivity analysis on different influential factors was carried out to identify the significant biofilm biochemical mechanisms for water quality improvement.

4. Conclusions

Based on experimental investigation of oyster-shell- and gravel-bedded CW systems on wastewater treatment efficiency, economic cost and numerical modeling of water quality, the present study has led to following conclusions,

(1) The four study SSF CWs showed a significant difference in the waste removal quantity of BOD, DO, NO_3 , and SS. The waste removal quantity of the horizontal SSF oyster-shell-bedded CW (HA) was significantly higher than the horizontal SSF gravel-bedded CW (HB) but similar to the vertical SSF oyster-shell CW (VB). Comparison between bagged (VA) and scattered (VB) arrangement oyster-shell-bedded CWs indicated that the waste removal quantity and treatment efficiency between these two wetlands were generally similar. However, VA wetland demonstrated significantly highest BOD removal capacity among all study sites but also showing the lowest cost per mass BOD removed (6.56 US$/kg) as compared to other three CWs (10.88–25.01 US$/kg). Therefore, VA was determined as the best option for SFF CW in terms of waste treatment efficiency and cost-effectiveness.

(2) The total costs of the four study CWs ranged from 2,737 (VB) to 2,869 (HB) US$/yr in 20-year annuity whereas they were between 10,711 (HA) and 13,586 (HB) US$ for only 55-day operation period. Also, the relative importance of capital costs to the total costs of all CWs for long-term operation (20-year annuity) was only one fifth of that for 55 days' operation. Therefore, results of the cost-effectiveness analysis highlighted that the economic returns of CWs would be higher for long-term operation.

(3) The average waste removal quantity of most wastewater parameters increased slightly from 35-day to 55-day-periods but the average treatment efficiency of all wastewater parameters remained fairly constant between 35-day and 55-day-periods. Our findings suggested that establishment time could be critical for the success of CWs with respect to wastewater treatment efficiency.

(4) The results of our numerical water quality model demonstrated that, biofilm adsorption played the most essential role in the wastewater treatment processes in oyster-shell-bedded CWs but biochemical degradation was the most significant mechanism in gravel-bedded CW.

(5) The feasible range of each water quality parameter in oyster-shell bedded wetlands was identified in the present study, and it was obtained by a regression model using the field monitoring data. These feasible ranges could be used for water quality simulations in the CWs

and this could help characterizing different CWs by determining the quantitative importance of different biochemical treatment processes in SSF CWs.

Therefore, our study confirmed that oyster shells were an effective adsorption medium in SSF CWs because of its lower cost and better wastewater treatment performance as compared to the conventional gravel-bedded SSF CW. However, the hydraulic design and arrangement of oyster shells could be important in determining the waste removal efficiency and cost-effectiveness of CWs. Data from the present study would then be used in future investigation of its effects in the vegetated CWs to enhance our understanding on the vegetation influence in the waste treatment efficiency in the oyster-shell-bedded CWs. We will further extend the study period in order to confirm the waste treatment efficiency of the four types of study CWs in long-term operation and provide more field data for the simulated model instead of the literature values. Also, the environmental impacts during the construction of operation period of the oyster-bedded CWs will be evaluated to provide information for developing this type of sustainable natural waste treatment system, e.g., [52].

Acknowledgements

This research was supported by the National Science Council ROC (Grant No.: NSC 100-2221-E-002-191-MY3 and 101-2625-M-002-006). The authors are grateful to the editors and anonymous reviewers for their detailed suggestions and helpful comments that improved this manuscript. We also thank YL Teng for her technical assistance of this work.

References

1. Economopoulou, M.A.; Tsihrintzis, V.A. Design methodology and area sensitivity analysis of horizontal subsurface flow constructed wetlands. *Water Resour. Manag.* **2003**, *17*, 147–174.
2. Vymazal, J.; Brix, H.; Cooper, P.F.; Haberl, R.; Perfler, R.; Laber, J. Removal Mechanisms and Types of Constructed Wetlands. In *Constructed Wetlands for Wastewater Treatment in Europe*; Backhuys: Leiden, The Netherland, 1998; pp. 17–66.
3. Vymazal, J. Constructed wetlands for wastewater treatment. *Water* **2010**, *2*, 530–549.
4. Teng, C.J.; Leu, S.Y.; Ko, C.H.; Fan, C.; Sheu, Y.S.; Hu, H.Y. Economic and environmental analysis of using constructed riparian wetlands to support urbanized municipal wastewater treatment. *Ecol. Eng.* **2012**, *44*, 249–258.
5. Chen, G.Q.; Shao, L.; Chen, Z.M.; Li, Z.; Zhang, B.; Chen, H.; Wu, Z. Low-carbon assessment for ecological wastewater treatment by a constructed wetland in Beijing. *Ecol. Eng.* **2011**, *37*, 622–628.
6. U.S. Environmental Protection Agency (USEPA). *Wastewater Technology Fact Sheet Wetlands: Subsurface Flow*; EPA 832-F-00-023, Office of Water, USEPA: Washington, DC, USA, 2000.
7. U.S. Environmental Protection Agency (USEPA). *Manual Constructed Wetlands Treatment of Municipal Wastewaters*; EPA 625-R-99-010, USEPA: Cincinnati, OH, USA, 2000.
8. Hsu, Y.F.; Pan, W.C. Study on the Production and Labor of Oyster-Cultivating Industry in Tong-Shih: A Periphery in the Commodity Chain. In *Nanhua University Policy Research*; Nanhua University: Chiayi, Taiwan, 2003; pp. 105–138.

9. Lin, Y.F.; Jing, S.R. *Recycling of Seafood Solid Waste as Substrate Material Used in Constructed Wetland for Wastewater Treatment*; National Sciences Council: Taipei, Taiwan, 2006.

10. Chen, B.; Chen, Z.M.; Zhou, Y.; Zhou, J.B.; Chen, G.Q. Emergy as embodied energy based assessment for local sustainability of a constructed wetland in Beijing. *Commun. Nonlinear Sci. Numer. Simul.* **2009**, *14*, 622–635.

11. Seo, D.C.; Cho, J.S.; Lee, H.J.; Heo, J.S. Phosphorus retention capacity of filter media for estimating the longevity of constructed wetland. *Water Res.* **2005**, *39*, 2445–2457.

12. Park, W.H.; Polprasert, C. Roles of oyster shells in an integrated constructed wetland system designed for P removal. *Ecol. Eng.* **2008**, *34*, 50–56.

13. Huett, D.O.; Morris, S.G.; Smith, G.; Hunt, N. Nitrogen and phosphorus removal from plant nursery runoff in vegetated and unvegetated subsurface flow wetlands. *Water Res.* **2005**, *39*, 3259–3272.

14. Reichert, P. *AQUASIM 2.0: Computer Program for the Identification and Simulation of Aquatic Systems*; Swiss Federal Institute for Environmental Science and Technology (EAWAG): Dübendorf, Switzerland, 1998.

15. Environmental Protection Administration, Republic of China (ROCEPA). *Environmental Water Quality Information*; ROCEPA: Taipei, Taiwan, 2010. Available online: http://wq.epa.gov.tw/WQEPA/Code/?Languages=en (accessed on 1 February 2013).

16. Chang, C.F. Economical Analysis of Oyster Shells Contact Bed in Wastewater Treatment. M.S. Thesis, National Taiwan University, Taipei, Taiwan, 2009.

17. APHA. *Standard Methods for Examination of Water and Wastewater*, 20th ed.; American Public Health Association, American Water Works Association, Water Environment Federation: Washington, DC, USA, 1998.

18. Henrichs, M.; Langergraber, G.; Uhl, M. Modelling of organic matter degradation in constructed wetlands for treatment of combined sewer overflow. *Sci. Total Environ.* **2007**, *380*, 196–209.

19. Yang, C.P.; Kuo, J.T.; Lung, W.S.; Lai, J.S.; Wu, J.T. Water quality and ecosystem modeling of tidal wetlands. *J. Environ. Eng.* **2007**, *133*, 711–721.

20. Downing, A.L. Population Dynamics in Biological System. In Proceedings of 3rd International Conference of Water Pollution Research, Munich, Germany, 1966; Volume 2, pp. 117–137.

21. Ottova, V.; Balcarova, J.; Vymazal, J. Microbial characteristics of constructed wetlands. *Water Sci. Tech.* **1997**, *35*, 117–123.

22. Fountoulakis, M.S.; Terzakis, S.; Chatzinotas, A.; Brix, H.; Kalogerakis, N.; Manios, T. Pilot-scale comparison of constructed wetlands operated under high hydraulic,loading rates and attached biofilm reactors for domestic wastewater treatment. *Sci. Total Environ.* **2009**, *407*, 2996–3003.

23. Lopes, J.F.; Silva, C. Temporal and spatial distribution of dissolved oxygen in the Ria de Aveiro lagoon. *Ecol. Model.* **2006**, *197*, 67–88.

24. Hull, V.; Parrella, L.; Falcucci, M. Modelling dissolved oxygen dynamics in coastal lagoons. *Ecol. Model.* **2008**, *211*, 468–480.

25. Lin, T.S. The Effect of Variation of Tank Volume on Heterotrophic/Nitrifying Species in TNCU3 Activated Sludge Process. M.S. Thesis, Chaoyang University of Technology, Taichung, Taiwan, 2005.

26. Wynn, T.M.; Liehr, S.K. Development of a constructed subsurface-flow wetland simulation model. *Ecol. Eng.* **2001**, *16*, 519–536.

27. Reichert, P.; Borchardt, D.; Henze, M.; Rauch, W.; Shanahan, P.; Somlyody, L.; Vanrolleghem, P. River water quality model No. 1 (RWQM1): II. Biochemical process equations. *Water Sci. Tech.* **2001**, *43*, 11–30.

28. Henze, M. *Wastewater Treatment: Biological and Chemical Processes*; Springer: Berlin, Germany, 2002.

29. Poirier, M.R. Minimum Velocity Required to Transport Solid Particles from the 2H-Evaporator to the Tank Farm. Westinghouse Savannah River Company: Aiken, SC, USA, 2000; WSRC-TR-2000-00263.

30. Polprasert, C.; Khatiwada, N.R. An integrated kinetic model for water hyacinth ponds used for wastewater treatment. *Water Res.* **1998**, *32*, 179–185.

31. Tchobanoglous, G.; Burton, F.L. *Wastewater Engineering: Treatment, Disposal, and Reuse*; McGraw-Hill: New York, NY, USA, 1991.

32. Young, D.F.; Munson, B.R.; Okiishi, T.H.; Huebsch, W.W. *A Brief Introduction to Fluid Mechanics*; John Wiley & Sons: New York, NY, USA, 2010.

33. Baca, R.G.; Arnett, R.C. *A Limnological Model for Eutrophic Lakes and Impoundments*; Battelle, Pacific Northwest Laboratories, for USEPA, Office of Research and Development, Richland, WA, USA, 1976.

34. Martin, J.F.; Reddy, K.R. Interaction and spatial distribution of wetland nitrogen processes. *Ecol. Model.* **1997**, *105*, 1–21.

35. Jørgensen, S.E.; Nielsen, S.N.; Jørgensen, L.A. *Handbook of Ecological Parameters and Ecotoxicology*; Elsevier: Amsterdam, The Netherland, 1991.

36. Mayo, A.W.; Bigambo, T. Nitrogen transformation in horizontal subsurface flow constructed wetlands I: Model development. *Phys. Chem. Earth.* **2005**, *30*, 658–667.

37. Charley, R.C.; Hooper, D.G.; McLee, A.G. Nitrification kinetics in activated-sludge at various temperatures and dissolved-oxygen concentrations. *Water Res.* **1980**, *14*, 1387–1396.

38. Ferrara, R.A.; Harleman, D.R.F. Dynamic nutrient cycle model for waste stabilization ponds. *J. Environ. Eng. ASCE.* **1980**, *106*, 37–54.

39. Thibodeaux, L.J. *Environmental Chemodynamics: Movement of Chemicals in Air, Water, and Soil*; John. Wiley & Sons: New York, NY, USA, 1996.

40. Hill, D. Diffusion-coefficients of nitrate, chloride, sulfate and water in cracked and uncracked chalk. *J. Soil Sci.* **1984**, *35*, 27–33.

41. Domingos, R.F.; Benedetti, M.F.; Croué, J.P.; Pinheiro, J.P. Electrochemical methodology to study labile trace metal/natural organic matter complexation at low concentration levels in natural waters. *Anal. Chim. Acta* **2004**, *521*, 77–86.

42. Henze, M, *Activated Sludge Models ASM1, ASM2, ASM2d and ASM3*; IWA Publishing: London, UK, 2000.

43. Reed, S.C.; Crites, R.W.; Middlebrooks, E.J. *Natural Systems for Waste Management and Treatment*; McGraw-Hill: New York, NY, USA, 1998.

44. Ghermandi, A.; Vandenberghe, V.; Benedetti, L.; Bauwens, W.; Vanrolleghem, P.A. Model-based assessment of shading effect by riparian vegetation on river water quality. *Ecol. Eng.* **2009**, *35*, 92–104.

45. Henze, M.; Gujer, W.; Mino, T.; Matsuo, T.; Wentzel, M.C.; Marais, G.V.R.; van Loosdrecht, M.C.M. Activated sludge model No.2d, ASM2d. *Water Sci. Tech.* **1999**, *39*, 165–182.

46. Wanner, O.; Reichert, P. Mathematical modeling of mixed-culture biofilms. *Biotechnol. Bioeng.* **1996**, *49*, 172–184.

47. Wang, Y.C.; Lin, Y.P.; Huang, C.W.; Chiang, L.C.; Chu, H.J.; Ou, W.S. A system dynamic model and sensitivity analysis for simulating domestic pollution removal in a free-water surface constructed wetland. *Water Air Soil Pollut.* **2012**, *223*, 2719–2742.

48. Her, J.J.; Huang, J.S. Influences of carbon surface and C/N ratio on nitrate nitrite denitrification and carbon breakthrough. *Bioresour. Technol.* **1995**, *54*, 45–51.

49. Luederitz, V.; Eckert, E.; Lange-Weber, M.; Lange, A.; Gersberg, R.M. Nutrient removal efficiency and resource economics of vertical flow and horizontal flow constructed wetlands. *Ecol. Eng.* **2001**, *18*, 157–171.

50. Bouwer, E.J. Theoretical investigation of particle deposition in biofilm systems. *Water Res.* **1987**, *21*, 1489–1498.

51. Tufenkji, N.; Elimelech, M. Correlation equation for predicting single-collector efficiency in physicochemical filtration in saturated porous media. *Environ. Sci. Technol.* **2004**, *38*, 529–536.

52. Dixon, A.; Simon, M.; Burkitt, T. Assessing the environmental impact of two options for small-scale wastewater treatment: Comparing a reedbed and an aerated biological filter using a life cycle approach. *Ecol. Eng.* **2003**, *20*, 297–308.

Water **2013**, *5*, 505-524; doi:10.3390/w5020505

ISSN 2073-4441
www.mdpi.com/journal/water

Article

Microbial Community Structure of a Leachfield Soil: Response to Intermittent Aeration and Tetracycline Addition

Janet A. Atoyan [1], Andrew M. Staroscik [2], David R. Nelson [2], Erika L. Patenaude [1], David A. Potts [3] and José A. Amador [1],*

[1] Laboratory of Soil Ecology & Microbiology, 024 Coastal Institute, University of Rhode Island, Kingston, RI 02881, USA; E-Mails: jatoyan@gmail.com (J.A.A.); erikap7108@gmail.com (E.L.P.)

[2] Department of Cell and Molecular Biology, University of Rhode Island, Kingston, RI 02881, USA; E-Mails: ams@staroscik.com (A.M.S.); dnelson@uri.edu (D.R.N.)

[3] Geomatrix, LLC, Old Saybrook, CT 06475, USA; E-Mail: dpotts@geomatrixllc.com

* Author to whom correspondence should be addressed; E-Mail: jamador@uri.edu; Tel.: +1-401-874-2902; Fax: +1-401-874-4561.

Received: 22 January 2013; in revised form: 3 April 2013 / Accepted: 15 April 2013 / Published: 25 April 2013

Abstract: Soil-based wastewater treatment systems, or leachfields, rely on microbial processes for improving the quality of wastewater before it reaches the groundwater. These processes are affected by physicochemical system properties, such as O_2 availability, and disturbances, such as the presence of antimicrobial compounds in wastewater. We examined the microbial community structure of leachfield mesocosms containing native soil and receiving domestic wastewater under intermittently-aerated (AIR) and unaerated (LEACH) conditions before and after dosing with tetracycline (TET). Community structure was assessed using phospholipid fatty acid analysis (PLFA), analysis of dominant phylotypes using polymerase chain reaction-denaturing gradient gel electrophoresis (PCR–DGGE), and cloning and sequencing of 16S rRNA genes. Prior to dosing, the same PLFA biomarkers were found in soil from AIR and LEACH treatments, although AIR soil had a larger active microbial population and higher concentrations for nine of 32 PLFA markers found. AIR soil also had a larger number of dominant phylotypes, most of them unique to this treatment. Dosing of mesocosms with TET had a more marked effect on AIR than LEACH soil, reducing the size of the microbial population and the number and concentration of PLFA markers. Dominant phylotypes decreased by ~15% in response to TET in both treatments, although the AIR treatment retained a higher number of phylotypes than the LEACH treatment. Fewer than 10% of clones were common to both

AIR and LEACH soil, and fewer than 25% of the clones from either treatment were homologous with isolates of known genus and species. These included human pathogens, as well as bacteria involved in biogeochemical transformations of C, N, S and metals, and biodegradation of various organic contaminants. Our results show that intermittent aeration has a marked effect on the size and structure of the microbial community that develops in a native leachfield soil. In addition, there is a differential response of the microbial communities of AIR and LEACH soil to tetracycline addition which may be linked to changes in function.

Keywords: PLFA; PCR-DGGE; domestic wastewater; intermittent aeration; tetracycline

---

1. Introduction

An understanding of how microbial communities respond to changes in physicochemical conditions and disturbances is necessary for effective development and management of innovative soil-based wastewater treatment systems. Although microorganisms are universally acknowledged as key components in the treatment of septic tank effluent (STE) in soil-based systems, information about the size, structure and function of these microbial communities—and their response to changes in environmental conditions—is scant. This is in contrast with biological processes in centralized wastewater treatment plants, to which state-of-the-art molecular techniques have been applied to elucidate the structure and function of the microbial communities involved in wastewater renovation for some time [1].

Early studies examining microbial populations of soil absorption systems employed culture-based methods [2,3]. Culture-based analyses of the microbial community, although a useful first step, provide limited information, since only a fraction of the community—that amenable to growth under the conditions provided – can be analyzed using this approach [4]. Culture-based analyses of microbial communities can lead to erroneous conclusions regarding the importance of particular organisms in treatment processes and thus ineffective or counterproductive recommendations for their optimization.

Amador *et al.* [5] employed molecular techniques to examine the microbial community structure of soil-based treatment systems using mesocosms filled with synthetic sand. Phospholipid fatty acid (PLFA) and polymerase chain reaction-denaturing gradient gel electrophoresis (PCR-DGGE) analyses indicated that intermittent aeration affected the size and structure of the microbial community. Proteobacteria and actinomycetes/sulfate-reducing bacteria constituted a higher proportion of the community in the aerated treatment, whereas anaerobic Gram-negative bacteria/firmicutes were more prominent in the unaerated treatment. In addition, higher species richness was found in the aerated treatment. The marked effects of intermittent aeration on community structure of soil-based treatment systems are likely linked with improvements in water quality (e.g., BOD, nutrient and pathogen removal) resulting from aeration [6]. More recently Tomaras *et al.* [7] used 16S rDNA gene sequence analysis to assess microbial community diversity in onsite wastewater treatment systems (OWTS). They reported strong differences in community composition among septic tank effluent, the biomat at the infiltrative surface, and soil that had not received STE. Furthermore, there was no overlap of sequences between STE and biomat communities, with considerably less phylogenetic diversity in the latter.

In the present study we describe the results of a mesocosm-scale study at an OWTS research facility using mesocosms filled with native soil to simulate conventional and intermittently aerated soil treatment areas. STE amended with tetracycline (TET) was used to regularly dose the lysimeters for a period of 10 days. Tetracycline was chosen as the antibiotic for evaluation because: (i) it has been shown to persist in the environment by adsorbing to soils [8,9]; (ii) it is a broad-spectrum antibiotic used in human medicine that is effective against both Gram-negative and Gram-positive bacteria [10]; and (iii) several of its degradation products also have antibiotic activity [11]. The soil microbial community was characterized using PLFA analysis, PCR-DGGE, and cloning followed by 16S rDNA gene sequence analysis. Differences in community structure were examined between aerated and unaerated soil before the addition of TET, and in response to TET addition for each treatment.

2. Materials and Methods

2.1. Experimental Facility

The study was conducted at a research facility in southeastern Connecticut, USA built adjacent to a two-family home fitted with a conventional septic system. Three to six people inhabited the home continuously during the study. A detailed description of the facility can be found in Potts *et al.* [6]. To the best of our knowledge, none of the residents was taking antibiotics during the course of our study. Septic tank effluent was diverted to a high-density polyethylene (HDPE) storage tank (1325 L) above the laboratory in a climate-controlled room (17–19 °C) (Figure 1). STE from the storage tank was pumped every 6 h (3:00 a.m., 9:00 a.m., 3:00 p.m. and 9:00 p.m.) to dosing tanks in the laboratory. Levels of dissolved organic carbon in STE ranged from 71 to 121 mg C L^{-1}. The dose flowed by gravity from these tanks into mesocosms consisting of stainless steel lysimeters (35.6 cm i.d., 61 cm height) filled with a mixture of B and C horizon soil from a sandy-skeletal, mixed, mesic Typic Udorthent (particle size distribution: 92% sand, 8% silt), representative of soil used in OWTS construction in the southern New England, USA region. The soil was homogenized using a cement mixer prior to use. The remaining space constituted the headspace. The dose was delivered to the soil surface through a horizontal PVC pipe in which holes were drilled. The bottom of the mesocosms was filled with 7.5 cm of No. 4 silica sand overlaid with 30 cm of native soil. The mesocosms began receiving wastewater on 13 August 2003 at a rate of 4 cm day^{-1}. On 22 June 2004, this rate was increased to 12 cm day^{-1}, remaining constant for the duration of the experiment.

2.2. Aeration

The headspace of mesocosms was either vented to the septic system leachfield of the house to simulate a conventional leachfield atmosphere (LEACH treatment) or was aerated intermittently with ambient air (AIR treatment) using a process that has been employed successfully to rejuvenate hydraulically-failed septic systems [12]. Each treatment was replicated three times. Air was pumped at regular intervals into the headspace of the AIR mesocosms to maintain O_2 levels close to atmospheric (~0.21 mol mol^{-1}) (Figure 1).

Figure 1. (**a**) Schematic diagram of laboratory facility and (**b**) leachfield mesocosms employed in this study. Drawings are not to scale (after Patenaude *et al.* [13])

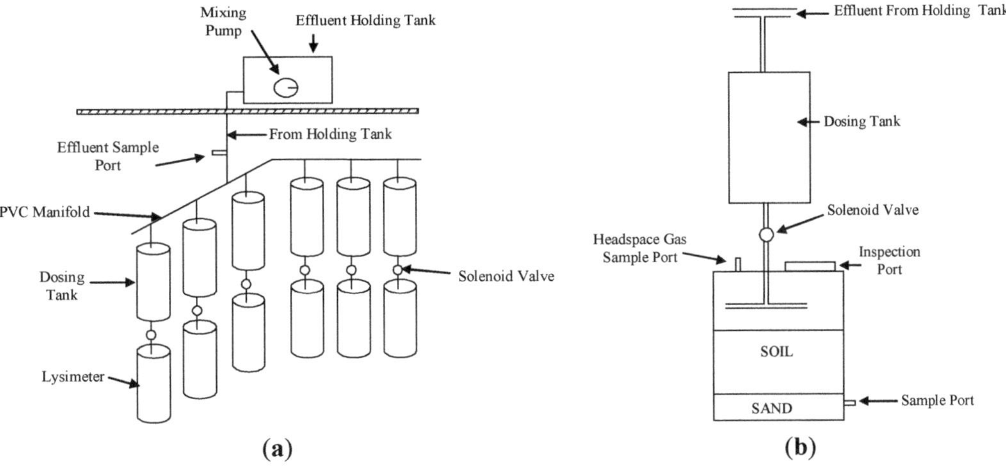

2.3. Antibiotic Dosing

Mesocosms were dosed with STE amended with tetracycline (final conc. = 5 mg L^{-1}) every 6 h for 10 days, beginning on 13 June 2005 at 3 p.m. (Day 0). The rationale for antibiotic dosing along with wastewater properties, are described in Patenaude *et al.* [13] and Atoyan *et al.* [14]. To amend the wastewater with TET, an aqueous stock solution (500 mg tetracycline HCl L^{-1}; CAS 64-75-5, Sigma Aldrich, Saint Louis, MO, USA) was prepared and kept at ~8 °C in an insulated container packed with ice and equipped with an IceProbe® thermoelectric water chiller (Coolworks®, San Rafael, CA, USA). A peristaltic pump (Thomas Scientific, Swedesboro, NJ, USA) was actuated by a solenoid valve to deliver ~28 mL of TET stock solution to the horizontal PVC pipe within the lysimeters (Figure 1) every 6 h, coincident with wastewater dosing. This mixed the antibiotic stock solution with the wastewater as it flowed into the lysimeters.

2.4. Soil Sampling

Soil samples (4-cm deep) were collected on Days 0 and 11. Approximately 4 h prior to the 3 p.m. dosing event the access port was opened, and STE on the soil surface of the LEACH mesocosms was removed by siphoning and stored. No STE had accumulated on the soil surface of AIR mesocosms, thus there was no need for removal. Five soil cores (2.75-cm dia., 4-cm height) were taken aseptically from each mesocosm using cut-off, 60-mL plastic syringes. STE was returned to the mesocosms after soil sampling. Soil cores were placed in sterile Whirl-Pak® bags and kept on ice during transport to the laboratory. Immediately upon returning to the laboratory, 50 g of homogenized soil from each mesocosm was shipped on ice by overnight courier to Microbial Insights, Inc. (Rockford, TN, USA) for PLFA analysis. The remaining soil was stored at −80 °C for subsequent analysis.

2.5. Phospholipid Fatty Acid Analysis

PLFAs were extracted using a modification [15] of the method of Bligh and Dyer [16], with one soil sample analyzed per mesocosm. Fatty acid methyl esters were separated by gas chromatography and identified by retention time and mass spectrometry as described by Tunlid *et al.* [17]. The detection limit was 7 pmoles of PLFA. For the purpose of community structure analysis, PLFAs were divided into markers for six different microbial groups [18–21]: (i) firmicutes/anaerobic Gram-negative bacteria, (ii) proteobacteria, (iii) anaerobic metal reducers, (iv) sulfate-reducing bacteria (SRB)/actinomycetes, (v) general bacteria, and (vi) eukaryotes.

2.6. DNA Extraction from Soil

DNA was extracted from ~1 g homogenized soil from each mesocosm using the bead-beating UltraClean Soil DNA Isolation kit (MoBio, Carlsbad, CA, USA) per manufacturer's instructions. DNA was further purified by spin-column chromatography following the protocol for BD Chroma Spin + TE-100 columns (Clontech, Mountain View, CA, USA), and concentrated by ethanol precipitation and resuspension in 20 µL EB buffer.

2.7. PCR-DGGE

Extracted DNA was amplified by polymerase chain reaction (PCR) with the primers 518R (5'-ATT ACC GCG GCT GCT GG-3') and 357F-GC (5'-CCT ACG GGA GGC AGC AGC GCC CGC CGC GCG CGG CGG GCG GGG CGG GGG CAC GGG GGG-3') specific for the 16S rDNA gene of bacteria, modified from Marchesi *et al.* [22] by the addition of a GC clamp [23]. Four PCR reactions were performed for each replicate mesocosm. PCR was performed using the Taq PCR Master Mix kit (Qiagen, Valencia, CA, USA) following the manufacturer's protocol with 10 ng of template DNA per 50 µL reaction. PCR was performed in a GeneAmp thermocycler (Applied Biosystems, Foster City, CA, USA) under the following conditions: initial denaturation at 94 °C for 5 min, followed by 30 cycles of 94 °C for 30 s, 55 °C for 30 s, and 72 °C for 1 min 30 s, and a final extension at 72 °C for 7 min. PCR products were purified and concentrated using the Qiaquick PCR Purification kit (Qiagen). The products from all four PCR reactions from a mesocosm were applied to one column and quantified using an Ultrospec 4000 spectrophotometer (Pharmacia Biotech, Piscataway, NJ, USA).

Approximately 200 ng of PCR product per lane was loaded onto a polyacrylamide gel for generation of community profiles. Electrophoresis was run as described by Muyzer *et al.* [24] using a CBS Scientific DGGE system (Del Mar, CA, USA) on a 0.75-mm thick, 8% (*w/v*) polyacrylamide gel with a gradient from 60% to 40% denaturant, where 100% denaturant had a concentration of 7 M urea and 40% (*v/v*) formamide. The gel was run in 0.5 × TAE buffer for 16 h at 200 V and 60 °C and stained for 30 min in SYBR Green dye. The gel was visualized using a Typhoon 9410 variable mode imager. Bands were identified using ImageJ software [25] with rolling ball subtraction (r = 10).

2.8. Clone Libraries

Extracted DNA was amplified by PCR with primers B27f (5'-AGA GTT TGA TCC TGG CTC AG-3') and 1387R (5'-GGG CGG WGT GTA CAA GGC-3'), specific for the 16S rDNA of bacteria [22]. Four

PCR reactions were performed for each replicate mesocosm. PCR, amplicon purification, and quantification were performed as for PCR-DGGE analysis. Four clone libraries were constructed: one per treatment—AIR and LEACH—for Day 0 and Day 11. Cloning reactions were performed following the standard protocol for the TOPO TA Cloning Kit for Sequencing (Invitrogen, Chicago, IL, USA) using mixed PCR product from each of the three replicates per treatment weighted by the concentration of DNA in each replicate. Approximately 100 colonies were then chosen randomly for sequencing on a Beckman Coulter CEQ 8000 using the primer B27f. Clone library sequences were aligned and chimeric sequences were removed using the NAST alignment tool and Bellerophon [26]. Clones were analyzed for phylogenetic similarity using the Greengenes DNA maximum likelihood (DNAML) classification tool.

2.9. Data Analysis

The Dice similarity coefficient, Cs, was calculated as described by Amador *et al.* [5]. Indices of richness (S) were calculated based on Staddon *et al.* [27]. Paired *t*-tests were used to compare the responses of this variable to TET addition (Day 0 *vs.* Day 11) within a particular treatment. The *p* value for all analyses was <0.05. Principal component analysis was performed on PLFA concentration (expressed as nmoles g^{-1} soil) and the DGGE presence/absence matrix using XLSTAT (Version 2008.1; Addinsoft, New York, NY, USA).

3. Results

3.1. Effects of Intermittent Aeration

3.1.1. PLFA Analysis

A total of 37 different PLFAs were detected on Day 0 from all AIR and LEACH treatments, of which 32 were common to all six mesocosms (data not shown). The active microbial biomass—represented by the total concentration of PLFA in a sample—prior to the addition of tetracycline was approximately twice as high in AIR as in LEACH soil (Table 1) and was significantly different. The main group contributing to total PLFA in both treatments was Proteobacteria, which accounted for a significantly larger proportion of the community in AIR (64%) than in LEACH soil (54%). In addition, the contribution of anaerobic metal reducers to total PLFA was significantly higher in the AIR treatment. General markers for bacteria, SRB/Actinomycetes and Firmicutes/anaerobic Gram-negative bacteria made up a significantly higher fraction of total PLFA in soil from the LEACH treatment. Eukaryotes constituted approximately 3% of the total PLFA in both treatments.

The Dice similarity coefficient (Cs)—computed from a presence/absence matrix of individual PLFAs—was 0.97, indicating a high degree of similarity between AIR and LEACH treatments. When principal component analysis was performed based on the concentration of individual PLFAs, there was clear separation between AIR and LEACH treatments along PC1 and PC2, which explained 96.8% and 2.2% of the variability, respectively (Figure 2).

Table 1. Active microbial biomass and relative amounts of PLFA for different microbial groups in AIR and LEACH soils before (Day 0) and after (Day 11) the tetracycline dosing period. Values are means (*n* = 3).

					Community structure			
Tmt	Day	Total PLFA concentration [a] (nmol g^{-1} soil)	Firmicutes/ Anaerobic G$^-$ bacteria	Proteobacteria	Anaerobic metal reducers	Actinomycetes /SRB	General	Eukaryotes
					% of total PLFA			
AIR	0	**117,673**	**9.3**	**63.5**	**2.4**	**0.6**	**21.1**	3.1
	11	55,305	8.5	61.2	2.4	0.8	21.5	5.5
LEACH	0	58,599	13.3	54.2	1.9	1.3	26.8	2.6
	11	53,819	11.9	55.6	1.9	1.3	26.5	2.9

Note: [a] Significant differences between AIR and LEACH treatments on Day 0 are indicated in **bold**.

Figure 2. (**a**) Principal component analysis based on PLFA concentration and (**b**) dominant phylotypes in soil from replicates of intermittently-aerated (AIR; A1, A2, A3) and unaerated (LEACH; L1, L2, L3) leachfield mesocosms before (Day0) and after (Day11) dosing with tetracycline.

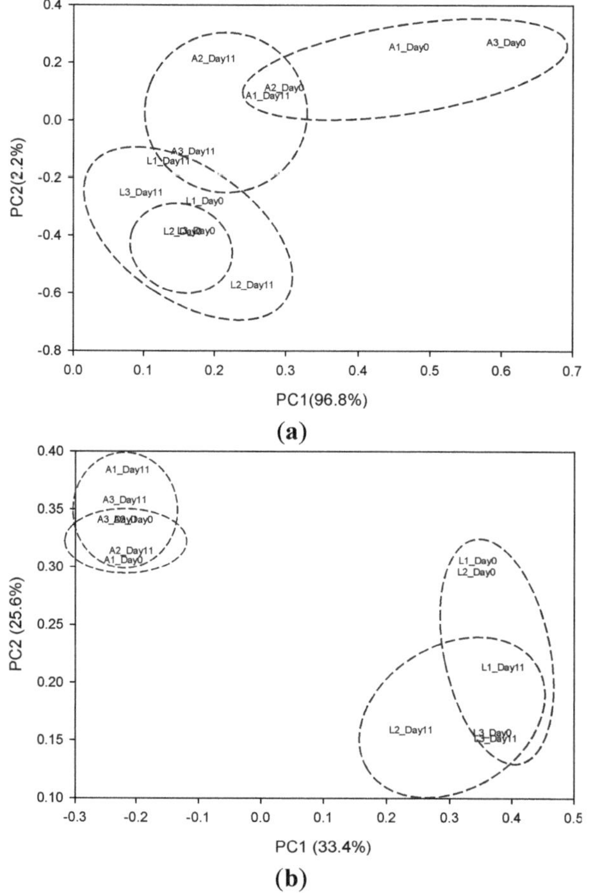

3.1.2. PCR-DGGE Analysis

A total of 10 DGGE bands—or dominant phylotypes—were common to all six mesocosms from both treatments (data not shown). An average of 51 bands was detected in AIR soil, of which 49 were common to all three replicates in the AIR treatment. Soil from the LEACH treatment had an average of 27 bands, of which only 16 were common to all three replicates, indicating greater variability in the composition of the microbial community among replicate LEACH mesocosms. Of all the bands detected in all replicates within a treatment, 20 were unique to the AIR treatment and four were unique to the LEACH treatment. Species richness—based on the number of bands detected—was significantly higher in AIR soil (Table 2). The Dice similarity coefficient computed from the PCR-DGGE presence/absence data showed clear differences between soil from the LEACH and AIR treatments, with a Cs of 0.78. Similarly, principal component analysis based on DGGE data clearly separated AIR and LEACH treatments along PC1 and PC2, which explained 33.4 and 24.5% of the variation between treatments, respectively (Figure 2).

Table 2. Richness (S) index based on PCR-DGGE data for intermittently aerated (AIR) and unaerated (LEACH) soil from leachfield mesocosms before (Day 0) and after (Day 11) dosing with tetracycline.

Treatment	Day 0	Day 11
AIR	**50.7**	<u>44.0</u>
LEACH	27.0	<u>23.0</u>

Notes: Significant differences between AIR and LEACH treatments on Day 0 are indicated in **bold**; significant differences between Day 0 and Day 11 within a treatment are indicated by <u>underlining</u>.

3.1.3. Clone Libraries

Analysis of clone libraries also indicated that there were differences in community composition between treatments. Of all the clones obtained, a total of 87 and 82 were sequenced from AIR and LEACH soil, of which 70 and 69 were free of chimeras and subjected to matching. Within these sequences, there were 42 and 48 unique operational taxonomic units (OTUs) in the AIR and LEACH soil, respectively. Bacteria from 10 different phyla were detected in both treatments (7 in AIR and 8 in LEACH soil) (Figure 3). Of these, five were common to both treatments (Acidobacteria, Actinobacteria, Bacterioidetes, Firmicutes, Proteobacteria), with two phyla unique to AIR soil (Cyanobacteria and Nitrospirae) and three unique to LEACH soil (Planctomycetes, Spirochaetes and Verrucomicrobia). As was the case for PLFA analysis, the soil microbial community from both treatments was dominated by Proteobacteria, which accounted for 77% and 45% of all clones in AIR and LEACH soil, respectively (Figure 3). Within this phylum, the class α-Proteobacteria accounted for 29% and 35% of all clones in AIR and LEACH soil, respectively.

Figure 3. Relative distribution of clones in different phyla in soil from intermittently aerated (AIR) and unaerated (LEACH) leachfield mesocosms before (Day 0) and after (Day 11) dosing with tetracycline.

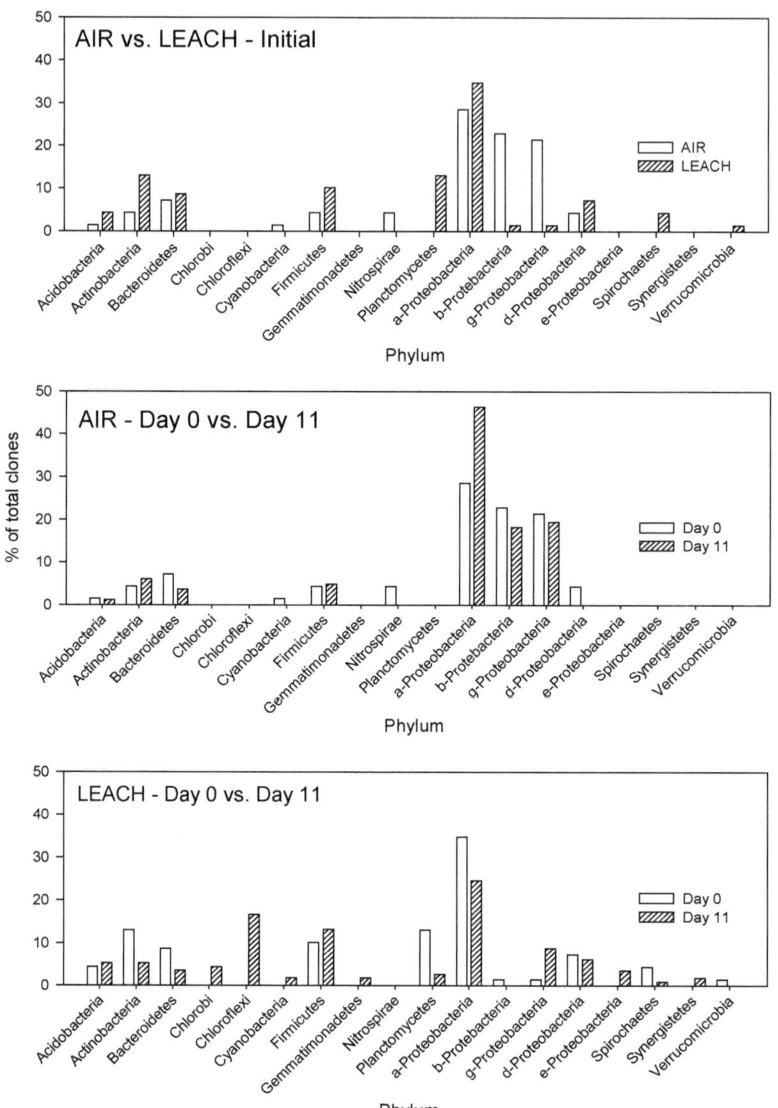

Only four OTUs were common to both treatments, belonging to the β-Proteobacteria, γ-Proteobacteria, Firmicutes, and Acidobacteria. None of these common OTUs met the 97% similarity threshold for identification. Homology with an isolate of known genus and species was observed for 21% of the OTUs from the AIR treatment and 22% of those from the LEACH treatment. Of the clones analyzed from AIR soil, 17 were identified with a particular genus or genus and species (applying a 97% similarity threshold for identification), whereas nine clones from LEACH soil were identified with a genus or genus and species (Table 3).

Table 3. Phylum, genus and species (closest match; similarity $\geq$ 97%) and potential function for OTUs from intermittently aerated (AIR) and unaerated (LEACH) soil from leachfield mesocosms before (Day 0) and after (Day 11) dosing with tetracycline. Dark squares indicate the presence of an OTU in a treatment.

Phylum	Genus and species	AIR Day 0	AIR Day 11	LEACH Day 0	LEACH Day 11	Potential function
Acidobacteria	*Terriglobus roseus*		■		■	Extracellular polysaccharide production [28]
Actinobacteria	*Leucobacter komagatae*		■			Biosurfactant production [29]
	Mycobacterium arupense		■			Pathogen [30]
	Mycobacterium sp.	■		■	■	Pathogen; PAH degradation [30,31]
	Rhodococcus coprophilus		■			Phenol degradation [32]
Bacteroidetes	*Flavobacterium succinicans*	■				Cellulose & polysaccharide degradation [33]
Firmicutes	*Bacillus* sp.			■	■	Pathogen; various
	Clostridium sp.			■	■	Pathogen; various
Nitrospirae	*Nitrospira* sp.	■				NO$_2^-$ oxidation [34]
α-Proteobacteria	*Caulobacter* sp.				■	Unknown [35]
	Phenylobacterium sp.	■	■		■	Degradation of chlorinated N-heterocyclics & linear alkylbenzenesulfonates [36]
	Beijerinckia sp.			■		Non-symbiotic N fixation; degradation of aromatic compounds [37]
	Afipia sp.		■	■	■	Pathogen [38]
	Bradyrhizobium elkanii		■	■	■	Symbiotic N fixation [39]
	Nitrobacter vulgaris		■	■	■	NO$_2^-$ oxidation [40]
	Methylocystis parvus	■				CH$_4$ oxidation [41]
	Methylocystis sp.			■		CH$_4$ oxidation [41]
	Labrys sp.		■			Unknown
	Erythrobacter sp.	■	■			Aerobic phototrophic bacteria
	Sphingobium sp.		■			Degradation of phenolic compounds [42]
	Sphingopyxis sp.	■				Degradation of polyvinyl alcohols [42]
β-Proteobacteria	*Acidovorax defluvii*		■			Denitrification [43]
	Acidovorax facilis	■				Degradation of polyhydroxyalkanoates [44]
	Thiobacillus sp.		■			Fe, S & S^{2-} oxidation
	Dechloromonas sp.	■	■			Perchlorate reduction [45]
	Rhodocyclus tenuis	■	■			Purple, non-S photosynthetic bacteria; methanol & formate oxidation
	Zoogloea ramigera	■				Extracellular polysaccharide production
δ-Proteobacteria	*Desulfovibrio desulfuricans*			■		SO$_4^{2-}$ & NO$_3^-$ reduction
γ-Proteobacteria	*Legionella pneumophila*				■	Pathogen [46]
	Methylosarcina sp.				■	Methane oxidation [47]
	Pseudomonas stutzeri	■				Pathogen; denitrification; degradation of CCl$_4$ [48–50]
	Pseudomonas umsongensis	■				Various [51]
	Pseudomonas sp.	■				Various
	Luteibacter rhizovicinus	■			■	Chitin degradation [52]
	Lysobacter sp.	■	■		■	Glucan & chitin degradation [53]
	Thermomonas sp.	■	■			Fe^{2+} oxidation; NO$_3^-$ reduction [54]

3.2. Effects of Tetracycline

3.2.1. PLFA Analysis

The mass of PLFA in the AIR soil declined to 55,305 nmol PLFA g^{-1} soil in response to TET addition, nearly 50% of the value on Day 0. By contrast, total mass of PLFA in LEACH soil declined by only 8% (Table 1). These effects were not statistically significant for either treatment. Total PLFA values were similar for AIR and LEACH treatments after TET dosing. The relative contribution of different microbial groups to total PLFA in soil from the LEACH treatment was not significantly affected by the addition of tetracycline (Table 1).

The total number of PLFAs detected in LEACH soil declined from 37 on Day 0 to 34 after TET dosing. Four previously present PLFA general bacteria markers were absent on Day 11. In addition, one previously absent marker for eukaryotes was present following TET dosing. TET dosing had no significant effect on the concentration the PLFA markers present in LEACH soil on both Day 0 and Day 11, nor did it affect the relative contribution of different microbial groups to total PLFA.

The total number of PLFAs detected in AIR soil declined from 36 on Day 0 to 32 after TET dosing (Table 3). Four previously present general markers for bacteria in AIR soil were absent following TET dosing—these were the same markers lost in response to TET dosing in soil from the LEACH treatment. TET dosing had no significant effect on species richness (Table 2) in the AIR treatment. The contribution of different microbial groups to total PLFA in AIR soil was minimally affected by TET dosing, with only the contribution of Proteobacteria decreasing significantly from 64% on Day 0 to 61% on Day 11 (Table 1).

Principal component analysis performed on individual PLFA concentrations showed separation between Day 0 and Day 11 for the AIR treatment along PC1, which accounted for 96.8% of the variability (Figure 2), but no separation was observed for the LEACH treatment. PC2, which explained 2.2% of the variability, did not separate Day 0 and Day 11 for either treatment.

3.2.2. PCR-DGGE Analysis

The number of dominant phylotypes common to all replicates in both treatments declined from 10 to 4 (data not shown) after TET dosing. An average of 44 bands was present in AIR mesocosms (a decline of ~13%), of which 35 were common to all three replicates. A total of 36 phylotypes persisted in soil from all AIR replicates following TET dosing. One phylotype absent on Day 0 was detected in soil from all three replicates in the AIR treatment on Day 11. The average number of DGGE bands in soil from the LEACH treatment decreased to 23 in response to TET (a decline of ~15%, from an average of 27 on Day 0). Of these, 10 were common to all replicates and no new phylotypes were detected. Species richness based on number of OTUs was significantly lower in both treatments following TET dosing (Table 2). Principal component analysis based on PCR-DGGE data did not separate pre- and post-TET dosing communities in either treatment (Figure 2).

3.2.3. Clone Libraries

The total number of clones sequenced from AIR and LEACH soil after TET addition was 82 and 84, respectively. Of these, 45 unique OTUs were identified in the AIR treatment and 62 in the LEACH treatment. The number of unique phyla in the AIR treatment declined from seven before TET dosing to five after, with Cyanobacteria and Nitrospirae absent following TET dosing (Figure 3). Proteobacteria continued to dominate the distribution of phyla after TET dosing, accounting for 85% and 46% of all clones in AIR and LEACH soil, respectively. Eight different OTUs persisted in soil from the AIR treatment after TET dosing: one Firmicute and seven Proteobacteria (Table 3).

Eight phyla were represented in LEACH soil before TET addition, whereas 12 phyla were present after dosing with antibiotic (Figure 3). Chlorobi, Chloroflexi, Cyanobacteria, Gemmatimonadetes, and Synergistetes were newly detected, whereas Verrucomicrobiales were lost from the community following TET addition. The microbial community of LEACH soil was dominated by Proteobacteria before and after TET dosing, accounting for 45% and 43% of total clones on Day 0 and Day 11, respectively. A total of 18 OTUs persisted after TET addition, belonging to six phyla: Acidobacteria, Actinobacteria, Bacteroidetes, Firmicutes, Planktomycetes and Proteobacteria. The persistent OTUs included *Mycobacterium* sp., *Bacillus* sp., *Clostridium* sp., *Afipia* sp., *Bradyrhizobium elkanii*, and *Nitrobacter vulgaris* (Table 3).

4. Discussion

4.1. Effects of Intermittent Aeration

LEACH mesocosms have elevated levels of CH_4, H_2S, and CO_2, and levels of O_2 that are considerably below ambient. In addition, dissolved oxygen (DO) levels in drainage water are low and levels of Fe^{2+} are high [6,13]. By contrast, aerobic conditions prevail in AIR mesocosms, evidenced by ambient levels of O_2 in the headspace, near saturation levels of DO in drainage water, and the absence of Fe^{2+} in drainage water [6,13]. The pH of soil and drainage water of LEACH mesocosms is near-neutral, whereas in AIR mesocosms it is acidic [13]. In addition, levels of dissolved organic carbon in drainage water are consistently higher in LEACH (65 to 105 mg C L^{-1}) than in AIR (6 to 20 mg C L^{-1}) mesocosms [13]. These differences in physicochemical properties and carbon availability within LEACH and AIR mesocosms argue for divergence in community composition, which we observed and discuss below. However, the presence of PLFA markers for the same groups of organisms, as well as shared phylotypes and OTUs found in both treatments, indicates that there is a fraction of the microbial community that is present under both sets of environmental conditions. The disparate conditions under which these organisms are found suggest that many of these are facultative anaerobes capable of tolerating a wide range of pH values and high levels of H_2S, and the presence and absence of O_2. Furthermore, the PLFA markers common to both treatments represent a wide range of active prokaryotic and eukaryotic organisms, and the common OTUs represent three different prokaryotic phyla, suggesting that this tolerance is present across a broad range of taxa.

Beyond the fraction of the microbial community shared by both treatments, there was considerable divergence among these communities in terms of size, richness and diversity. The size of the active microbial population in AIR soil was larger (Table 1), the relative amounts of PLFA contributed by microbial groups were different (Table 1), and a number of individual PLFA markers were present at higher levels in the AIR treatment. The two communities were clearly separated based on the concentration of PLFA markers and presence/absence of dominant phylotypes by principal component analysis (Figure 2). The AIR soil had a larger number of dominant phylotypes and, of the phylotypes present in all replicates within a treatment, there were 5× more that were unique to the AIR soil community. In addition, only 4.8% of all unique OTUs were common to both treatments. The fact that the soil used in our mesocosms and the STE inputs were the same for both treatments, suggests that differences in microbial community structure are being driven by intermittent aeration.

The larger community size and greater species richness in AIR mesocosms are consistent with the expectations for ecosystems with few physicochemical constraints [55]. In a previous study at the same experimental facility on the effects of intermittent aeration in leachfield mesocosms filled with synthetic silica sand, Amador *et al.* [5] observed differences between AIR and LEACH treatments using PLFA and PCR-DGGE analysis similar to those observed in the present study using mesocosms filled with native soil. The similarities in response to aeration for mesocosms filled with media with such different physical, chemical and biological properties (synthetic sand *vs.* native soil) further suggest that intermittent aeration exerts an important control on the structure of the leachfield microbial communities that develop.

Species accumulation curves indicated that the full diversity of these soils was not covered by the number of clones sequenced (data not shown). Thus we are unable to quantitatively evaluate differences in species composition between AIR and LEACH treatments. Nevertheless, the genus (and in some instances, species) of bacteria found in soil from AIR and LEACH treatments prior to TET addition provide us with a qualitative picture of the presence of pathogens as well as bacteria that may be involved in biogeochemical transformations and metabolism of organic pollutants (Table 3). AIR soil had bacteria in the genus *Mycobacterium* and the species *Pseudomonas stutzeri*, and, in LEACH soil, bacteria in the genus *Mycobacterium*, *Bacillus*, *Clostridium* and *Afipia* were present. All of these genera include species known to be human pathogens. Among bacteria with the capacity to be involved in biogeochemical processes in AIR soil we found *Nitrospira* (nitrite oxidation), *Methylocystis parvus* (methane oxidation), *Flavobacterium succinicans* (cellulose, polysaccharide degradation), *Erythrobacter* (aerobic phototrophic bacteria); *Rhodocyclus tenuis* (purple non-sulfur photosynthetic bacterium; methanol, formate oxidation), *Zooglea ramigera* (extracellular polysaccharide production), *Pseudomonas stutzeri* (denitrification), *Luteibacter rhizvicinus* (chitin degradation), *Lysobacter* sp. (glucan, chitin degradation), and *Thermomonas* (iron oxidation, nitrate reduction). Bacteria involved in biogeochemical processes found in LEACH soil include *Beijerinckia* (non-symbiotic nitrogen fixation), *Bradyrhizobium elkanii* (symbiotic nitrogen fixation), *Nitrobacter vulgaris* (chemoautotrophic nitrite oxidation), *Methylocystis* (methane oxidation), and *Desulfovibrio desulfuricans* (sulfate, nitrate reduction). Bacteria with potential for metabolism of organic contaminants found in AIR soil include *Phenylobacterium* (degradation of N-heterocyclic chlorinated compounds), *Sphingopyxis* (degradation of polyvinyl alcohols), *Acidovorax facilis* (degradation of polyhydroxyalkanoates), *Dechloromonas* (perchlorate reduction), *P. stutzeri* (carbon tetrachloride degradation), as well as *Nitrospira*, *Nitrobacter*, and *Methylocystis*, known to oxidize a variety of aromatic and low-molecular weight halogenated alkanes.

4.2. Effects of Tetracycline

Dosing of mesocosms with TET for 10 days caused a decrease in microbial biomass in AIR mesocosms (to the level observed in the LEACH treatment), whereas TET had no effect on biomass in the latter. The differential effect of TET is likely associated with the physiological state of the microbial community in AIR mesocosms and the mode of action of the antibiotic. AIR soil has been shown to have population densities of bacteriovores (protozoa and nematodes) that are orders of magnitude larger than LEACH mesocosms, and their grazing activities are expected to keep the microbial community in a continuous state of growth [5]. Tetracycline is a bacteriostatic agent—it does not directly kill bacteria but rather prevents protein synthesis, thereby inhibiting their growth [10]. Grazing of bacteria by protozoa and nematodes in AIR soil likely lowers the biomass, and tetracycline prevents bacterial replication, resulting in a greater impact on the active microbial biomass in AIR. By contrast, there is less grazing pressure in the LEACH soil, where protozoa and nematode numbers are lower and bacteria are less likely to be in the growth phase, thus this treatment was less affected by TET dosing.

Species richness generally decreased in both treatments in response to antibiotic dosing, with some of the effects of tetracycline addition on community composition shared by both treatments. For instance, four PLFA biomarkers for general bacteria that were present in soil from both treatments prior to dosing were absent in both treatments following tetracycline addition. In addition, of the 10 dominant phylotypes shared by all replicates in both treatments, six were absent after dosing with tetracycline. Thus, there is a fraction of the microbial community present in both treatments that is susceptible to the effects of tetracycline. However, analysis of dominant phylotypes indicates that a large proportion of the microbial community persists following TET dosing, as indicated by the persistence of ~70% and ~90% of previously present bands in the AIR and LEACH treatments, respectively, following TET dosing.

Beyond the shared responses, there were a number of differences in community structure in response to TET dosing. Whereas dosing had little effect on the relative contribution of different microbial groups to total PLFA in LEACH soil, in the AIR soil it resulted in a significantly lower contribution of Proteobacteria. Furthermore, there were lower concentrations of biomarkers for anaerobic Gram-negative/Firmicutes, anaerobic metal reducers, and general bacteria. PLFA biomarkers whose concentration declined likely represent those organisms that were actively growing in soil. These results also suggest that TET affects most of the groups that make up this community, as expected for a broad spectrum antibiotic. The overall effects of TET on AIR soil communities—as measured by PLFA analysis—are likely the result of shared susceptibility to the antibiotic and/or indirect effects of TET, such as selection for resistant bacteria.

The detection of OTUs and PLFAs only after TET dosing in soil from both treatments suggests that some of the effects of the antibiotic on these microbial communities are indirect. For example, TET dosing may have suppressed competing organisms, allowing otherwise less competitive—but TET-resistant—organisms to grow in numbers. Alternatively, TET may be used as a carbon source by some bacteria, as has been shown for a number of other antibiotics in soil [56], selecting for organisms capable of this function. These interpretations must be tempered by the limitations of the PCR-based methods used, which tend to result in a picture of the bacteria community that is skewed towards the most numerous organisms. Thus, lack of detection of an OTU prior to TET addition may not be due to its absence from soil, but rather to its low population density. Independent of mechanism, the eleven

OTUs that were detected only after TET addition to AIR soil (Table 3) were associated with a variety of potential functions, including pathogens (*Mycobacterium arupense, Afipia* sp.), degradation of aromatic compounds (*Rhodococcus coprophilus, Sphingobium* sp.), production of surfactants and polysaccharides (*Terriglobus roseus, Leucobacter komagatae*), nitrogen cycling (*Bradyrhizobium elkanii, Nitrobacter vulgaris, Acidovorax defluvii*), and iron and sulfur transformations (*Thiobacillus* sp.). The seven OTUs found only in LEACH soil after TET dosing (Table 3) also represented a variety of potential functions, including pathogens (*Legionella pneumophila*), extracellular polysaccharide production (*Terriglobus roseus*), degradation of heterocyclic compounds (*Phenylobacterium*), methane oxidation (*Methylosarcina* sp.), and degradation of chitin and glucans (*Luteibacter rhizovicinus, Lysobacter* sp.).

The differential effects of TET dosing on the community structure of AIR and LEACH soil would be expected to affect the community function in these ecosystems. For example, our data for the AIR mesocosms—although limited in terms of genus and species identified with a particular function (Table 3)—suggest that a number of processes in this treatment may be unaffected by TET dosing (e.g., Fe oxidation, NO_3 reduction), whereas some may diminish (e.g., degradation of polyhydroxyalkanoates), and others may be enhanced (e.g. phenol degradation). In a companion study Patenaude *et al.* [13] reported lower concentrations of Fe^{2+} and SO_4^{2-} in drainage water and higher levels of H_2S and CH_4 in the headspace of LEACH mesocosms dosed with TET. Effects on iron and sulfate concentrations were apparent for at least six weeks after antibiotic additions ceased, whereas gas levels returned to pre-dosing conditions shortly after dosing stopped. Some of the organisms that disappeared in response to TET dosing in LEACH mesocosms may represent iron-reducing and/or sulfur-oxidizing bacteria susceptible to TET. Changes in H_2S and CH_4 levels suggest that some of the absent organisms were also associated with sulfide- and methane-oxidizing bacteria sensitive to TET, with the transient nature of the effect suggesting eventual recovery of these populations. Our results lend qualitative support to this interpretation, as suggested by the loss of *Methylocystis* sp. and *Desulfovibrio desulfuricans* from the LEACH soil following TET dosing (Table 3). Within AIR mesocosms, a transient decrease in N removal capacity was observed by Patenaude *et al.* [13] in response to TET dosing, which was ascribed to inhibitory effects on nitrification (Patenaude *et al.* [13]). We observed the disappearance of *Nitrospira* sp., which carries out nitrite oxidation, in response to TET dosing of AIR mesocosms (Table 3). In addition, diminished N removal may also be associated with effects on denitrifiers, which could be reflected in the lower concentrations of various PLFAs observed in response to antibiotic dosing, since the capacity to denitrify is associated with a wide range of bacteria [50]. The relatively small effect of TET dosing on the water quality functions of AIR mesocosms [13] is in contrast with the various negative effects of TET on microbial community structure observed in the present study. This disparity may be the result of greater functional redundancy and/or prevalence of TET resistance within the microbial community of AIR soil, which may make OWTS that incorporate this technology more resilient to environmental disturbances.

5. Conclusions

Our results suggest that the microbial communities of intermittently aerated and unaerated leachfield native soil can differ markedly with respect to size and structure. Leachfield soil under intermittent aeration has a larger active microbial biomass and significantly higher richness and diversity of taxa, as indicated by data from PLFA and PCR-DGGE analysis. Qualitative analysis of community function based on sequencing of OTUs suggests that there may also be differences in the presence or absence of pathogenic bacteria and bacteria involved in elemental cycling and degradation of organic contaminants. Tetracycline dosing appears to have a differential effect on the leachfield communities, with intermittently aerated soil exhibiting greater loss of active microbial biomass and a higher proportional loss of richness and diversity relative to unaerated soil. These data provide evidence that the size, structure and function of the microbial community of leachfield soil can be manipulated by the introduction of air. Furthermore, the introduction of air can also affect the response of the community to disturbances such as short-term exposure to antibiotics relative to unaerated soil.

Acknowledgments

This research was supported by funds from the Rhode Island Agricultural Experiment Station (Contribution No. 5334) to J.A. Amador, by a grant from the University of Rhode Island's Office of the Provost to J.A. Atoyan, and by a USDA Cooperative State Research, Education and Extension Service Award (grant no. 2006-34438-17306) to D.R. Nelson. We are grateful to RI-INBRE for the use of its Research Core Facility, supported by Grant # P20 RR16457 from NCRR and NIH. This research was based in part upon work conducted using the Rhode Island Genomics and Sequencing Center, which is supported in part by the National Science Foundation under EPSCoR Grants Nos. 0554548 and EPS-1004057.

References

1. Sanz, J.L.; Kochling, T. Molecular biology techniques used in wastewater treatment: An overview. *Process Biochem.* **2007**, *42*, 119–133.
2. Calaway, W.T.; Carroll, W.R.; Long, S.K. Heterotrophic bacteria encountered in intermittent sand filtration of sewage. *Sew. Ind. Wastes* **1952**, *24*, 642–653.
3. Pell, M.; Nyberg, F. Infiltration of wastewater in a newly started pilot sand-filter system: II. Development and distribution of the bacterial populations. *J. Environ. Qual.* **1989**, *18*, 457–462.
4. Amann, R.I.; Ludwig, W.; Schleifer, K.H. Phylogenetic identification and *in situ* detection of individual microbial cells without cultivation. *Microbiol. Rev.* **1995**, *59*, 143–169.
5. Amador, J.A.; Potts, D.A.; Savin, M.C.; Tomlinson, P.; Görres, J.H.; Nicosia, E.L. Mesocosm-scale evaluation of faunal and microbial communities of aerated and unaerated leachfield soil. *J. Environ. Qual.* **2006**, *35*, 1160–1169.
6. Potts, D.A.; Görres, J.H.; Nicosia, E.L.; Amador, J.A. Effects of aeration on water quality from septic system leachfields. *J. Environ. Qual.* **2004**, *33*, 1828–1838.
7. Tomaras, J.; Sahl, J.W.; Siegrist, R.L.; Spear, J.R. Microbial diversity of septic tank effluent and a soil biomat. *Appl. Environ. Microbiol.* **2009**, *75*, 3348–3351.
8. Rabølle, M.; Spliid, N.H. Sorption and mobility of metronidazole, olaquindox, oxytetracycline, and tylosin in soil. *Chemosphere* **2000**, *40*, 715–722.

9. Loke, M.J.T.; Halling-Sørensen, B. Determination of the distribution coefficient (logKd) of oxytetracycline, tylosin A, olaquindox and metronidazole in manure. *Chemosphere* **2002**, *48*, 351–361.

10. Chopra, I.; Roberts, M. Tetracycline antibiotics: Mode of action, applications, molecular biology, and epidemiology of bacterial resistance. *Microbiol. Mol. Biol. Rev.* **2001**, *65*, 232–260.

11. Halling-Sørensen, B.; Sengelov, G.; Tjornelund, J. Toxicity of tetracyclines and tetracycline degradation products to environmentally-relevant bacteria, including selected tetracycline-resistant bacteria. *Arch. Environ. Contam. Toxicol.* **2002**, *42*, 263–271.

12. Amador, J.A.; Potts, D.A.; Loomis, G.W.; Kalen, D.V., Patenaude, E.L.; Görres, J.H. Improvement of hydraulic and water quality renovation functions by intermittent aeration of soil treatment areas in onsite wastewater treatment systems. *Water* **2010**, *2*, 886–903.

13. Patenaude, E.L.; Atoyan, J.A.; Potts, D.A.; Amador, J.A. Effects of tetracycline on water quality, soil and gases in aerated and unaerated leachfield mesocosms. *J. Environ. Sci. Health Pt. A* **2008**, *43*, 1054–1063.

14. Atoyan, J.A.; Patenaude, E.L.; Potts, D.A.; Amador, J.A. Effects of tetracycline on antibiotic resistance and removal of fecal indicator bacteria in aerated and unaerated leachfield mesocosms. *J. Environ. Sci. Health Part A* **2007**, *42*, 1571–1578.

15. White, D.C.; Davis, W.M.; Nickels, J.S.; King, J.D.; Bobbie, R.J. Determination of the sedimentary microbial biomass by extractable lipid phosphate. *Oecologia* **1979**, *40*, 51–62.

16. Bligh, E.G.; Dyer, W.J. A rapid method of total lipid extraction and purification. *Can. J. Biochem. Physiol.* **1959**, *37*, 911–917.

17. Tunlid, A.; Hoitink, J.A.; Low, C.; White, D.C. Characterization of bacteria that suppress *Rhizoctonia* damping-off in bark compost media by analysis of fatty acid biomarkers. *Appl. Environ. Microbiol.* **1989**, *55*, 1368–1374.

18. Dowling, N.J.E.; Widdel, F.; White, D.C. Phospholipid ester-linked fatty acid biomarkers of acetate-oxidizing sulfate reducers and other sulfide forming bacteria. *J. Gen. Microbiol.* **1986**, *132*, 1815–1825.

19. Edlund, A.; Nichols, P.D.; Roffey, R.; White, D.C. Extractable and lipopolysaccharide fatty acid and hydroxyl acid profiles from *Desulfovibrio* species. *J. Lipid Res.* **1985**, *26*, 982–988.

20. White, D.C.; Pinkart, H.C.; Ringelberg, D.B. Biomass measurements: Biochemical approaches. In *Manual of Environmental Microbiology*; Hurst, C.J., Knudsen, G.R., McInerney, M.J., Stetzenbach, L.D., Walter, M.V., Eds.; ASM Press: Washington, DC, USA, 1997; pp. 91–101.

21. White, D.C.; Stair, J.O.; Ringelberg, D.B. Quantitative comparisons of *in situ* microbial biodiversity by signature biomarker analysis. *J. Ind. Microbiol.* **1996**, *17*, 185–196.

22. Marchesi, J.R.; Sato, T.; Weightman, A.J.; Martin, T.A.; Fry, J.C.; Hiom, S.J.; Wade, W.G. Design and evaluation of useful bacterium-specific PCR primers that amplify genes coding for bacterial 16S rRNA. *Appl. Environ. Microbiol.* **1998**, *64*, 795–799.

23. Ferris, M.J.; Muyzer, G.; Ward, D.M. Denaturing gradient gel electrophoresis profiles of 16S rRNA-defined populations inhabiting a hot spring microbial mat community. *Appl. Environ. Microbiol.* **1996**, *62*, 340–346.

24. Muyzer, G.; de Waal, E.C.; Uitterlinden, A.G. Profiling of complex microbial populations by denaturing gradient gel electrophoresis analysis of polymerase chain reaction-amplified genes coding for 16S rRNA. *Appl. Environ. Microbiol.* **1993**, *59*, 695–700.

25. Rasband, W.S. *ImageJ*, Version 1.38; U.S National Institutes of Health: Bethesda, MD, USA, 1997–2007. Available online: http://rsb.info.nih.gov/ij/ (accessed on 16 April 2013).

26. DeSantis, T.Z.; Hugenholtz, P.; Keller, K.; Brodie, E.L.; Larsen, N.; Piceno, Y.M.; Phan, R.; Andersen, G.L. NAST: A multiple sequence alignment server for comparative analysis of 16S rRNA genes. *Nucleic Acids Res.* **2006**, *34*, W394–W399.

27. Staddon, W.J.; Duchesne, L.C.; Trevors, J.T. Microbial diversity and community structure of postdisturbance forest soils as determined by sole-carbon-source utilization patterns. *Microb. Ecol.* **1997**, *34*, 125–130.

28. Eichorst, S.A.; Breznak, J.A.; Schmidt, T.M. Isolation and characterization of soil bacteria that define *Terriglobus* gen. nov., in the Phylum Acidobacteria. *Appl. Environ. Microbiol.* **2007**, *73*, 2708–2717.

29. Takeuchi, M.; Weiss, N.; Schumann, P.; Yokota, A. *Leucobacter komagatae* gen. nov., sp. nov., a new aerobic Gram-positive, nonsporulating rod with 2,4-diaminobutyric acid in the cell wall. *Int. J. Syst. Bacteriol.* **1996**, *46*, 967–971.

30. Tortoli, E. The new mycobacteria: An update. *FEMS Immunol. Med. Microbiol.* **2006**, *48*, 159–178.

31. Leys, N.M.; Ryngaert, A.; Bastiaens, L.; Wattiau, P.; Top, E.M.; Verstraete, W.; Springael, D. Occurrence and community composition of fast-growing Mycobacterium in soils contaminated with polycyclic aromatic hydrocarbons. *FEMS Microbiol. Ecol.* **2005**, *51*, 375–388.

32. Mara, D.D.; Oragui, J.I. Occurrence of *Rhodococcus coprophilus* and associated actinomycetes in feces, sewage, and freshwater. *Appl. Environ. Microbiol.* **1981**, *42*, 1037–1042.

33. Anderson, R.L.; Ordal, E.J. *Cytophaga succinicans* sp. n., a facultatively anaerobic, aquatic myxobacterium. *J. Bacteriol.* **1961**, *81*, 130–138.

34. Koops, H.-P.; Pommerening-Röser, A. Distribution and ecophysiology of the nitrifying bacteria emphasizing cultured species. *FEMS Microbiol. Ecol.* **2001**, *37*, 1–9.

35. Nierman, W.C.; Feldblyum, T.V.; Laub, M.T.; Paulsen, I.T.; Nelson, K.E.; Eisen, J.; Heidelberg, J.F.; Alley, M.R.K.; Ohta, N.; Maddock, J.R.; *et al.* Complete genome sequence of *Caulobacter crescentus*. *Proc. Natl. Acad. Sci. USA* **2001**, *98*, 4136–4141.

36. Ke, N.; Xiao, C.; Ying, Q.; Ji, S. A new species of the genus *Phenylobacterium* for the degradation of LAS (linear alkylbenzene sulfonate) [in Chinese]. *Wei Sheng Wu Xue Bao* **2003**, *43*, 1–7.

37. Gibson, D.T. *Beijerinckia* sp strain B1: A strain by any other name. *J. Ind. Microbiol.* **1999**, *23*, 284–293.

38. Brenner, D.J.; Hollis, D.G.; Moss, C.W.; English, C.K.; Hall, G.S.; Vincent, J.; Radosevic, J.; Birkness, K.A.; Bibb, W.F.; Quinn, F.D. Proposal of *Afipia* gen. nov., with *Afipia felis* sp. nov. (formerly the cat scratch disease bacillus), *Afipia clevelandensis* sp. nov. (formerly the Cleveland Clinic Foundation strain), *Afipia broomeae* sp. nov., and three unnamed genospecies. *J. Clin. Microbiol.* **1991**, *29*, 2450–2460.

39. Rumjanek, N.G.; Dobert, R.C.; van Berkum, P.; Triplett, E.W. Common soybean inoculant strains in Brazil are members of *Bradyrhizobium elkanii*. *Appl. Environ. Microbiol.* **1993**, *59*, 4371–4373.

40. Bock, E.; Koops, H.-P.; Möller, U.C.; Rudert, M. A new facultatively nitrite oxidizing bacterium, *Nitrobacter vulgaris* sp. nov. *Arch. Microbiol.* **1990**, *153*, 105–110.

41. Hou, C.T.; Laskin, A.I.; Patel, R.N. Growth and polysaccharide production by *Methylocystis parvus* OBBP on methanol. *Appl. Environ. Microbiol.* **1979**, *37*, 800–804.

42. Takeuchi, M.; Hamana, K.; Hiraishi, A. Proposal of the genus *Sphingomonas* sensu stricto and three new genera, *Sphingobium*, *Novosphingobium* and *Sphingopyxis*, on the basis of phylogenetic and chemotaxonomic analyses. *Int. J. Syst. Evol. Microbiol.* **2001**, *51*, 1405–1417.

43. Schulze, R.; Spring, S.; Amann, R.; Huber, I.; Ludwig, W.; Schleifer, K.H.; Kämpfer, P. Genotypic diversity of *Acidovorax* strains isolated from activated sludge and description of *Acidovorax defluvii* sp. nov. *Syst. Appl. Microbiol.* **1999**, *22*, 205–214.

44. Mergaert, J.; Webb, A.; Anderson, C.; Wouters, A.; Swings, J. Microbial degradation of poly(3-hydroxybutyrate) and poly(3-hydroxybutyrate-co-3-hydroxyvalerate) in soils. *Appl. Environ. Microbiol.* **1993**, *59*, 3233–3238.

45. Achenbach, L.A.; Michaelidou, U.; Bruce, R.A.; Fryman, J.; Coates, J.D. *Dechloromonas agitata* gen. nov., sp. nov. and *Dechlorosoma suillum* gen. nov., sp. nov., two novel environmentally dominant (per)chlorate-reducing bacteria and their phylogenetic position. *Int. J. Syst. Evol. Microbiol.* **2001**, *51*, 527–533.

46. Friedman, H.; Yamamoto, Y.; Klein, T.W. *Legionella pneumophila* pathogenesis and immunity. *Sem. Pediatr. Infect. Dis.* **2002**, *13*, 273–279.

47. Kalyuzhnaya, M.G.; Stolyar, S.M.; Auman, A.J.; Lara, J.C.; Lidstrom, M.E.; Chistoserdova, L. *Methylosarcina lacus* sp. nov., a methanotroph from Lake Washington, Seattle, USA, and emended description of the genus *Methylosarcina*. *Int. J. Syst. Evol. Microbiol.* **2005**, *55*, 2345–2350.

48. Lalucat, J.; Bennasar, A.; Bosch, R.; García-Valdés, E.; Palleroni, N.J. Biology of *Pseudomonas stutzeri*. *Microbiol. Molec. Biol. Rev.* **2006**, *70*, 510–547.

49. Baggi, G.; Barbieri, P.; Galli, E.; Tollari, S. Isolation of a *Pseudomonas stutzeri* strain that degrades o-xylene. *Appl. Environ. Microbiol.* **1987**, *53*, 2129–2132.

50. Zumft, W.G. The denitrifying prokaryotes. In *The Prokaryotes*; Balows, A., Ed.; Springer-Verlag: New York, NY, USA, 1992.

51. Kwon, S.W.; Kim, J.S.; Park, I.C.; Yoon, S.H.; Park, D.H.; Lim, C.K.; Go, S.J. *Pseudomonas koreensis* sp. nov., *Pseudomonas umsongensis* sp. nov. and *Pseudomonas jinjuensis* sp. nov., novel species from farm soils in Korea. *Int. J. Syst. Evol. Microbiol.* **2003**, *53*, 21–27.

52. Johansen, J.E.; Binnerup, S.J.; Kroer, N.; Mølbak, L. *Luteibacter rhizovicinus* gen. nov., sp. nov., a yellow-pigmented gammaproteobacterium isolated from the rhizosphere of barley (*Hordeum vulgare* L.). *Int. J. Syst. Evol. Microbiol.* **2005**, *55*, 2285–2291.

53. Christensen, P.; Cook, F.D. *Lysobacter*, a new genus of nonfruiting, gliding bacteria with a high base ratio. *Int. J. Syst. Bacteriol.* **1978**, *28*, 367–393.

54. Mergaert, J.; Cnockaert, M.C.; Swings, J. *Thermomonas fusca* sp. nov. and *Thermomonas brevis* sp. nov., two mesophilic species isolated from a denitrification reactor with poly(epsilon-caprolactone) plastic granules as fixed bed, and emended description of the genus *Thermomonas*. *Int. J. Syst. Evol. Microbiol.* **2003**, *53*, 1961–1966.

55. Atlas, R.M.; Bartha, R. *Microbial Ecology: Fundamentals and Applications*, 4th ed.; Benjamin Cummings: Menlo Park, CA, USA, 1998.
56. Dantas, G.; Sommer, M.O.A.; Oluwasegun, R.D.; Church, G.M. Bacteria subsisting on antibiotics. *Science* **2008**, *320*, 100–103.

Water **2013**, *5*, 394-404; doi:10.3390/w5020394

OPEN ACCESS

water

ISSN 2073-4441
www.mdpi.com/journal/water

Article

CO_2 Emission Factor for Rainwater and Reclaimed Water Used in Buildings in Japan

Yasutoshi Shimizu *, Satoshi Dejima and Kanako Toyosada

ESG Promotion Department, TOTO LTD, 2-1-1, Nakashima, Kokurakita-ku, Kitakyushu 802-8601, Japan; E-Mails: satoshi.dejima@jp.toto.com (S.D.); kanako.toyosada@jp.toto.com (K.T.)

* Author to whom correspondence should be addressed; E-Mail: yasutoshi.shimizu@jp.toto.com; Tel.: +81-93-952-3315; Fax: +81-93-952-3468.

Received: 24 January 2013; in revised form: 26 February 2013 / Accepted: 19 March 2013 / Published: 8 April 2013

Abstract: From the standpoint of the preservation of water resources, rainwater and reclaimed water have been widely used in buildings in many countries. However, the CO_2 emission factors of these two waters factors that determine their environmental impacts—have not been calculated. In a previous study, the CO_2 emission factor of water for waterworks and sewer systems was determined. In this paper, we evaluate the emission factors of rainwater and reclaimed water in the same manner. First, the emission factor for pumping water in buildings is determined using published values for operating performances. About half of the residential dwellings in Japan are multistory apartments, and these apartments use pumps for the delivery of water. The emission factor of pumping is calculated as 0.69 kg CO_2/m^3, which adds 16% to the emission factor of waterworks and sewer systems. Next, the CO_2 emission factors of rainwater and reclaimed water are calculated for different water delivery cases in buildings. As a result, it is found that the use of reclaimed water increases CO_2 emissions by 62%, compared to the use of ordinary water.

Keywords: CO_2 emission factor; energy consumption rate; environmental impact; rainwater; reclaimed water

1. Introduction

In recent years, studies associating water use with CO_2 emissions have been performed around the world [1–3]. In Japan, research linking the water-saving performance of bathroom fixtures, such as toilets and showers, with a reduction in CO_2 emissions has progressed, and the fact that the widespread use of water-saving fixtures can be effective in cutting CO_2 emissions has been recognized [4,5]. As a result, a carbon credit project based on the adoption of water-saving apparatuses has been launched in Japan. In addition, feasibility studies in China and Vietnam have been carried out as Bilateral Offset Credit Mechanism projects, which the Japanese government is promoting [6,7].

Carbon credits are calculated by measuring the amount of water saved by replacing conventional equipment with energy-saving or water-saving products and multiplying this value by the CO_2 emission factor to convert the values into the amount of CO_2. The CO_2 emission factor of water has been calculated by using the energy consumption values for waterworks and sewer systems.

A report on Hong Kong, where buildings consists of skyscrapers, states that energy consumption related to pumping water in a building accounts for 45% of the total energy consumption of water usage [8]. In a previous paper, the latest value for the CO_2 emission factor of water in Japanese waterworks and sewer systems was determined [9]. In this research, the emission factor from pumping water in Japanese buildings was calculated.

From the standpoint of preserving water resources, rainwater and reclaimed water have been widely used in buildings in Japan. According to "Water Resources of Japan", published by the Ministry of Land, Infrastructure, Transport and Tourism, rainwater and reclaimed water systems had been introduced by about 3400 institutions by 2008 [10]. Comparisons of the different environmental impacts resulting from the use of ordinary water *versus* the use of rainwater or reclaimed water in buildings have been studied by many researchers [11–14]. However, as neither the calculation boundaries nor the calculation conditions—such as an energy coefficient—used are unified, the evaluation results cannot be compared. The most ambiguous point of existing research is the CO_2 emission factor of the water for waterworks and sewer systems used as a standard value in evaluations. The emission factors in Japan are 0.59 kg CO_2/m^3, as presented by the Ministry of the Environment in 1996, and 2.011 kg CO_2/m^3, as given in "LCA Guideline Buildings, 1999", published by the Society of Architecture of Japan. Considering that electricity accounts for more than 90% of the energy required for the operation of waterworks and sewer systems and that the CO_2 emission factor for electricity changes annually depending on the composition ratio of the type of power-generation processes—such as nuclear and thermal power generation—the CO_2 emission factor derived from energy consumption should be reexamined every year [15]. When comparing the environmental impacts of rainwater and reclaimed water use with the value of ordinary water use, the adoption of the same energy coefficient is vital. Thus, by using the reported values for the energy consumptions and treated water volumes for each process, the environmental impacts were calculated as CO_2 emission factors using the same evaluation boundary and energy coefficient.

2. Analysis

2.1. Establishing the Evaluation Boundary and Model

The energy consumption rates and CO_2 emission factors of rainwater and reclaimed water used for toilet flushing and watering in buildings were calculated and compared with ordinary water use. The evaluation boundary was established as the series of processes from water generation to wastewater treatment. Only energy consumption in the operation phase was targeted in the evaluation. Rainwater and reclaimed water are used in large buildings of a multistory structure, such as public facilities. In such buildings, water has to be pumped. Therefore, the water-pumping process in buildings is taken into consideration in this research. The evaluation boundary and model are shown in Figure 1.

Figure 1. Evaluation boundary and model.

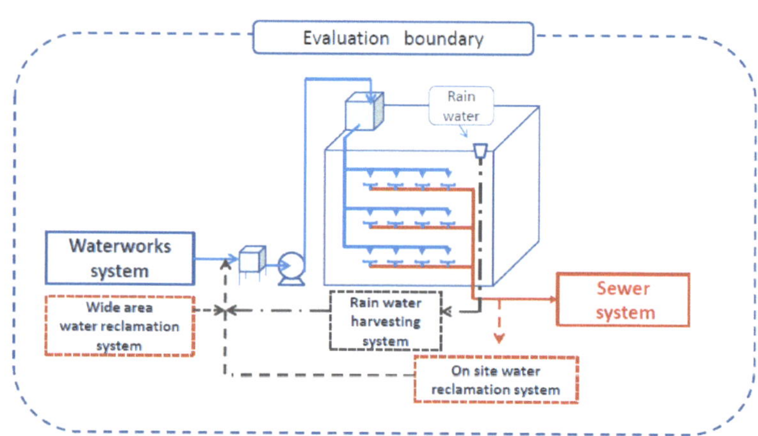

2.1.1. Water-Generation Process

In the water-generation process, waterworks, rainwater harvesting and the water-reclamation process were evaluated. The energy consumption rate and CO_2 emission factor for waterworks were taken from previous research. The reported values were the weighted averages of data for all Japanese facilities composited from raw water intake, purification and the delivery process. For the evaluation of large areas and on-site water-reclamation systems, the reported values of treated water amounts and consumed energy values were used for analysis [16–20]. According to a Ministry of Land, Infrastructure, Transport and Tourism report, "Rainwater and Reclaimed Water Use Institution Survey", a common rainwater-harvesting system is processed by using potential energy [21]. Rain is collected on the roof of a building and flows down through sedimentation, filtration and disinfection by chlorine processes toward a basement tank. Therefore, it was thought that the energy consumption used in the operation process for rainwater harvesting was very small. Therefore, the energy for water supply in a building was taken into consideration in rainwater use.

2.1.2. Water-Pumping Process in Buildings

For water delivery in buildings, elevated tank systems were commonly used until 1994. However, it became widely known as a result of TV reports and other sources that there were buildings where the management of water in the elevated tanks had not been undertaken to a satisfactory sanitary level and that improvement was required. Therefore, the booster pump system, which is a more sanitary water supply system, is used in buildings in Japan built since that time. An outline of the system is shown in Figure 2. Based on the reported values [22] for the energy consumption of pumps, the energy consumption rates and CO_2 emission factors were calculated. The average values of the energy consumption rate and emission factor for a Japanese residence were calculated by considering the adoption rate for each water distribution system.

Figure 2. Water delivery systems in buildings.

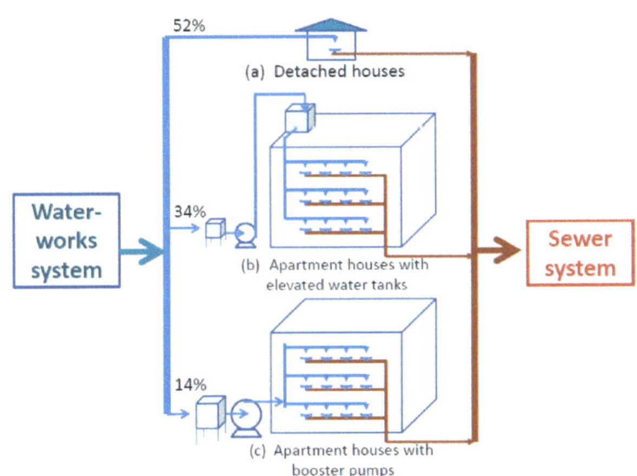

2.1.3. Wastewater-Treatment Process

Drainage from each water system is handled by sewers, and the energy consumption rate and CO_2 emission factor were quoted from the previous research. The water-reclamation system removes the pollution in wastewater discharged from buildings. Therefore, the pollution load of sewer systems decreases with the spread of the water-reclamation system. However, reclaimed water usage was only 259 million m^3/year [10]—about 1.8% of the 14,440 million m^3/year of the annual sewer treatment amount [23]—and since it is small, this effect can be ignored.

2.2. Calculation

The energy consumption and CO_2 emission per cubic meter of water were calculated by using Equations (1) and (2) for each of the above-mentioned processes. In the calculation, the energy consumption for operation was positioned as the object of evaluation, and the emission factor of electricity was adopted as the average value for all power sources in Japan at both the receiving and generating ends after credit compensation by assuming its application in Clean Development Mechanism and the Bilateral Offset Credit Mechanism [6,7].

$$CEw = \sum Ew(i) / Qw \tag{1}$$

$$CFw = \sum \{Ew(i) \cdot CFe(i)\} / Qw \tag{2}$$

where *CEw* is the energy consumption rate of water (MJ/m^3), *Ew(i)* is the energy consumption of each energy source (MJ/year), *Qw* is the volume of water treated by the system (m^3/year), *CFw* is the CO_2 emission factor for water (kg CO_2/m^3) and *CFe(i)* is the CO_2 emission factor for each energy source (kg CO_2/MJ).

3. Results and Discussion

3.1. Energy Consumption Rate and CO_2 Emission Factor for Pumping in Buildings

Statistical data on non-residential buildings, such as the number of buildings, their purpose and the year of construction, is not compiled in Japan. Thus, the energy consumption rate and CO_2 emission factor for pumping was calculated for residential buildings, for which ample data for analysis is available. In detached houses in Japan, water is supplied by tap water pressure, and the use of individual water tanks and pumps is not common. Apartment buildings, however, have water-delivery systems that use elevated tanks or booster pumps.

The number of detached houses and apartment buildings constructed every fiscal year is recorded in the annual report on construction statistics published by the Ministry of Land, Infrastructure, Transport and Tourism. An example is shown in Figure 3. The water-delivery systems of apartment buildings have changed since 1994, as mentioned earlier. However, there are no statistics values, such as numbers classified by the water-supply system of buildings. Thus, based on the hearing from building designers and owners, buildings constructed before 1994 were assumed to have elevated tanks, and for buildings constructed after 1994, 80% of them use booster pumps, 20% use elevated tanks and the building ratio of each system was calculated. The result is shown in Figure 2. The ratio of detached houses without pumping, apartment buildings with elevated tanks and apartment buildings with booster pumps was estimated as 52%, 23% and 14%, respectively.

Figure 3. Changes in the numbers of newly built houses in Japan.

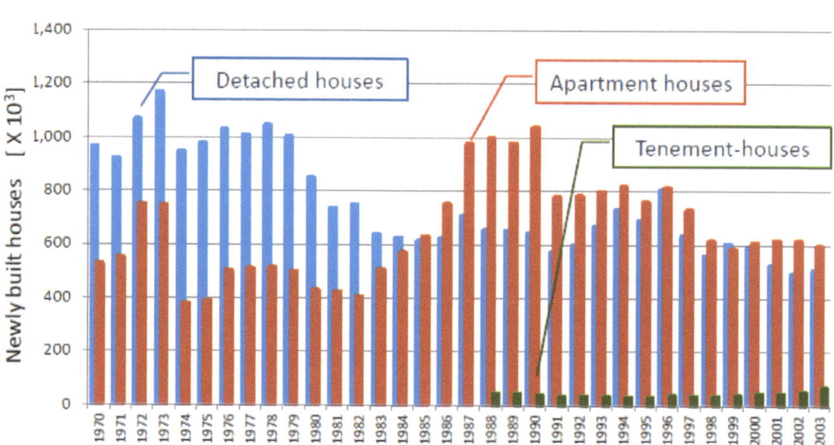

Next, data on the power consumption of water pumps for houses were extracted from pump manufacturers' report [22], as shown in Table 1. The energy consumption rate of pump to elevated tank was 0.8 MJ/m^3, according to a major pump manufacturer. The calculated value, 1.00 MJ/m^3, was about the same as the manufacturer's value, and it was decided that the calculated value was appropriate. For an elevated tank system, pressurized water of around 0.2 MJ is delivered to each apartment using potential energy (gravity). In a booster pump system, the piping pressure is maintained at around 0.2 MJ by pump operation. Therefore, a booster pump system consumes more energy than an elevated tank system. For the operation of booster pumps, there are control systems with pump inverters and systems that combine the operation of two or more pumps. The energy efficiency of such systems was calculated as 1.2–4.4 MJ/m^3, and an average value of 2.52 MJ/m^3 was adopted in this research.

Table 1. Energy consumption for housing pump system in Japan.

Type	Pump system	Delivered water (m^3/year)	Consumed energy (kWh/year)	Energy consumption rate (MJ/m^3)
Apartment house with elevated water tank	T-405X5S M3.7	12,410 *	3,444	0.999
Apartment houses with booster pumps	KDP2-40A2.2A	12,410 *	8,616	2.499
	50KNV325P2.2	12,410 *	15,000	4.351
	KF2-32P1.9	12,410 *	8,376	2.430
	KF2-50R3-3.7	36,500 **	12,396	1.223
	100KNV505R3-3	36,500 **	21,384	2.109

* 17-story apartment (34 houses; pumping height: 50 m); ** 13-story apartment (100 houses; pumping height: 39 m).

Although this calculation regarded the water supply for apartment buildings, the authors studied the water supply for office buildings using these findings. Office buildings of a size suitable for the application of data on water-delivery systems for apartment buildings were selected for this study. According to "Plumbing sanitary planning/designing know-how", published by the Society of Heating, Air-Conditioning and Sanitary Engineering of Japan, water consumptions by men and women in an office are 50 L and 100 L/(person·day), respectively [24]. When the man-to-woman ratio of an office is set at 75:25, the average consumption is 62.5 L/(person·day). From this, it was presumed that the water-delivery system for apartment buildings mentioned above was equivalent to an office for 540–1600 people. Therefore, it was considered that the environmental impact of the water-delivery system for houses, as shown in Tables 2 and 3, is applicable for a general office building.

Table 2. Energy consumption rates of Japanese water system for housing (MJ/m^3).

Classification of buildings	Waterworks system	Water supply in buildings	Sewer system	Total
(a) Detached houses		0		4.20
(b) Apartment houses with elevated water tanks	1.98	1.00	2.22	5.20
(c) Apartment houses with booster pumps		2.52		6.72
Average values	1.98	0.69	2.22	4.89

Table 3. CO_2 emission factor of Japanese water system for housing (kg-CO_2/m^3).

Classification of buildings	Calculated with generating end electricity (0.335 kg CO_2/kWh)	Calculated with receiving end electricity (0.373 kg CO_2/kWh)
(a) Detached houses	0.376	0.415
(b) Apartment houses with elevated water tanks	0.469	0.519
(c) Apartment houses with booster pumps	0.611	0.676
Average values	0.441	0.487

3.2. Energy Consumption Rates and CO$_2$ Emission Factors of Rainwater and Reclaimed Water

Although the use of reclaimed water has progressed in public facilities and evaluation research into the energy consumption rate has been conducted in Japan, the boundary of evaluation is not uniform, and as mentioned above, the results cannot be compared. Therefore, the energy consumption and CO_2 emissions for ordinary water, as well as rainwater and reclaimed water use, in buildings were evaluated under the same conditions. The operation data for the energy consumption for a water-reclamation system was extracted from reports, as shown in Table 4. The correlation between energy consumption and the processing scale, processing system, *etc.*, was not observed.

From these results, rainwater and reclaimed water use were compared with ordinary water use, water for waterworks and sewer systems. The results are shown in Figure 4. In the evaluation of ordinary water use, the energy consumption for the water-delivery process in buildings was added. As for the water-purification process, the rainwater-harvesting system exploited potential energy after collection from roofs, and since it was filtered and stored, the energy consumption presupposed that it can be ignored; we counted only the energy consumption for the water-delivery process in buildings from storage tanks. In a water reclamation system, water treatment and pumping energies were contained in the extracted operation data. In addition, for all systems, wastewater treatment after use was needed, and the energy required was added. As a result, energy consumption is in the order: rainwater use < waterworks and sewerage use < reclaimed water use in a building; and reclaimed water use was found to consume 1.4-times more energy than ordinary water use. The emission factor for water was determined for each water resource, as shown in Table 5.

Figure 4. Energy consumption rates of water systems for non-residential buildings with booster pumps (MJ/m^3).

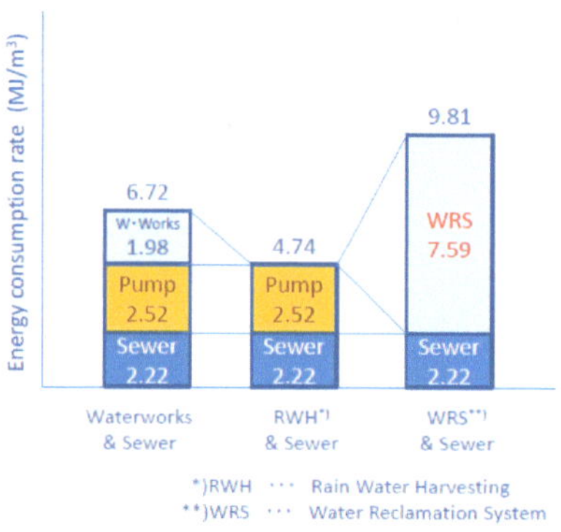

Table 4. Energy consumption for grey water treatment system in Japan.

Treatment method	Facility	Treatment method						Treated water volume (m³/year)	Energy consumption (kWh/year)	Energy consumption rate (MJ/m³)	Source
		B/T	M/F	S/F	SE	O₃	AC				
Wide area water reclamation system	Facility A	○			○	○		52,484	276,930	19.00	[15]
	Facility B			○	○	○		1,917,964	2,509,400	4.71	
	Facility C	○	○	○		○		1,267,332	2,215,000	6.29	
	Facility D			○		○		823,116	553,662	2.42	
	Shibayama Housing Complex	○				○		55,896	110,507	7.12	[17]
	Heijou New-town	○		○		○		20,955	76,500	13.14	
On-site water reclamation system	Facility F			○				92,407	59,260	2.31	[15]
	Facility G	○		○			○	28,244	43,281	5.52	
	TV Center (Tokyo)	○					○	58,400	111,325	6.86	[16,19]
	Office Building (Tokyo)	○		○				146,000	251,685	6.21	[18]
	Apartment House (Fukuoka)	○				○		9,855	16,584	6.06	
	Factory (Mie)	○		○			○	25,550	80,899	11.40	

B/T: bio-treatment; M/F: membrane filtration; S/F: sand filtration; SE: sedimentation; O₃: ozone treatment; AC: activated carbon treatment.

Table 5. CO_2 emission factors of Japanese water system (kg CO_2/m³).

Classification of buildings		Calculated with Generating end electricity (0.335 kg CO₂/kWh)	Calculated with receiving end electricity (0.373 kg CO₂/kWh)
Detached houses	Waterworks and sewer system	0.376	0.415
Apartment houses/buildings with elevated water tanks	Waterworks and sewer system	0.469	0.519
	RWH and sewer system	0.288	0.318
	WRS and sewer system	0.901	1.003
Apartment houses/buildings with booster pumps	Waterworks and sewer system	0.611	0.676
	RWH and sewer system	0.429	0.476
	WRS and sewer system	0.901	1.003

RWH: rain water harvesting; WRS: water reclamation system.

The above mentioned evaluation is the case for Japan, a country with comparatively abundant water resources.

It is known that the water-transfer process accounts for 80 percent of the energy consumption of a waterworks system [9]. One report states that in California, where the water source area and the consuming area are separated, waterworks require a larger amount of water-transfer energy than reclaimed water generation [25]. It is necessary to carry out evaluation of the environmental impact of water in consideration of the water situation for each country.

Moreover, this research was an evaluation that paid attention only to energy consumed in the operation of the water system. To carry out an entire lifecycle assessment, which would include every step, from facility construction to abandonment, for rainwater harvesting through to water reclamation, the installation of exclusive water-piping systems and water-purification facilities must be taken into consideration. In that case, the environmental impact of rainwater and reclaimed water use increases. Moreover, as it is thought that various substances are dissolved and become mixed in with rainwater and reclaimed water, the risk to maintenance, such as piping and the degradation of pump speeds, must be considered. Further study along these lines will be necessary in the future.

4. Conclusions

Through this research, the energy consumption rate and the CO_2 emission factor of water, required for the environment assessment of a multistory building, have been established. As a result of the establishment of this value, the carbon credit program established for detached houses, achieved by the spread of the water-saving apparatus, will expand to multistory buildings, such as apartment buildings, hotels and public facilities.

In addition, the energy consumption rate and the CO_2 emission factor of rainwater and reclaimed water used in buildings were also determined. Water reclamation systems introduced mostly for public facilities in many countries were found to have an environmental impact that was larger than ordinary water use. Therefore, it was judged that the meaning of the environmental impact reduction by water saving in a water reclamation system was large. From now on, a carbon credit program resulting from the adoption of water-saving apparatus spread can be developed, ranging from houses using ordinary water, to buildings using rainwater and reclaimed water. This is expected to contribute to measures against global warming.

References

1. Kenwey, S.J.; Priestly, A.; Cook, S.; Inman, M.; Gregory, A.; Hall, M. *Energy Use in the Provision and Consumption of Urban Water in Australia and New Zealand*; Water Services Association of Australia (WSAA): Sydney, Australia, 2008. Available online: http://www.csiro.au/files/files/pntk.pdf (accessed on 23 January 2013).
2. Hackett, M.J.; Gray, N.F. Carbon dioxide emission savings potential of household water use reduction in the UK. *J. Sustain. Dev.* **2009**, *2*, 36–43.
3. Walker, G. Water and energy use efficiency are increasingly linked. *Energy World* **2009**, *369*, 18–19.
4. Yasutoshi, S.; Kanako, T. The economic and environmental impact of remodeling in the case of a water-saving toilet bowl (in Japanese). *J. Soc. Heat. Air-Cond. Sanit. Eng. Jpn.* **2009**, *152*, 9–14.
5. Yasutoshi, S.; Kanako, T.; Kiyoshi, N. Prediction of CO_2 emission associated with residential plumbing equipment (in Japanese). *J. Soc. Heat. Air-Cond. Sanit. Eng. Jpn.* **2010**, *163*, 11–18.

6.   Domestic Clean Development Mechanism (in Japanese). Available online: http://jcdm.jp/process/methodology.html (accessed on 23 January 2013).

7.   New Mechanism Information Platform (in Japanese). Available online: http://www.mmechanisms.org/initiatives/index.html (accessed on 23 January 2013).

8.   Wong, L.T.; Mui, K.W. Energy efficiency of elevated water supply tanks for high-rise buildings. *Appl. Energy* **2013**, *103*, 658–691.

9.   Yasutoshi, S.; Satoshi, D.; Kanako, T. The CO_2 emission factor of water in Japan. *Water* **2012**, *4*, 759–769.

10.   *Water Resources of Japan* (in Japanese); Ministry of Land, Infrastructure, Transport and Tourism, Tokyo, Japan, 2011.

11.   Yasuhiko, W.; Hiroyuki, M.; Ritsuo, T.; Taira, O. Utilization of unused water resources in urban areas for the establishment of a water conservation city (in Japanese). *J. Jpn. Soc. Civ. Eng.* **1999**, *662*, 59–71.

12.   Tsutomu, N.; The recent usage of rainwater systems in buildings (in Japanese). *Build. Eng. Equip.* **2004**, *1*, 152–154.

13.   Masashi, O.; Shunsuke, N. Quantitative evaluation of CO_2 reduction by using reclaimed water (in Japanese). In Annual Meeting Proceedings of Japan Society on Water Environment; Japan Society on Water Environment: Tokyo, Japan, 2011.

14.   Hideki, Y.; Ken, M.; Toshiharu, I.; Shin, H.; kazuhito, O. A study on reduction of environmental impact caused by building facilities and urban development (in Japanese). In Proceedings of the Society of heating, Air-Conditioning and Sanitary Engineers of Japan; The Society of Heating, Air-Conditioning and Sanitary Engineers of Japan: Tokyo, Japan, 1992; pp. 1237–1240.

15.   Keidanren (Japan Business Federation). *Voluntary Action Plan on the Environment, the Electricity Emission Factor in Global Warming Countermeasure* (in Japanese); Japan Business Federation: Tokyo, Japan, 2012. Available online: http://www.keidanren.or.jp/japanese/policy/2011/113/honbun.pdf (accessed on 23 January 2013).

16.   Sewerage Treatment Research Facility, National Institute for Land and Infrastructure Management. *About the CO_2 Emission by Using Reclaimed Water (The Results of the CO_2 Emission Survey from Operating the Reuse and Supply Facilities)* (in Japanese); Ministry of Land, Infrastructure, Transport and Tourism: Tokyo, Japan, 2009.

17.   Yusuke, F. The waste water recycling facility of the head office building of Nippon television network corporation, the Society of heating (in Japanese). *Air-Cond. Sanit. Eng. Jpn.* **1985**, *59*, 45–57.

18.   Japan Housing Equipment System Federation. *The Waste Water Recycling System in Buildings* (in Japanese); Japan Housing Equipment System Federation: Tokyo, Japan, 1984.

19.   Housing and Urban Development Corporation. *Basic Research on Integrated Water Circulating System in Housing Estate 2* (in Japanese); Housing and Urban Development Corporation: Tokyo, Japan, 1987.

20.   Fresh Water Generation Technology. *Challenging the Limits of Biological Wastewater Treatment. (Water for Miscellaneous Use at Tokyo NTV Building)* (in Japanese); Fresh Water Generation Technology: Tokyo, Japan, 1985; pp. 28–32.

21. Ministry of Land, Infrastructure, Transport and Tourism. *Rain Water/Recycled Water Usage Research Facility* (*Case Study*) (in Japanese); Ministry of Land, Infrastructure, Transport and Tourism: Tokyo, Japan, 2002.

22. Kawamoto Pump Mfg. Co. Ltd. *Kawamoto Technical Report* (in Japanese); Kawamoto Pump Mfg. Co. Ltd.: Nagoya, Japan, 2013; Available online: http://www.kawamoto.co.jp/catalog/documents/81128d14.pdf (accessed on 23 January 2013).

23. Japan Sewage Works Association. *Sewerage Statistics* 2008 (in Japanese); Japan Sewage Works Association: Tokyo, Japan, 2008.

24. The Society of Heating, Air-Conditioning and Sanitary Engineers of Japan. *Plumbing Sanitary Planning/Designing Know-How* (in Japanese); The Society of Heating, Air-Conditioning and Sanitary Engineers of Japan, Tokyo, Japan, 2001.

25. Taffler, D.; Lesley, D.; Zelenka, A. Hidden potential, recycled water and the water-enegy-carbon nexus. *Water Environ. Tech.* **2008**, *11*, 34–41.

Water **2013**, *5*, 243-261; doi:10.3390/w5010243

OPEN ACCESS

water

ISSN 2073-4441
www.mdpi.com/journal/water

Article

Fecal Coliform and *E. coli* Concentrations in Effluent-Dominated Streams of the Upper Santa Cruz Watershed

Emily C. Sanders, Yongping Yuan * and Ann Pitchford

Landscape Ecology Branch, Environmental Sciences Division, National Exposure Research Laboratory, United States Environmental Protection Agency Office of Research and Development, 944 East Harmon Avenue, Las Vegas, NV 89119, USA; E-Mails: sanders.emily@epamail.epa.gov (E.C.S.); pitchford.ann@epamail.epa.gov (A.P.)

* Author to whom correspondence should be addressed; E-Mail: yuan.yongping@epamail.cpa.gov; Tel.: +1-702-798-2112; Fax: +1-702-798-2208.

Received: 4 January 2013; in revised form: 26 February 2013 / Accepted: 26 February 2013 / Published: 11 March 2013

Abstract: This study assesses the water quality of the Upper Santa Cruz Watershed in southern Arizona in terms of fecal coliform and *Escherichia coli* (*E. coli*) bacteria concentrations discharged as treated effluent and from nonpoint sources into the Santa Cruz River and surrounding tributaries. The objectives were to (1) assess the water quality in the Upper Santa Cruz Watershed in terms of fecal coliform and *E. coli* by comparing the available data to the water quality criteria established by Arizona, (2) to provide insights into fecal indicator bacteria (FIB) response to the hydrology of the watershed and (3) to identify if point sources or nonpoint sources are the major contributors of FIB in the stream. Assessment of the available wastewater treatment plant treated effluent data and in-stream sampling data indicate that water quality criteria for *E. coli* and fecal coliform in recreational waters are exceeded at all locations of the Santa Cruz River. For the wastewater discharge, 13%–15% of sample concentrations exceeded the 800 colony forming units (cfu) per 100 mL sample maximum for fecal coliform and 29% of samples exceeded the full body contact standard of 235 cfu/100 mL established for *E. coli*; while for the in-stream grab samples, 16%–34% of sample concentrations exceeded the 800 cfu/100 mL sample maximum for fecal coliforms and 34%–75% of samples exceeded the full body contact standard of 235 cfu/100 mL established for *E. coli*. Elevated fecal coliform and *E. coli* concentrations were positively correlated with periods of increased streamflow from rainfall. FIB concentrations observed in-stream are significantly greater (*p*-value < 0.0002) than wastewater treatment plants effluent concentrations; therefore, water quality managers should focus on nonpoint sources to reduce overall fecal indicator loads. Findings indicate that fecal coliform and *E. coli* concentrations are highly variable, especially along urban streams and generally increase with streamflow and precipitation events. Occurrences of peaks in FIB concentrations during baseflow conditions indicate that further assessment of ecological factors such as interaction with sediment, regrowth, and source tracking are important to watershed management.

Keywords: fecal indicator bacteria; *Escherichia coli*; fecal coliforms; water reuse; Santa Cruz River; Upper Santa Cruz watershed; effluent-dominated

---

1. Introduction

In the semi-arid southwest, rapid urbanization and population growth have led to increased use of treated effluent to augment and maintain hydrologic conditions in the watershed resulting in both positive and negative consequences in terms of overall watershed quality [1,2]. Planned water reuse is a common occurrence globally and began as early as 1918 in California and Arizona in order to provide irrigation water for crops [3]. Discharge of treated effluent into stream channels recharges the groundwater aquifers, supports riparian habitation, enhances ecosystem services, and is commonly implemented by state agencies for these reasons [4,5]. For example, natural perennial and ephemeral flows in the Upper Santa Cruz River are artificially augmented by treated effluent from the cities of Nogales and Tucson where, historically, portions of the Santa Cruz River near the city of Tucson were pumped dry as early as 1910 [6].

However, reliance on treated effluent for perennial streamflow potentially endangers human health due to recreational exposure and possible contamination of domestic water supplies by increased microbial pathogen concentrations in surface and ground waters [4,7–9]. Common sources of potential pathogenic contamination in surface waters include storm runoff from urban and agricultural landscapes, wild animal wastes, wastewater treatment plant discharges, and failing septic system drainage [8,10,11]. Monitoring river networks for all potential pathogenic agents is expensive and not feasible; therefore, methodologies for monitoring fecal indicator bacteria (FIB) and determining acceptable risk have been established [12–15]. Current ambient water quality criteria for FIB in fresh waters are aimed to protect human health from gastroenteritis due to pathogenic exposure based on the estimated relative risk of 8 cases of gastroenteritis per 1000 swimmers [12]. The appropriateness of the methods used and FIB capability for correlating and identifying human health risk from pathogens has been debated in the literature [16–19]. Despite the ongoing debate, most states monitor for total coliforms, fecal coliforms, *Escherichia coli* (*E. coli*), fecal streptococci, or enterococci as indicators of potential pathogens in water resources. In Arizona, *E. coli* has replaced fecal coliform as the preferred FIB in stream networks [20,21].

To minimize the potential risk of wastewater to public health and the environment, state agencies regulate and permit planned wastewater reclamation and reuse facilities [3]. In many cases, these facilities, regardless if the intended reuse is for recharge or irrigation, achieve a high degree of consistent water quality, and the removal of microbial and other contaminants associated with human waste are of paramount concern [22,23]. As this case study will show, additional research and assessment of the fate and transport of pollutants released indirectly into effluent-dominated and/or effluent dependent stream networks are critical to controlling overall FIB loading in the watershed. The objectives of this study are (1) to assess the water quality in the Upper Santa Cruz Watershed in terms of FIB by comparing the available data to the water quality criteria established by Arizona, (2) to provide insights into FIB response to the hydrology of a semi-arid watershed and (3) to identify major FIB contributors (point sources *versus* nonpoint sources) to the stream.

2. Study Location: Santa Cruz Watershed

The entire Santa Cruz Watershed is composed of approximately 28,749 km^2, roughly 10% of the state of Arizona; land ownership is approximately 40% tribal, 25% federal, 20% private and 15% state [20]. The Santa Cruz River has its headwaters in Arizona's San Rafael Valley, which is in the southeast/central part of the state. The river flows south and makes a 40 km loop through Mexico before returning to the United States (U.S.) about eight kilometers east of Nogales, Arizona. The river then flows north from the U.S.-Mexico border and converges with the Gila River, just southwest of Phoenix. According to the Arizona Department of Environmental Quality (ADEQ), grazing is the dominant land use while irrigated crop production is limited to areas near streams, but restricted land uses have been established near several wilderness areas, national forests, and national monuments. In addition, mining operations, both active and abandoned, are located throughout the watershed [20]. Annual precipitation ranges from 280 to 860 mm (valley to mountain, respectively). This study focuses on the sub watersheds containing the Santa Cruz River south of Tucson, Arizona.

Most of the population in the Upper Santa Cruz Watershed is found in the city of Tucson (population 530,000), the state's second largest city after Phoenix [24]. There is also a population of 370,000 located on the U.S.-Mexico border in the sister border cities of Heroica Nogales, Sonora, Mexico and Nogales, Arizona, U.S. According to the U.S. Census Bureau (2005), the population in the state of Arizona is projected to increase by approximately 52% over 30 years from 2000 to 2030 which is expected to increase the urban water demand by approximately 45% despite sustainable development efforts [25,26]. The growth in Sonora, Mexico is expected to increase at an even higher rate which is anticipated to increase the urban water demand by 18% by 2030 [25]. As more demand from urban growth and land use is placed on the system, understanding the fate and transport of pollutants released and how treated effluent impacts the overall water quality, especially water supplies designated for human consumption, is necessary.

Water quantity and quality issues in the Upper Santa Cruz watershed are confounded by the quality of waters flowing from areas of Mexico which have less regulated infrastructure to handle wastewater treatment [27]. Continuous efforts are being made by both countries to provide wastewater service in rural areas and to enhance wastewater treatment and reclamation infrastructure to meet future needs [28]. The Groundwater Storage, Savings, and Replenishment Program managed by the Arizona Department of Water Resources (ADWR) permits groundwater and surface water recharge facilities to discharge reclaimed waters into infiltration basins and, in some cases, directly into the Santa Cruz River [5]. The ADEQ permits 22 facilities, each issued an Arizona Pollution Discharge Elimination System (AZPDES) permit, to discharge treated effluent into the Santa Cruz River and its tributaries [29]. These facilities, not all of which are actively discharging, include wastewater treatment plants (WWTP), wastewater reclamation facilities, and water pollution control facilities. The Central Arizona Project (CAP) canal allocates 563,947,056 m^3 of Colorado River water per year to Pima, Pinal, and Maricopa counties to supplement domestic water supplies and also to maintain aquifer levels [30]. In 2010, Pima County, Arizona produced approximately 84,860,000 m^3 of treated effluent of which about 76,720,000 m^3 was discharged from facilities located in Tuscon, Arizona [31]. In Santa Cruz County, Arizona, the newly expanded Nogales International Wastewater Treatment Plant (NIWTP) (see Figure 1 Map ID C) treats more than 56,781 m^3/day, approximately 20,720,000 m^3 annually, of wastewater from both Nogales, Arizona and Heroica Nogales, Sonora and discharges it to the Santa Cruz River after advanced biological treatment [32].

Figure 1. The Santa Cruz Watershed, fecal coliform and *E. coli* sampling locations, weather stations, and United States Geological Survey (USGS) streamflow stations.

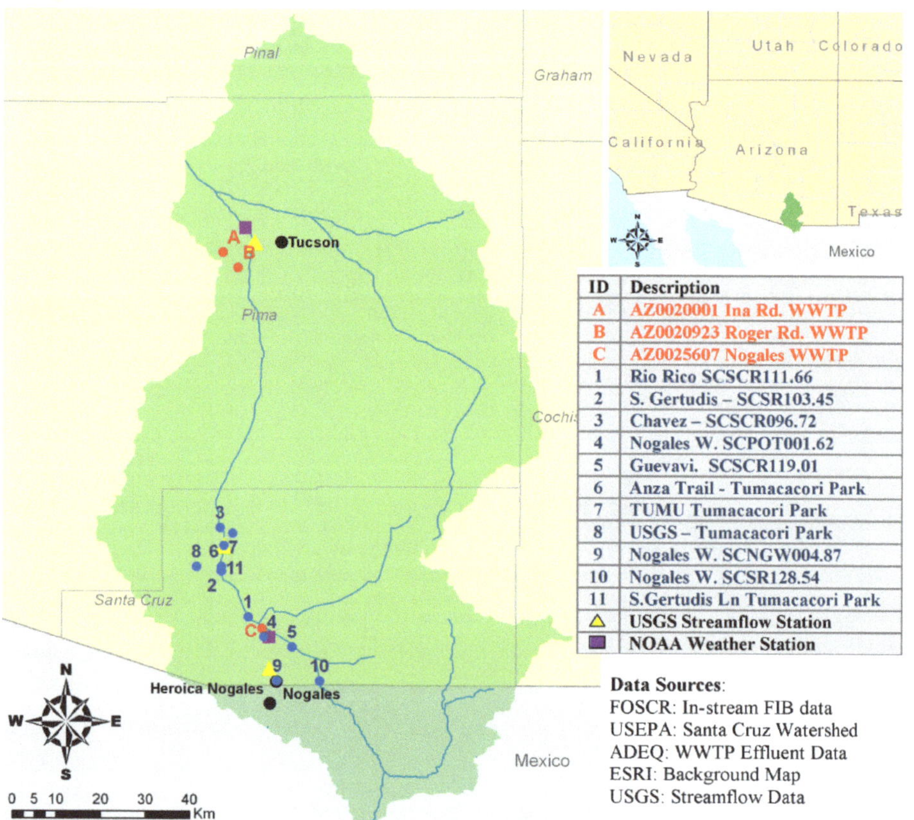

ID	Description
A	AZ0020001 Ina Rd. WWTP
B	AZ0020923 Roger Rd. WWTP
C	AZ0025607 Nogales WWTP
1	Rio Rico SCSCR111.66
2	S. Gertudis – SCSR103.45
3	Chavez – SCSCR096.72
4	Nogales W. SCPOT001.62
5	Guevavi. SCSCR119.01
6	Anza Trail - Tumacacori Park
7	TUMU Tumacacori Park
8	USGS – Tumacacori Park
9	Nogales W. SCNGW004.87
10	Nogales W. SCSR128.54
11	S.Gertudis Ln Tumacacori Park
△	USGS Streamflow Station
■	NOAA Weather Station

Data Sources:
FOSCR: In-stream FIB data
USEPA: Santa Cruz Watershed
ADEQ: WWTP Effluent Data
ESRI: Background Map
USGS: Streamflow Data

3. Data Collection and Analysis

Monthly *E. coli* and fecal coliform monitoring data from both point sources such as WWTP discharge pipes and nonpoint sources from numerous stream segments throughout the Upper Santa Cruz Watershed as shown in Figure 1 were used in this study. The *E. coli* and fecal coliform data used in this study are from numerous sampling records including ADEQ in conjunction with Friends of the Santa

Cruz River (FOSCR), National Park Service at Tumacacori National Historical Park and Sonoran Desert Network, Sonoran Institute, United States Geological Survey (USGS), and U.S. Environmental Protection Agency (EPA) Envirofacts permit compliance system (PCS) database. For the point sources data, a custom search on the Envirofacts PCS database was preformed to assess indicator bacteria concentrations from WWTP monthly discharge monitoring reports (DMR) prepared by AZPDES permitted facilities which discharge treated effluent into the Santa Cruz River and surrounding washes and tributaries (Figure 1, Map ID A and B) [29]. These grab samples show a snap shot in time and space of the FIB activity for a given location and were collected to either fulfill the AZPDES monitoring requirements or for water quality assessment purposes. The available data for the watershed are organized by location and vary in regard to sample frequency, period of record, sampling method, and FIB assessed (fecal coliforms or *E. coli*). The WWTP DMRs data collected were summarized into a monthly report. For nonpoint source data, in-stream samples were collected primarily on a quarterly or monthly basis unless no sample could be obtained due to low or no streamflow conditions; several gaps in the sampling record exist at each location. The geometric mean and sample maximum for each WWTP DMR and each in-stream sampling location available are summarized in the results section below. Variations in the targeted FIB disallow direct comparison of each sampling location for the entirety of the sampling record and the reported concentrations have differences in terms of method quantification limits and the lab methods used. The lab method reported for *E. coli* samples is listed as SM9223B and fecal coliform concentrations were determined using direct plating methods (SM9222E) or the Most Probable Number (MPN) method [13,15]. For the raw in-stream sampling data, a geometric mean and maximum concentration are calculated for the FIB reported at each location. The results are presented in the Tables 2–5 below.

The available data at each sampling location are compared to regulatory water quality criteria for FIB established in Arizona as summarized in Table 1. According to the regulatory standards listed in Table 1, wastewater dischargers report bacteria concentrations as a geometric mean of all the test results obtained during a reporting period, which is helpful when analyzing bacteria concentrations that may vary anywhere from 10 to 10,000 fold over a given period. The single sample maximum value is also needed to ensure that public health is protected from unusually high microbial loads.

Average daily baseflow conditions were determined using the Web Based Hydrograph Analysis Tool (WHAT) and the local minimum method for daily streamflow from 1 March 1996 to 30 April 2008 at two USGS stations (09481740 and 09480500) within close proximity of the sampling locations [33]. Since the local minimum method generally overestimates baseflow during storm events, the WHAT results were compared to precipitation data for a better estimation of actual baseflow conditions. Then, the correlation between streamflow/precipitation and in-stream fecal coliform/*E. coli* concentrations was analyzed to identify potential factors impacting the in-stream fecal coliform/*E. coli* concentrations. Precipitation data was obtained from weather stations maintained by the National Oceanic and Atmospheric Administration (NOAA). Streamflow data was collected from gage stations maintained by the USGS. Finally, data collected from point source WWTPs were compared with nonpoint in-stream grab samples and statistical tests were performed to see if fecal coliform/*E. coli* concentrations were significantly different between WWTPs and nonpoint sources. In instances where the sample value was reported as greater than the upper method detection limit or less than the lower method detection limit, the detection limit was used in the statistical comparison.

Table 1. Water quality standards for *E. coli* and fecal coliforms. Units are colony forming units (cfu)/100 mL.

	E. coli [a]	
Water Quality Criteria	***FBC*** [d]	***PBC*** [e]
Geometric Mean [c]	126	126
Single sample maximum	235	575
	Fecal Coliform [b]	
Water Quality Criteria	***FBC*** [d]	***Other Designated Uses*** [f]
Geometric Mean [c]	200	1000
10% of samples over 30 days	400	2000
Single Sample Maximum	800	4000

Notes: [a] Source: Bacterial Water Quality Standards for Recreational Waters: Status Report (EPA-823-R-03-008) [14]; [b] Source: Pathogen TMDL in Slide Rock State Park, Oak Creek Canyon, Arizona [34]; [c] Minimum of four samples in 30 days [35]; [d] "Full-body contact (FBC)" means the use of a surface water for swimming or other recreational activity that causes the human body to come into direct contact with the water to the point of complete submergence [35]; [e] "Partial-body contact (PBC)" means the recreational use of a surface water that may cause the human body to come into direct contact with the water, but normally not to the point of complete submergence (for example, wading or boating) [35]; [f] "other designated uses" may include fish consumption, aquatic and wildlife, agricultural irrigation or livestock watering [35].

4. Results

4.1. Fecal Coliform and E. coli *Concentrations from Point Source WWTP Effluent*

Consistent concentration data was found for three permitted locations (Map ID A–C in Figure 1) in the Upper Santa Cruz watershed from approximately 1988 to 2008 for fecal coliform and approximately 2008 to 2011 for *E. coli*. The values represented in Tables 2 and 3 were obtained from the DMRs filed with the USEPA as required by the AZPDES permit for each facility. It is important to note that the following tables reflect the number of reported average and maximum values for all reported monitoring periods for each facility and not the actual number of grab samples collected at each facility location. Table 2 summarizes the maximum grab sample value reported in each DMR period and represents the "worst case" fecal coliform concentrations released from these facilities into the Santa Cruz River and its tributaries. Table 3 summarizes the averaged values reported for each DMR period for each facility. The values were then compared to the current water quality standards shown in Table 1 for fecal coliform and *E. coli*.

Table 2 shows instances in which maximum DMR values exceed the maximum allowable concentration of 800 colony forming units (cfu) per 100 mL for fecal coliform for the facilities with available data from about 1988 to 2008. 13% of the DMR periods at Pima County Rd WWTP and 15% of the DMR periods at Roger Road WWTP contained fecal coliform concentrations which exceeded the 800 cfu/100 mL single sample maximum standard. These facilities are located near Tucson where surface water withdrawals are used for municipal water supplies. At the Nogales International WWTP, *E. coli* levels in the treated effluent exceed the maximum concentration of 235 cfu/100 mL for FBC associated with recreational use in 29% of the DMR periods. The single sample maximum of 575 cfu/100 mL for PBC was exceeded in 18% of the maximum concentrations reported for each

DMR period. Table 3 indicates that the mean concentration values for the monitoring periods are below the WQ standards for fecal coliforms. The geometric mean of 126 cfu/100 mL for *E. coli* is exceeded in 11% of the monitoring periods available for assessment from the Nogales International WWTP. The treated effluent from WWTP facilities appears to have a minor contribution to the fecal coliform and *E. coli* concentrations found within the watershed.

Table 2. Summary of the maximum concentrations reported for discharge monitoring reports (DMRs) period compared to the fecal coliform maximum standard of 800 cfu/100 mL for a single sample value or to the *E. coli* full body contact (FBC) maximum standard of 235 cfu/100 mL and to the 575 cfu/100 mL for partial body contact (PBC) for a single sample value.

Facility Name Permit ID	# of Reporting Periods [a]	The highest value of Maximum concentrations reported by the facility during DMRs period	Mean of the Maximum Concentrations reported during DMRs period	Reporting Periods >800 cfu/100 mL (Fecal)	Reporting Periods >235 cfu/100 mL (FBC *E. coli*)	Reporting Periods >575 cfu/100 mL (PBC *E. coli*)
Pima County Ina Road WWTP AZ0020001	94	1600	231	13%	----	----
Roger Road WWTP AZ0020923	98	1600	269	15%	----	----
Nogales International WWTP AZ0025607	27	2400	330	----	29%	18%

Notes: [a] # of reporting periods represent the number of DMRs submitted and not the actual number of raw sample data collected at the facility. DMRs represent monthly data.

Table 3. Summary of the averaged concentrations reported during each DMRs period compared to the fecal coliform geometric mean standard of 200 cfu/100 mL or the *E. coli* geometric mean standard of 126 cfu/100 mL for FBC and PBC.

Facility Name Permit ID	# of Reporting Periods [a]	The highest value of Average Concentrations [b] reported by the facility during DMRs period	Mean of the Average Concentrations [b] reported by the facility during DMRs period	Reporting Periods >200 cfu/100 mL (Fecal)	Reporting Periods >126 cfu/100 mL (*E. coli*)
Pima County Ina Road WWTP AZ0020001	94	79	16.2	0	---
Roger Road WWTP AZ0020923	98	104	17.4	0	---
Nogales International WWTP AZ0025607	27	229	41.6	----	11%

Notes: [a] # of reporting periods represent the number of DMRs submitted and not the actual number of raw sample data collected at the facility. DMRs represent monthly data; [b] Average concentration represents the value reported on the Discharge Monitoring Report (DMR) as the geometric mean grab sample value for the given monitoring period.

4.2. In-Stream Fecal Coliform and E. coli Data Analysis

4.2.1. Fecal Coliform and *E. coli* Concentrations from Nonpoint In-Stream Sources

Data used in this study from in-stream monitoring locations (Map ID 1–11 in Figure 1) for the Upper Santa Cruz River was obtained primarily *via* coordination between ADEQ and nonprofit organizations such as the FOSCR. Fecal coliform grab sampling results were organized by location; the geometric mean and sample maximum for each location for the entire period of record available was summarized in Table 4. An extremely large range of individual sample values exists for all locations; however, the geometric mean standard of 200 cfu/100 mL for fecal coliform was not exceeded at any location. The single sample maximum of 800 cfu/100 mL for fecal coliform is exceeded during several sampling events at each location as shown in the last column of Table 4.

Table 4. Fecal coliform concentration (cfu/100 mL) summary from in-stream sampling locations in the Upper Santa Cruz Watershed.

Reach ID ADEQ ID	# of samples	Start Date	End Date	Single Sample Max	Geometric Mean	% > 800 (Fecal)
Rio Rico SCSCR111.66 ADEQ 100238	112	3/1988	12/2008	139,000	161	19%
S. Gertudis SCSCR103.45 ADEQ 100247	98	2/1993	12/2008	27,100	149	21%
Chavez SCSCR096.72 ADEQ 100244	89	11/1992	12/2008	49,200	99	15%
Nogales W. (Portero Creek) SCPOT001.62 ADEQ 100571	70	3/1996	12/2008	24,000	146	24%
Nogales Guevavi SCSCR119.01 ADEQ 100246	32	11/1992	7/2001	79,000	39	13%

E. coli grab sampling results were organized by location; the geometric mean and sample maximum for each location for the entire period of record available was summarized into Table 5. *E. coli* concentrations at all in-stream sampling locations indicate the geometric mean standard of 126 cfu/100 mL is exceeded by more than double at all sampling locations. In addition, the maximum standards for a single sample value (235 cfu/100 mL for partial body contact and 575 cfu/100 mL for full body contact) are also exceeded at every location in at least 33% and up to 75% of the samples evaluated. The *E. coli* concentrations reported consistently exceed those concentration reported for fecal coliforms, which is likely due to differences in the methods of analysis for the specific indicator species targeted [17,18].

Tables 4 and 5 show that in-stream concentrations of *E. coli* and fecal coliform are much higher than that observed in the point source effluent discharges. The in-stream data available for assessment was limited to stream segments along the Santa Cruz River except in two locations at Nogales W. Portero Creek and USGS Tumacacori Park (Map ID 4 and 8 in Figure 1, respectively). Samples collected from these tributary washes at Portero Creek and Tumacacori Park exceeded the FBC water quality standards for *E. coli* in approximately 61% and 75% of samples collected, respectively (see Table 5). Additional sampling from contributing effluent-dominated washes and tributaries would allow better estimates of the true fecal coliform and *E. coli* indicator concentrations in the Santa Cruz River from point and nonpoint sources.

Table 5. *E. coli* concentration (cfu/100 mL) summary from in-stream sampling locations in the Upper Santa Cruz Watershed.

Reaches ID ADEQ ID	# of samples	Start Date	End Date	MAX	Geometric Mean	% > 235 (FBC *E. coli*)	% > 575 (PBC *E. coli*)
Santa Gertudis Lane Tubac Basin Tumacacori Park (NPS)	159	6/2007	9/2010	547,500	668	61%	45%
Anza Trail River Crossing Tubac Basin Tumacacori Park (NPS)	64	6/2007	9/2010	173,290	316	53%	33%
TUMA Educational Site Tubac Basin Tumacacori Park (NPS)	88	7/2007	9/2010	241,960	609	57%	42%
Rio Rico SCSCR111.66 ADEQ 100238	29	2/2008	5/2011	241,920 [a]	306	34%	24%
S. Gertudis SCSCR103.45 ADEQ 100247	22	2/2008	5/2011	241,920 [a]	367	41%	18%
Chavez SCSCR096.72 ADEQ 100244	19	2/2008	4/2011	141,300	491	52%	26%
Nogales W. (Portero Creek) SCPOT001.62 ADEQ 100571	21	2/2008	5/2011	241,920 [a]	792	61%	38%
USGS Tumacacori Tubac	16	6/2/2010	9/8/2010	210,000	2265	75%	56%

Note: [a] Laboratory reported value is greater than the method quantification level (Method SM9223B).

4.2.2. Correlation of In-Stream Fecal Coliform and *E. coli* Concentrations to Streamflow and Precipitation

Daily streamflow and baseflow vary significantly in this watershed and are often near zero during low flow periods. For USGS station 09481740 near Tubac, Arizona, average baseflow is approximately 0.40 m^3/s and between September 1995 to 2012, a zero average daily flow was recorded on 152 days predominantly in the months of June and July. Further upstream at USGS station 09480500 near Nogales, Arizona average baseflow is approximately 0.02 m^3/s and experienced zero average daily flow on 4052 days and in all months of the year. Based on the sampling location and baseflow estimates, 25% to 60% of the fecal coliform samples which exceeded the 800 cfu/100 mL standard in Table 4 and zero to 12% of the *E. coli* samples which exceeded the 235 cfu/100 mL standard in Table 5 were collected during periods of above average baseflow. From this comparison, exceedances typically occur during average baseflow or lower than average streamflow; however, approximately 85% of all in-stream samples were collected during less than average streamflow conditions.

In-stream fecal coliform and *E. coli* concentrations fluctuate based on seasonal streamflow and precipitation trends with the greatest concentrations experienced predominantly during the summer months. In-stream fecal coliform and *E. coli* concentrations generally increase in response to increased streamflow as shown in Figures 2 and 3, respectively. The range of the raw data set is 0 to 76,000 cfu/100 mL for fecal coliform sampled between March 1996 and August 2001 and 0 to 241,920 cfu/100 mL for *E. coli* sampled between February 2008 and September 2010. The daily mean in-stream fecal coliform concentrations for all locations collected on the same day was compared to the average daily streamflow from USGS gage station 9481740 corresponding to that sample date, as shown graphically in Figure 2. The range of the mean data included in Figure 2 is 0 to 37,366 cfu/100 mL and includes the same locations listed in Table 4. In Figure 3, the daily mean in-stream *E. coli* concentration for all *E. coli* sampling locations was compared to the average daily streamflow recorded on that date from USGS gage station 9481740, which is located in the mid to southern portion of the watershed near Tubac, Arizona. The range of the mean data included in Figure 3 is 28 to 118,470 cfu/100 mL, and no month had zero *E. coli* concentration simultaneously at all locations. The sampling location data included in Figure 3 are those listed in Table 5 and additional *E. coli* data from Nogales Wash SCNGW004.87 and Nogales Wash at Johnsons Ranch SCSCR128.54 (these locations were not included in Table 5 due to limited sample availability). No samples were collected on days of zero streamflow thus daily streamflow shown in the below figures does not reflect the periods of no flow conditions.

In Figure 4, the in-stream fecal coliform concentrations from multiple locations are graphically compared to monthly accumulated rainfall for the years 1996 to 2001. Weather Station 025924 (Nogales 6N) had the most complete record of precipitation data for comparison to the fecal coliform data. In-stream fecal coliform loads fluctuate in response to precipitation amount. An overall increase in fecal coliform concentrations occurs during increased periods of precipitation.

Figure 2. In-stream fecal coliform concentrations along the Upper Santa Cruz River compared to average daily streamflow at USGS station 9481740 from March 1996 to August 2001. The Arizona WQ standard for fecal coliforms is 800 cfu/100 mL for a single sample maximum.

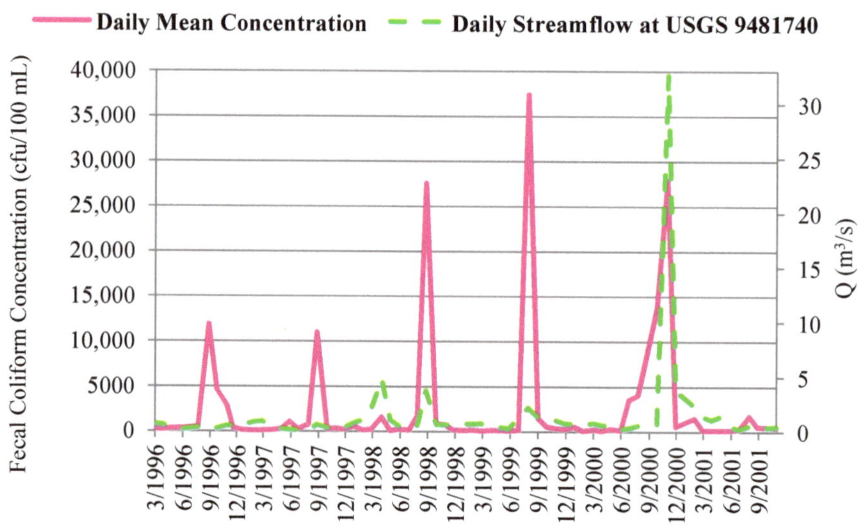

Figure 3. Mean in-stream *E. coli* concentrations in the Upper Santa Cruz River compared to daily streamflow at USGS station 9481740 from February 2008 to September 2010. The Arizona WQ standard for *E. coli* is 235 cfu/100 mL (FBC) and 575 cfu/100 mL (PBC) for a single sample maximum.

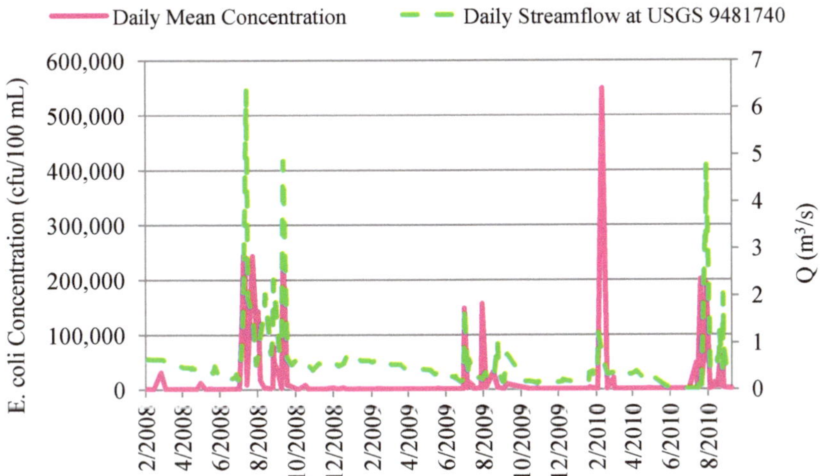

Figure 4. Impact of monthly accumulated rainfall on in-stream fecal coliform concentrations from March 1996 to December 2001.

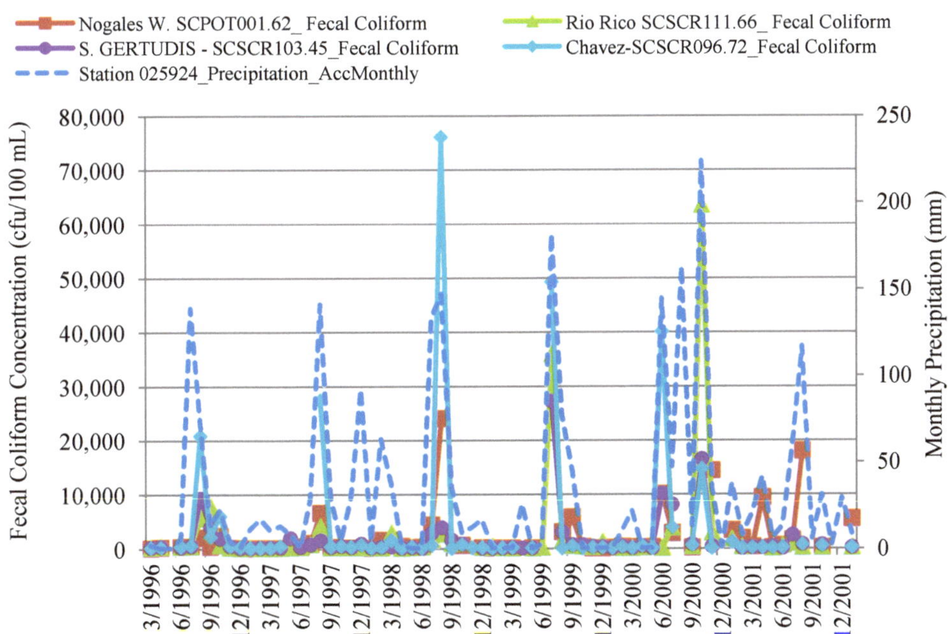

The strength of the positive correlation observed between the response of in-stream *E. coli* concentrations and streamflow (Figure 3); and in-stream fecal coliform concentrations and streamflow

(Figure 2)/precipitation (Figure 4) was tested using linear regression. The resulting *R*-square (R^2) values were 0.31 and 0.32 for correlation of *E. coli* to daily streamflow and fecal coliform to daily streamflow, respectively. The R^2 value for fecal coliform concentration correlation to monthly accumulated rainfall was 0.43. While a correlation exists between streamflow and FIB concentrations, the relationship is convoluted by other factors. Since many hydrological and ecological processes [36] would affect the relationship, the degree of correlation is dependent on factors such as antecedent soil moisture conditions, seasonal changes, sediment loads, proximity of point and nonpoint runoff sources, microbial life cycles.

4.3. In-Stream Concentrations versus WWTP Effluent Concentrations

The in-stream fecal coliform concentrations range from <1.0 to 2519 and the WTTP effluent fecal coliform DMR maximums range from 3 to 1600; in-stream *E. coli* concentrations range from <1.0 to 139,000 and WTTP effluent *E. coli* concentrations range from < 1.0 to 2400. As shown in Figures 5 and 6, the nonpoint source in-stream fecal coliform and *E. coli* concentrations are compared to the maximum concentration reported in each point source WWTP DMR period. The maximum concentration was used because it represents the "worse case" situation during that period of measure. In-stream sampling locations have mean concentrations that are significantly different than the WWTP effluent maximum DMR grab sample values at the 0.05 alpha level of significance as shown in Table 6. Figures 5 and 6 and the statistical summary in Table 6 show that the in-stream fecal coliform and *E. coli* concentrations are significantly greater than the concentrations found in WTTP effluent. Regardless of sample location or type, a high degree of variability occurs in all data sets. Table 6 also shows the range of the data in each category for the entire period of record.

Figure 5. Comparison of fecal coliform maximum WWTP effluent discharge concentrations to in-stream sampling locations. Only values exceeding the 800 cfu/100 mL standard are shown.

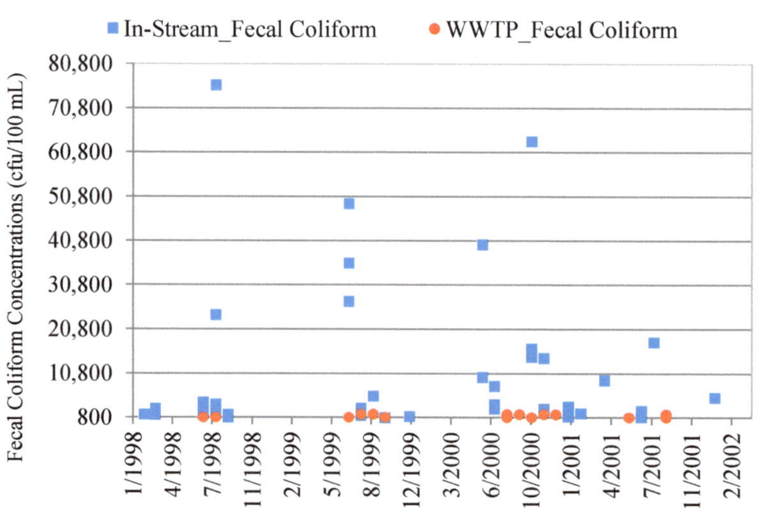

Figure 6. Comparison of *E. coli* from WWTP effluent discharge to in-stream sampling locations. Only values exceeding the 235 cfu/100 mL standard are shown.

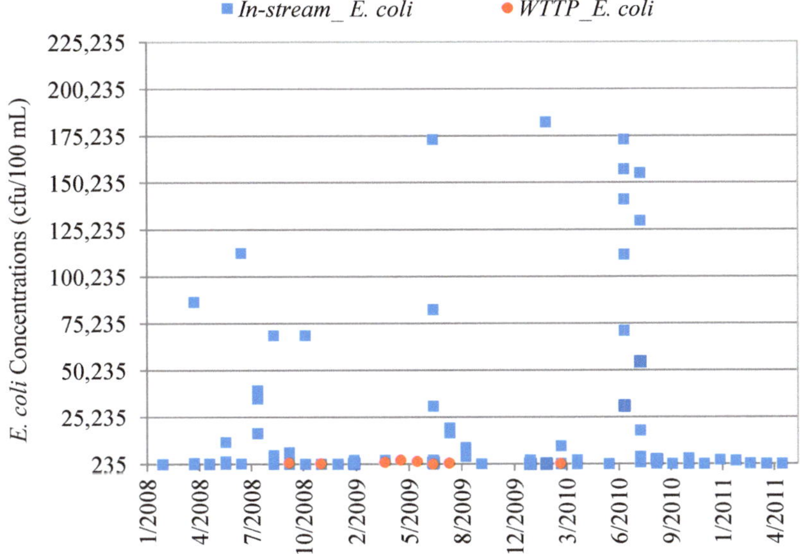

Table 6. Statistical summary of in-stream and WWTP effluent fecal coliform and *E. coli* concentrations. All units are cfu/100 mL.

Data Set	Mean Concentration (cfu/100 mL)	Minimum Concentration	Maximum Concentration	*p*-value *
WWTP Effluent *E. coli*	330	<1.0	2400	0.000002
In-stream *E. coli*	1,745	<1.0	139,000	
WWTP Effluent Fecal Coliform (DMR Maximums **)	285	3	1,600	0.0002
In-stream Fecal Coliform	2,519	<1.0	139,000	

Notes: * statistical test used: two tailed T-test, unequal variance; ** Maximum values reported from each DMR period reflect "worse case" concentrations.

5. Discussion

As this study verifies, significant surface water impairment is a result of nonpoint source pollution in Arizona. In-stream concentrations of fecal coliform and *E. coli* are significantly greater than those concentrations discharged in the treated effluent from WWTPs, as shown in Figures 5 and 6. Nonpoint sources such as faulty septic systems, agricultural and urban runoff, unregulated discharges to stream washes, land use practices, and in-stream fate and transport processes contribute a significant portion of the pollution load to the Santa Cruz River; the statistical data reported in Table 6 supports this finding. According to the ADEQ 2006/2008 statewide summary report, point source contributions to stream pollution impacted 46 miles of streams while nonpoint sources contributed to pollution to 3245 miles of the statewide stream network [20]. The data presented in this study indicate all sampling locations assessed in the Upper Santa Cruz watershed, both point and nonpoint, exceed the water quality criteria established by Arizona to protect human and aquatic health. DMRs submitted to regulatory agencies have several occurrences of FIB concentrations in the treated effluent exceeding the established water quality criteria. Depending on the specifics of the facility permit and wastewater class, these exceedances may be acceptable in some cases.

Studies have shown that FIB survival in surface waters varies from hours to days or even months if protected by sediments which make identifying the source of the FIB concentrations difficult [37,38]. The decay rate of FIB in surface water is a function of many ecological influences; therefore, water quality management, best management plan (BMP) development, watershed modeling, and risk assessment practices need to incorporate better methods as to how FIB interact with the environment, and furthermore, how well FIB accurately model true pathogenic concentrations in the watershed [16,17]. Researchers and regulators continuously question which pathogen indicators are appropriate to determine safe exposure levels in recreational waters. USEPA has approved several detection tests for evaluating FIB in water samples, and comparisons of these methods indicate high variability in sample results [17]. Field *et al.* [19] evaluates the application of fecal source tracking as a better method for human health risk assessments and managing water quality compared to current reliance on FIB criteria. Litton *et al.* [39] further identifies fecal markers and source tracking tools which could vastly change the approach to FIB monitoring and regulation. These studies and the one presented here provide data on FIB concentrations in selected streams with respect to concentration, relationship to recreational water-quality standards, and influence of environmental factors such as streamflow, rainfall, sediment, and runoff [36]. Findings indicate that FIB concentrations are highly variable, especially along urban streams even in the absence of significant rainfall. Though FIB generally increase with streamflow and precipitation events as shown in Figures 2–4, there are occurrences of peaks in FIB concentrations during baseflow conditions.

In Figures 2–4, it is important to provide insight into the data to reach sound conclusions. Overall, trends and correlations show that increased fecal coliform and *E. coli* concentration generally correspond to increased streamflow from rainfall and concentrations are generally higher in the summer months as shown in other similar studies [36]. However, there were instances of increased fecal coliform or *E. coli* concentrations observed during months of little precipitation or streamflow. The data also show that in months of little to no streamflow, several locations were noted as "no sample collected due to no-flow conditions" on the day of sampling. Opportunities for consistent sample collection are limited due to the ephemeral nature of the streamflow, especially at tributary locations. It is likely that in-stream sample collection was done during periods of higher streamflow than average during little or zero baseflow conditions; however, most sample collection was done during low flow conditions and not as a result of precipitation events. As shown in Figure 4, peaks in fecal coliform concentrations positively correlate (R^2 = 0.43) to months of high rainfall. The data compiled for this study provides insight into the water quality conditions related to pathogen indicators in the watershed; however, the underlying conditions, which could affect the grab sample concentrations—such as the sample collection and analysis method, agricultural activity, grazing activity, seasonal hydrology, and stream ecology—were not always clear in this assessment. The variation of the analysis methods and the FIB of interest disallow direct comparison of each sampling location for the entirety of the sampling record and may over or under estimate the actual value.

Efforts to mitigate nonpoint sources are mostly voluntary yet very active across the nation. Watershed managers encourage stakeholders to participate in watershed management groups, volunteer monitoring programs, BMP development and implementation, and education. Examples of successful BMPs for FIB mitigation in effluent dominated systems include engineered wetlands, bioretention areas, and filter strips [40,41]. In addition, improvements in watershed modeling capabilities allow better fate and transport for remediation studies and TMDL development [42,43]. In Arizona, the ADEQ adopted a suspended sediment concentration (SSC) standard of 80 mg/L in 2002 to replace its turbidity standard [20] which is closely linked to FIB concentrations released into the surface waters. Suspended sediment reduction is a priority in many watersheds in order to enhance water quality and to protect fish and aquatic communities. Hindering this progress is the lack of monitoring data in many watersheds which delays efforts to develop, implement, and assess the effectiveness of watershed control strategies such as the SSC standard.

6. Conclusions and Recommendations

Like much of the southwest, Arizona uses recycled waters for groundwater and surface water recharge to balance the supply and demand of a growing population. However, continuous monitoring of the fate and transport of FIB and their associated pathogens is an area needing further assessment. To fully assess the water quality in the Upper Santa Cruz watershed, a detailed analysis is needed which allows for FIB monitoring, source tracking, and reduction of nonpoint sources of pollution. This study assesses the influence of WWTP discharges and nonpoint sources on the indicator bacteria concentrations in the Santa Cruz River and surrounding tributaries. The results of this assessment find that the Upper Santa Cruz watershed is impaired with fecal coliform and *E. coli* at levels, which exceed the established water quality criteria in Arizona. This assessment indicates that a risk to human health exists especially during the summer months when concentration trends increase and water contact is most likely to occur. Fecal coliform and *E. coli* levels from the WWTP effluent assessed in this study are significantly lower than the in-stream samples assessed which indicates that nonpoint sources play a significant role in the water quality conditions. Regardless of the sample type (effluent or in-stream), all sampled locations with available data exceeded the water quality criteria for fecal coliform and *E. coli* indicators. Water quality issues in the Upper Santa Cruz watershed are confounded by the quality of waters flowing from urbanized areas of Mexico with less regulated infrastructure to handle wastewater treatment.

Using natural vegetation filters, stabilization of stream banks, improvement of riparian zones, and urban runoff reduction in order to reduce erosion and sedimentation, are effective watershed control strategies. Updating septic systems is another method of source reduction of potential pathogens to the aquatic environment. Sediment is linked to pollutants such as pathogens and nutrients, and suspended sediment reduction should be a priority in this watershed. Management practices aimed to reduce urban runoff and thus sediment could markedly reduce nonpoint sources of FIB in the stream network. Though likely a more expensive option, infrastructure improvements that eliminate faulty septic systems and combined sewer overflows would also reduce FIB concentrations released into the stream system. Advanced treatment of wastewater effluent and industrial discharges is another option to consider for reducing FIB concentrations within the watershed; the state of the art wastewater treatment at the Nogales plant is a good example of the current and ongoing efforts to achieve such objectives in Arizona. These recommendations could only be truly beneficial to the managers and regulators once TMDL values are established for impaired waterways and more data has been collected to assess how pathogens cycle through the entire watershed. As urbanization and population growth continues in the Santa Cruz watershed, water regulators, managers, and development planners will have to assess the impact of effluent-dominated stream sections in order to meet not only water quantity objectives, but also to maintain water quality standards.

Acknowledgments

The authors acknowledge and thank Claire Zugmeyer, ecological research specialist with the Sonoran Institute, for compiling and sharing the data and the following groups for collection of the in-stream sampling data used in this assessment: National Park Service at Tumacácori, National Historical Park; Friends of the Santa Cruz River, Riverwatch program; U.S. Geological Survey; National Park Service Sonoran Desert Network; and Arizona Department of Environmental Quality. The authors are also grateful for Juanita Francis-Begay who provided many valuable comments to improve the manuscript. Although this work was reviewed by U.S. EPA and approved for publication, it may not necessarily reflect official Agency policy.

References

1. Fayer, R.; Speer, C.A.; Dubey, J.P. *Cryptosporidium and Cryptosporidiosis*, 2nd ed.; CRC Press: Boca Raton, FL, USA, 1997; pp. 1–42.
2. Gaffield, S.J.; Goo, R.L.; Richards, L.A.; Jackson, R.J., Public health effects of inadequately managed stormwater runoff. *Am. J. Public Health* **2003**, *93*, 1527–1533.
3. Asano, T.; Levine, A.D. Wastewater reclamation, recycling and reuse: Past, present, and future. *Water Sci. Technol.* **1996**, *33*, 1–14.
4. Asano, T.; Cotruvo, J.A. Groundwater recharge with reclaimed municipal wastewater: Health and regulatory considerations. *Water Res.* **2004**, *38*, 1941–1951.
5. Arizona Department of Water Resources (ADWR) Web Page. Recharge. Available online: http://www.azwater.gov/AzDWR/WaterManagement/Recharge/default.htm (accessed on 6 July 2011).
6. Schladweiler, J. Tracking Down the Roots of Our Sanitary Sewers. Available online: https://www.sewerhistory.org (accessed on 8 August 2011).
7. Simon, T. Reuse of effluent water—Benefits and risks. *Agric. Water Manag.* **2006**, *80*, 147–159.
8. Brooks, B.; Riley, T.; Taylor, R. Water quality of effluent-dominated ecosystems: Ecotoxicological, hydrological, and management considerations. *Hydrobiologia* **2006**, *556*, 365–379.
9. Gannon, J.; Busse, M. *E. coli* and enterococci levels in urban stormwater, river water and chlorinated treatment plant effluent. *Water Res.* **1989**, *23*, 1167–1176.
10. U.S. Environmental Protection Agency (USEPA) Web Page. 5.11 Fecal Bacteria. Available online: http://water.epa.gov/type/rsl/monitoring/vms511.cfm (accessed on 26 July 2011).

11. U.S. Environmental Protection Agency (USEPA). *Review of Published Studies to Characterize Relative Risks From Different Sources of Fecal Contamination in Recreational Water*; EPA 822-R-09-001; U.S. Environmental Protection Agency: Washington, DC, USA, 2009.

12. U.S. Environmental Protection Agency (USEPA). *Ambient Water Quality Criteria for Bacteria—1986*; EPA 440/5-84-002; U.S. Environmental Protection Agency: Washington, DC, USA, 1986.

13. U.S. Environmental Protection Agency (USEPA). *Test Methods for* Escherichia coli *and* enterococci *in Water by the Membrane Filter Procedure (Method #1103.1)*; EPA 600/4-85-076; U.S. Environmental Protection Agency, Environmental Monitoring and Support Laboratory: Cincinatti, OH, USA, 1985.

14. U.S. Environmental Protection Agency (USEPA). *Bacterial Water Quality Standards for Recreational Waters (Freshwater and Marine Waters)*; EPA-823-R-03-008; U.S. Environmental Protection Agency: Washington, DC, USA, 2003.

15. American Public Health Association (APHA). *Standard Methods for the Examination of Water and Wastewater*, 18th ed.; American Public Health Association: Washington, DC, USA, 1992.

16. Gronewold, A.D.; Borsuk, M.E.; Wolpert, R.L.; Reckhow, K.H. An assessment of fecal indicator bacteria-based water quality standards. *Environ. Sci. Technol.* **2008**, *42*, 4676–4682.

17. Harwood, V.J.; Levine, A.D.; Scott, T.M.; Chivukula, V.; Lukasik, J.; Farrah, S.R.; Rose, J.B. Validity of the indicator organism paradigm for pathogen reduction in reclaimed water and public health protection. *Appl. Environ. Microbiol.* **2005**, *71*, 3163–3170.

18. Leclerc, H.; Mossel, D.A.A.; Edberg, S.C.; Struijk, C.B. Advances in the bacteriology of the Coliform Group: Their suitability as markers of microbial water safety. *Annu. Rev. Microbiol.* **2001**, *55*, 201–234.

19. Field, K.G.; Samadpour, M. Fecal source tracking, the indicator paradigm, and managing water quality. *Water Res.* **2007**, *41*, 3517–3538.

20. Arizona Department of Environmental Quality (ADEQ). 2006/2008 Status of Ambient Surface Water Quality in Arizona: Arizona's Integrated 305(b) Assessment and 303(d) Listing Report. Available online: http://www.azdeq.gov/environ/water/assessment/assess.html (accessed on 26 July 2011).

21. U.S. Environmental Protection Agency (USEPA). *Bacteriological criteria for those states not complying with Clean Water Act section 303(i)(1)(A)Title 40: Part 131.41 Arizona*; U.S. Environmental Protection Agency: Washington, DC, USA, 2010.

22. Marino, R.; Gannon, J. Survival of fecal coliforms and fecal streptococci in storm drain sediment. *Water Res.* **1991**, *25*, 1089–1098.

23. National Research Council. *Issues in Potable Reuse—The Viabilty of Augmenting Drinking Water Supplies with Reclaimed Water*; National Academy Press: Washington, DC, USA, 1998.

24. U.S. Environmental Protection Agency (USEPA) Web Page. Watershed Priorities: Santa Cruz River Watershed, AZ. Available online: http://www.epa.gov/region9/water/watershed/santacruz.html (accessed on 15 September 2011).

25. Scott, C.; Pasqualetti, M.; Hoover, J.; Garfin, G.; Varady, R.; Guhathakurta, S. *Water and Energy Sustainability with Rapid Growth and Climate Change in the Airzona-Sonora Border Region*; A Report to the Arizona Water Institute: Temple, AZ, USA, 2009.

26. U.S. Census Bureau. Table A1: Interim Projections of the Total Population for the United States and States: April 1, 2000 to July 1, 2030. Available online: http://wonder.cdc.gov/wonder/help/populations/population-projections/SummaryTabA1.pdf (accessed on 14 September 2011).

27. Sprouse, T.W. *Water Issues on the Arizona–Mexico Border: The Santa Cruz, San Pedro, and Colorado Rivers*; Water Resources Research Center, The University of Arizona: Tucson, AZ, USA, 2005.

28. U.S. Environmental Protection Agency (USEPA) Web Page. US-Mexico Border 2012. Available online: http://www.epa.gov/usmexicoborder/index.html (accessed on 30 September 2011).

29. U.S. Environmental Protection Agency (USEPA) Web Page. Envirofacts Database. Available online: http://www.epa.gov/enviro/ (accessed on 7 September 2011).

30. Arizona Department of Water Resources (ADWR) Web Page. Active Management Area Water Supply—Central Arizona Project Water. Available online: http://www.azwater.gov/azdwr/StatewidePlanning/WaterAtlas/ActiveManagementAreas/PlanningAreaOverview/WaterSupply.htm (accessed on 30 September 2011).

31. Pima County Regional Wastewater Reclamation Department (RWRD). *RWRD's 2010 Effluent Generation and Utilization Report*; Pima County Regional Wastewater Reclamation Department: Tucson, AZ, USA, 2011.

32. CH2MHILL. *Nogales International Wastewater Treatment Plant Maximum Allowable Headworks Loading Development*; CH2MHILL: El Paso, TX, USA, 2009.

33. Lim, K.J.; Engel, B.A.; Tang, Z.; Choi, J.; Kim, K.-S.; Muthukrishnan, S.; Tripathy, D. Automated web GIS based hydrograph analysis tool (WHAT). *JAWRA J. Am. Water Resour. Assoc.* **2005**, *41*, 1407–1416.

34. Arizona Department of Environmental Quality (ADEQ). *Total Maximum Daily Load For: Oak Creek- Slide Rock State Park Parameters:* Escherichia coliform; Open File Report 09-08; Arizona Department of Environmental Quality: Phoenix, AZ, USA, 1999.

35. U.S. Environmental Protection Agency (USEPA). *Water Quality Standards Handbook: Second Edition*; EPA-823-B-12-002; United States Environmental Protection Agency: Washington, DC, USA, 2012.

36. Lipp, E.; Kurz, R.; Vincent, R.; Rodriguez-Palacios, C.; Farrah, S.; Rose, J. The effects of seasonal variability and weather on microbial fecal pollution and enteric pathogens in a subtropical estuary. *Estuaries Coasts* **2001**, *24*, 266–276.

37. Kinnaman, A.; Surbeck, C.Q.; Usner, D. Coliform bacteria: The effect of sediments on decay rates and on required detention times in stormwater BMPs. *J. Environ. Prot.* **2012**, *3*, 787–797.

38. Easton, J.H.; Gauthier, J.J.; Lalor, M.M.; Pitt, R.E. Die-off of pathogenic *E. coli* O157:H7 in sewage contaminated waters. *JAWRA J. Am. Water Resour. Assoc.* **2005**, *41*, 1187–1193.

39. Litton, R.M.; Ahn, J.H.; Sercu, B.; Holden, P.A.; Sedlak, D.L.; Grant, S.B. Evaluation of chemical, molecular, and traditional markers of fecal contamination in an effluent dominated urban stream. *Environ. Sci. Technol.* **2010**, *44*, 7369–7375.

40. Hunt, W.; Smith, J.; Jadlocki, S.; Hathaway, J.; Eubanks, P. Pollutant removal and peak flow mitigation by a bioretention cell in urban Charlotte, N.C. *J. Environ. Eng.* **2008**, *134*, 403–408.

41. Van der Valk, A.G.; Jolly, R.W. Recommendations for research to develop guidelines for the use of wetlands to control rural nonpoint source pollution. *Ecolog. Eng.* **1992**, *1*, 115–134.

42. Baffaut, C.; Sadeghi, A. Bacteria modeling with SWAT for assessment and remediation studies: A review. **2010**, *53*, 1585–1594.
43. Benham, B.L.; Baffaut, C.; Zeckoski, R.W.; Mankin, K.R.; Pachepsky, Y.A.; Sadeghi, A.M.; Brannan, K.M.; Soupir, M.L.; Habersack, M.J. Modeling bacteria fate and transport in watersheds to support TMDLs. *Transact. ASABE* **2006**, *49*, 987–1002.

Water **2013**, *5*, 342-355; doi:10.3390/w5020342

OPEN ACCESS

water

ISSN 2073-4441
www.mdpi.com/journal/water

Article

Modeling and Optimization of New Flocculant Dosage and pH for Flocculation: Removal of Pollutants from Wastewater

Ammar Salman Dawood [1,2] **and Yilian Li** [1,*]

[1] Environmental Engineering Department, School of Environmental Studies, China University of Geosciences, Wuhan 430074, China; E-Mail: ammar@cug.edu.cn

[2] College of Engineering, University of Basrah, Karmat Ali, Basrah 42001, Iraq

* Author to whom correspondence should be addressed; E-Mail: yl.li@cug.edu.cn; Tel.: +86-27-87436235; Fax: +86-27-87436235.

Received: 22 January 2013; in revised form: 8 March 2013 / Accepted: 19 March 2013 / Published: 26 March 2013

Abstract: In this paper, a new ferric chloride-(polyvinylpyrrolidone-grafted-polyacrylamide) hybrid copolymer was successfully synthesized by free radical polymerization in solution using ceric ammonium nitrate as redox initiator. The hybrid copolymer was characterized by Fourier transform infrared spectroscopy (FTIR) and scanning electron microscopy (SEM). Response surface methodology (RSM), involving central composite design (CCD) matrix with two of the most important operating variables in the flocculation process; hybrid copolymer dosage and pH were utilized for the study and for the optimization of the wastewater treatment process. Response surface analyses showed that the experimental data could be adequately fitted to quadratic polynomial models. Under the optimum conditions, the turbidity and chemical oxygen demand (COD) removal efficiencies were 96.4% and 83.5% according to RSM optimization, whereas the optimum removals based on the genetic algorithm (GA) were 96.56% and 83.54% for the turbidity and COD removal models. Based on these results, wastewater treatment using this novel hybrid copolymer has proved to be an effective alternative in the overseeing of turbidity and COD problems of municipal wastewater.

Keywords: ferric chloride; polyvinylpyrrolidone; polyacrylamide; hybrid copolymer; wastewater treatment; RSM; genetic algorithm; optimization

1. Introduction

Coagulation-flocculation is one of the chemical treatment processes commonly used for water and wastewater. It has a wide range of application in water and wastewater facilities because it is efficient and simple to operate [1,2]. Domestic wastewater usually contains pathogens, suspended solids, nutrients and some other organic materials [3]. The benefit of wastewater treatment is to satisfy the requirements of discharging treated water into the environment. Aluminum and iron salts are widely used as coagulants in the conventional coagulation/flocculation processes and their mode of action is usually explained by two mechanisms: charge neutralization of negatively charged colloids by cationic hydrolysis products and incorporation of impurities in an amorphous hydroxide precipitate [4]. The efficiency of the coagulation/flocculation process depends on the type and dosage of coagulants/flocculants, wastewater pH, and mixing speed.

The addition of inorganic salts to organic flocculants was suggested as the main method of preparing hybrid-flocculants [5,6]. In this method, the enhancement of flocculants by aggregation power increased the ratio of effective component and positive charge of the flocculants [7].

The traditional method of experimentation involves changing one factor at a time. This conventional method of experimentation requires many experiments, which are not only time-consuming, but also lead to low efficiency of optimization. To find a solution to this problem, design of experiment (DOE) has been employed to study the effect of variables and their responses using a minimum number of experiments. Response surface methodology (RSM) is a collection of statistical and mathematical methods which are useful for developing, improving, and optimizing processes [8,9].

Genetic algorithm (GA) is defined as a search technique used in computing to find out the exact or estimated solution in order to optimize and investigate the problem. GA-based optimization is a stochastic search method that involves random generation of positional design solutions which it systematically evaluates and refines until a stopping criterion is met [10]. Through genetic operators and natural selection as well as by mutation and crossover, the best fitness is found.

In this research, the hybrid copolymer ferric chloride-(polyvinylpyrrolidone-grafted-polyacrylamide) ($FeCl_3$-(PVP-g-PAM)) was successfully synthesized by free radical polymerization and characterization of the novel hybrid copolymer was carried out using Fourier transform infrared spectroscopy (FTIR) and scanning electron microscopy (SEM). Moreover, the possible effectiveness of utilizing the hybrid copolymer was investigated as an alternative floculent in wastewater treatment to remove turbidity and chemical oxygen demand (COD) from that water. This paper also targeted the use of a genetic algorithm with RSM to find the optimal parameters and to investigate the promoted effectiveness of predication for removal of pollutants from wastewater.

2. Materials and Methods

2.1. Materials

Acrylamide (AM) was purchased from Amresco (Solon, OH, USA). Ferric chloride ($FeCl_3$) was provided by Shanghai Chemicals Reagent Corp. (Shanghai, China). Polyvinylpyrrolidone (PVP) was obtained from Shanghai Zhanyun Chemical Co., Ltd (Shanghai, China). Ammonium cerium (IV) nitrate (CAN) was supplied by Sinopharm Chemicals Reagent Co. Ltd (Beijing, China). Acetone used was of analytical regent grade.

2.2. Preparation of Hybrid Copolymer

The hybrid copolymer was synthesized by a ceric ion-induced redox initiation method according to the following steps: One gram of polyvinylpyrrolidone (PVP) was dissolved in 100 mL of distilled water at ambient temperature in a 500 mL three-necked flask, stirred with a mechanical stirrer and equipped with a thermostatic water bath, nitrogen line, a reflex condenser, and a rubber septum gap. After that, the system was purged with nitrogen for 30 min to remove the dissolved oxygen from the solution. Then, 0.1 mol of acrylamide and 0.55 mmol of ammonium cerium (IV) nitrate were added to the polymerization system under atmospheric nitrogen. The polymerization reaction was carried out for 2 h at 60 °C. To this, a solution, 1 M of ferric chloride (prepared in 50 mL of distilled water) was added to the polymerization flask at 60 °C and mixed with constant stirring under a nitrogen atmosphere for 3 h. Finally, the produced gel was cooled to ambient temperature, precipitated in acetone, and dried in a vacuum oven at 60 °C to constant weight.

2.3. Characterization of Hybrid Copolymer

The Fourier transform infrared (FTIR) spectra were measured on a Nexus FTIR spectrophotometer (Thermo Fisher Scientific Inc., Waltham, MA, USA) using the KBr pellet method. The IR spectra were recorded within the range of 4000–400 cm^{-1}. The morphology of the hybrid copolymer surface was investigated using scanning electron microscope (SEM) images with different magnification obtained from a QUANTA 200 scanning electron microscope (FEI Company, Hillsboro, OR, USA).

2.4. Wastewater Source

The samples of wastewater in this study were collected from the sewer system network in the east campus at the China University of Geosciences (Wuhan). The wastewater samples were transported to the laboratory within 20 min and then characterized there for turbidity, COD and pH. The measured values of the wastewater sample for the flocculation experiments were as follows: Turbidity 305 NTU, COD 368 mg/L, and pH 7.4.

2.5. Wastewater Flocculation

The jar test used in our experiments was a programmable apparatus (TA6, Wuhan, China). It consisted of six paddles on a bench. The paddles were connected to each other by a gear mechanism, and all of these paddles were simultaneously rotated by the same motor at a controlled speed and time.

Wastewater samples of 1000 mL each, were transferred to the jars and then pH adjusted using 0.5 M HCl or 0.5 M NaOH solutions. The required dose of FeCl$_3$-(PVP-g-PAM) hybrid copolymer was added to each beaker. Directly after the addition of the hybrid copolymer dosage, the wastewater sample in the jar was stirred rapidly at a paddle speed of 120 rpm for 2 min then stirred slowly at a paddle speed of 30 rpm for 20 min, and finally, the treated wastewater was allowed to settle for 30 min.

The removal of the pollutants (COD and turbidity) was calculated according to the following formula:

$$\text{Removal Efficiency} = \left[\frac{C_i - C_f}{C_i} \right] \times 100 \tag{1}$$

where, C_i and C_f are the initial and final concentrations of the pollutants.

2.6. Response Surface Methodology

Response surface methodology is a statistical method frequently used in designing experimental, building models, for evaluating the effects of several factors and to find the optimum conditions for desirable responses as well as to reduce the number of experiments [11,12]. Design-expert software version 8, was used to optimize the major operating factors which were $FeCl_3$-(PVP-g-PAM) hybrid copolymer dosage and wastewater pH. In this study, RSM used the common form of center composite design (CCD) which is called center composite face design (CCFD) that consists of 2^k factorial points (k means factors $= 2$), $2k$ axial points and two replicated at the center point to provide estimation of the experimental error variance. The turbidity removal and COD removal were selected as the dependent variables, while the hybrid copolymer dosage and wastewater pH were selected as independent variables.

$$\text{Total number of experiments} = (2^k) + (2k) + 2 = 10 \text{ experiments}$$

The independent variables (factors) were hybrid copolymer dosage (denoted by X_1) and wastewater pH (denoted by X_2). These factors have three levels as follows: low level (-1), center level (0), and high level ($+1$) as shown in Table 1. The actual values of the coded levels for these factors were selected and based on preliminary experiments. These codes are also included in Table 1.

Table 1. Experimental factor levels for independent variables.

Variables (Factors)	Symbol	Real values of coded levels		
		Low level (-1)	Center level (0)	High level ($+1$)
Dose (mg/L)	X_1	50	100	150
pH	X_2	5	7	9

The second order polynomial equation is used to prove the relationship between the factors (X_1 and X_2) and the investigated response (Y).

$$Y = f(x) = \beta_o + \sum_{i=1}^{k} \beta_i X_i + \sum_{i=1}^{k} \beta_{ii} X_i^2 + \sum_{i=1}^{k-1} \sum_{j=i+1}^{k} \beta_{ij} X_i X_j \tag{2}$$

where Y is the response model (turbidity removal and COD removal); β_0 is the constant coefficient; β_i is the coefficient of the linear term; β_{ii} is the coefficient of the square term; β_{ij} is the coefficient of the quadratic term; k is the number of independent variables; X_i and X_j are the coded values of the independent variables.

3. Results and Discussion

3.1. FTIR Spectra

The IR spectra of the hybrid copolymer are shown in Figure 1. The spectra were characterized by the following bands: band at 3417 cm^{-1} attributed to OH; band at 1653 cm^{-1} attributed to amide II; band at 1552 cm^{-1} attributed to amide I; band at 1451 cm^{-1} attributed to CH; band at 1108 cm^{-1} corresponding to -C-NH$_2$; band at 594 cm^{-1} corresponding to C-Cl [13]. The above analysis results show that the new hybrid copolymer has inorganic and organic components, and hence it is an inorganic-organic complex.

Figure 1. Fourier transform infrared spectroscopy (FTIR) spectra of the ferric chloride-(polyvinylpyrrolidone-grafted-polyacrylamide)(FeCl$_3$-(PVP-g-PAM)) hybrid copolymer.

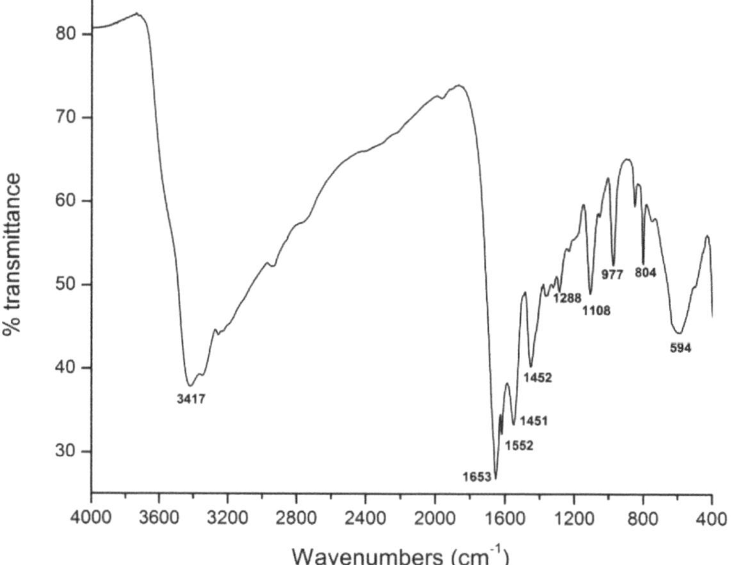

3.2. Morphological Analysis

The technique used for studying the surface morphology of the polymers is scanning electron microscopy (SEM). The SEM images obtained for the FeCl$_3$-(PVP-g-PAM) hybrid copolymer as shown in Figure 2 indicated that the surface of the hybrid copolymer has a porous surface and that this type of surface may have some influence on the flocculation process.

Figure 2. Scanning electron microscopy (SEM) images for FeCl$_3$-(PVP-g-PAM) at different magnification.

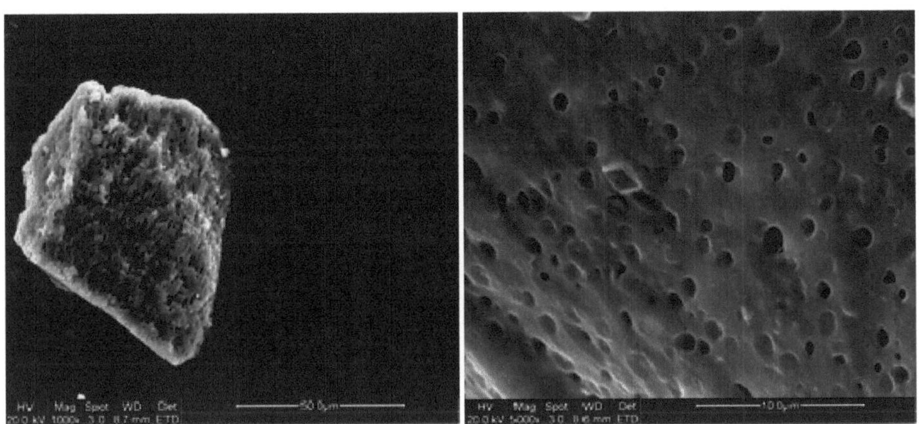

3.3. Statistical Analysis

Response surface methodology was used to determine the relationship between the flocculation process responses (turbidity and COD removals) with the most important variables (hybrid copolymer dosage and wastewater pH). A total of ten experiments was carried out as mentioned before and their results are shown in Table 2. There are several response models that can be derived, such as linear, interactive, quadratic and cubic models. These models may be correlated with the experimental data, but significant selection of the best model is required because the selected model correlates with the experimental data depending on the adequacy of that selected model. Thus, according to the experimental data, the quadratic model was suggested to represent the correlation between experimental data and all responses, because it has the lowest standard deviation and p value, as well as the highest coefficient of determination (R^2) , adjusted R^2 and predicted R^2 values. However, the cubic model was not recommended in this study because it had insufficient points to estimate the coefficients of the model.

Table 2. Experimental variables and results for wastewater flocculation.

Run No.	Coded variables		Real variables		Results	
	Dose	pH	Dose (mg/L)	pH	Turbidity Removal (%)	COD Removal (%)
1	1	−1	150	5	91	72
2	−1	1	50	9	78	50
3	0	0	100	7	93	83
4	−1	−1	50	5	83	54
5	0	−1	100	5	89	70
6	1	0	150	7	97	84
7	1	1	150	9	86	63
8	0	1	100	9	85	62
9	−1	0	50	7	88	60
10	0	0	100	7	95	80

3.4. Analysis of Variance (ANOVA)

Analysis of variance (ANOVA) with an alpha (α) level of 0.05 was employed to determine the statistical significance of all analyses. The final quadratic model for each response in terms of coded levels is shown (Equations 3 and 4) which represent the final quadratic model for the pollutants removal. These equations have some statistically non-significant terms containing the lowest F value. Therefore, it is necessary to eliminate these non-significant terms from the response equations as shown below:

$$Y_{\textit{Turbidity removal}} = 94.29 + 4.17\,X_1 - 2.33\,X_2 - 2.07\,X_1^2 - 7.57\,X_2^2 \tag{3}$$

$$Y_{\textit{COD removal}} = 80.57 + 9.17\,X_1 - 3.50\,X_2 - 7.64\,X_1^2 - 13.64\,X_2^2 \tag{4}$$

Table 3 shows the probability (p value) of the quadratic model for turbidity and COD removals that were 0.0005 and 0.0032 respectively. The p value in each pollutant removal model implies that the model is significant.

Table 3. Analysis of variance (ANOVA) results showing the terms in each response quadratic model.

Response	Source	Sum of Squares	df	F value	p value	Remark
Turbidity removal	X_1	104.17	1	128.68	0.0003	significant
	X_2	32.67	1	40.35	0.0031	significant
	X_1X_2	0.000	1	0.00	1.0000	not significant
	X_1X_1	10.01	1	12.37	0.0245	significant
	X_2X_2	133.76	1	165.24	0.0002	significant
COD removal	X_1	504.17	1	57.02	0.0016	significant
	X_2	73.50	1	8.31	0.0449	significant
	X_1X_2	6.25	1	0.71	0.4478	not significant
	X_1X_1	136.30	1	15.41	0.0172	significant
	X_2X_2	434.30	1	49.12	0.0022	significant

Notes: X_1: first variable, dose (mg/L); X_2: second variable, pH; df: degree of freedom.

To evaluate the quality of the model developed, the coefficient of determination (R^2) was used which gave the proportion of total variance in the response predicted by the model. The closer R^2 is to 1, the better the model predicts the response [14]. The coefficient of determination values for turbidity and COD removals were 0.9892 and 0.9726 respectively. This indicates that there is high dependence and correlation between the observed and predicted values of the response [15]. The adjusted R^2 for turbidity and COD removals were 0.9758, and 0.9383 respectively, so they are very close to the R^2 value in each of the response equations. Thus the predication of experimental data is considered to be satisfactory [16].

As shown in Table 4, the values of adjusted R^2 for the pollutants removal models suggested that the total variation was 98% and 97% for turbidity and COD removals respectively. This can be attributed to the independent variables and there is only about 2% and 3% of the total variation respectively that cannot be explained by these models.

Table 4. ANOVA results for response models.

Response	Probability	R^2	Adj. R^2	Pred. R^2	Adeq. precision	CV%
Turbidity removal	0.0005	0.9892	0.9758	0.9466	26.169	1.02
COD removal	0.0032	0.9726	0.9383	0.7469	14.859	4.39

Notes: R^2: coefficient of determination; Adj. R^2: adjusted R^2; Pred. R^2 : predicted R^2 ; Adeq. precision: Adequate precision; CV: coefficient variation.

The measures of the adequate precision for the response models were 26.169 and 14.859 for the turbidity and COD removal models. These values represent the measures of the signal to noise ratio [17]. A ratio greater than four is desirable. Hence, in this study the adeq. precision values for both the turbidity removal model and the COD removal model were more than four. This indicates the quadratic model equation can be used within the range of factors in the design space. The coefficient of variance (CV), represents the ratio of the standard error of estimate to the mean value of the observed model (represented as %). The model can be normally considered reproducible when its CV value is less than 10% [18]. The low CV values of the models of 1.02% and 4.39% indicate that the precision and reliability of the experiments is good. To judge the suitability of the model, a diagnostic plot *i.e.*, observed *vs.* predicted values was used (Figure 3). These diagnostic plots provide sufficient agreement between the experimental data and the values obtained from the models for turbidity and COD removals. To understand the effect of each factor on the final response of pollutants removal, Pareto graphics can be used. Pareto graphics have positive and negative bars (Figure 4). The positive bars suggest that by varying the factor the response increases. When it increases the X_1, response will also increase. While, the negative bars suggest that by varying the factor the response decreases.

Figure 3. (**a**) Predicted *vs.* observed values plot for turbidity removal model; and (**b**) Predicted *vs.* observed values plot for COD removal model.

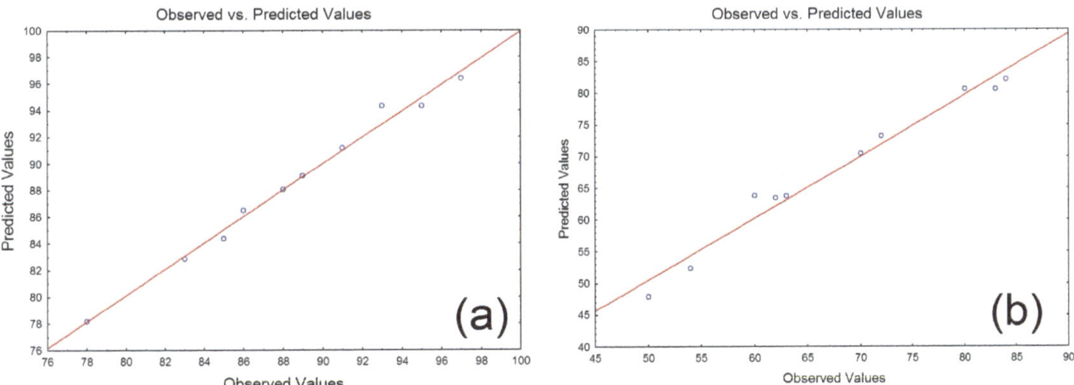

Figure 4. (**a**) Pareto graphics for hybrid copolymer dose and pH for turbidity removal model; and (**b**) Pareto graphics for hybrid copolymer dose and pH for COD removal model.

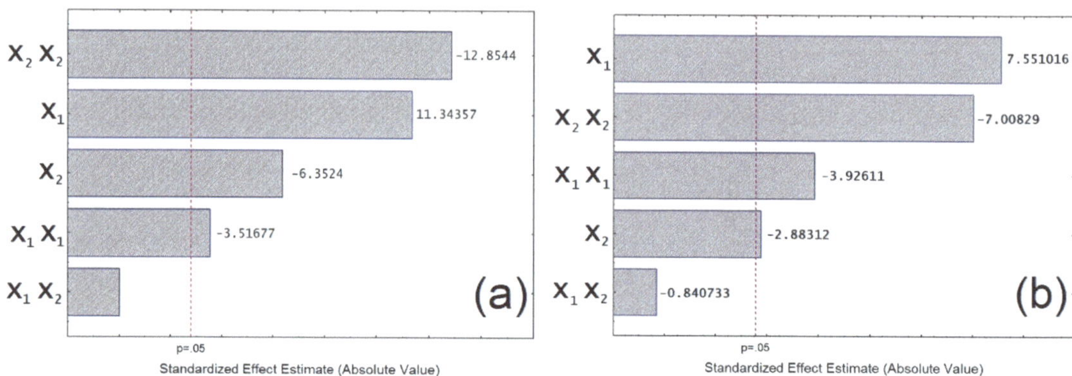

3.5. Analysis of Flocculation Process

The 3D surface plots for each model show the responses of experimental variables and these graphs can be used to identify the major interaction between the variables. The 3D surface plot and contour plot for the turbidity removal model (Figures 5a and 6a), show that a maximum turbidity removal of more than 95% occurs at pH range (6–7.5) with a hybrid copolymer dosage more than 110 mg/L. This maximum removal occurs due to the fact that the hybrid polymer $FeCl_3$-(PVP-g-PAM) becomes ionized, and the ionized Fe^{3+} can easily neutralize the residual charge on particles and expand the chain on the bridge. It is observed that an increase in pH beyond the maximum range will lead to a decrease in the flocculation process efficiency. This decrease in the removal process is due to the start $Fe(OH)_3$-(PVP-g-PAM) complexes in the alkaline region leading to adsorption of $Fe(OH)_3$-(PVP-g-PAM) onto wastewater particles.

Figure 5. 3D surface plot for (**a**) turbidity removal; and (**b**) for COD removal.

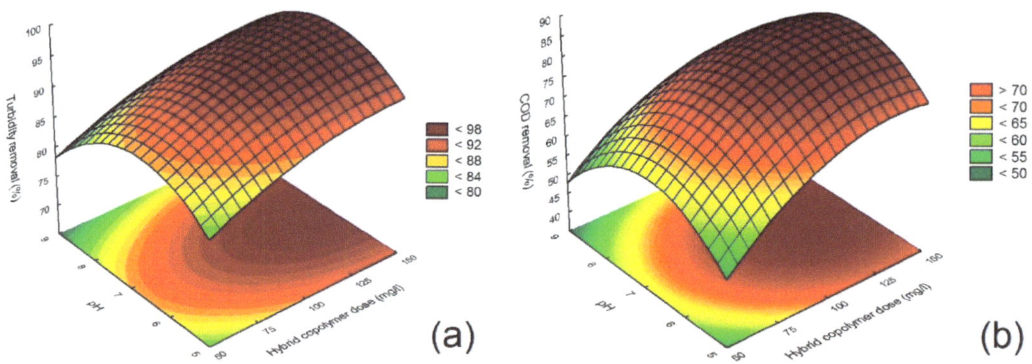

Figure 6. 2D contour plot for (a) turbidity removal; and (b) for COD removal.

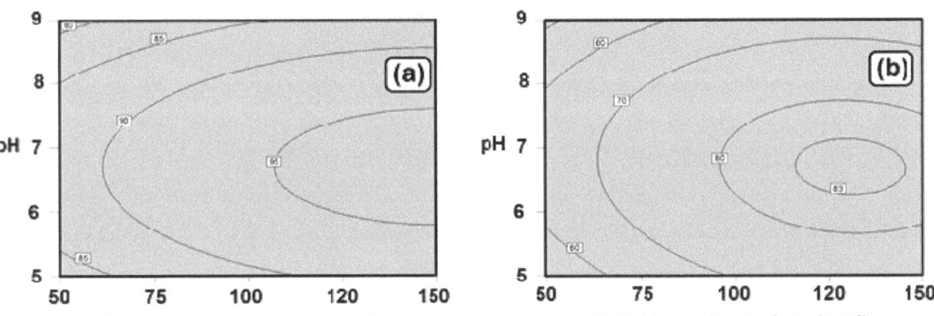

The common turbidity removal mechanism ensues by neutralizing the negative charge of particles and the positive charge of metal hydrolysis species followed by the aggregation of destabilized particles. There are other mechanisms for turbidity removal that take place by forming flocs composed of metal hydroxide precipitates accompanied or followed by sweep flocculation of colloidal particles [19].

The 3D surface plot and contour plot for the COD removal model (Figures 5b and 6b), show that the maximum COD removal was 83% in the pH range (6.3–7) with the hybrid copolymer dosage range (120–145 mg/L). As is shown in the figures, high dosage of the hybrid copolymer does not contribute to a noticeable increase in COD removal [20]. This phenomenon is due to the increase of $Fe(OH)_3$-(PVP-g-PAM) complexes that start to form because ferric hydroxide precipitates when alkaline and any increase in Fe^{3+} ion or OH^- ion also increases the solubility constant of the ferric hydroxide. The maximum removal of COD is at the pH value when almost all ferric ions are converted into perceptible hydroxide [21]. Beyond that optimum pH value, the COD removal decreases probably due to the increase in solubility of the ferric precipitate.

3.6. Optimization Conditions and Verification

The optimal conditions for maximum turbidity and COD removals were determined by the response model obtained from the experimental data. A desirable function was used to find the optimum condition for the two variables, of hybrid copolymer dosage and wastewater pH, in the study of the flocculation process of wastewater. In RSM, the desirability function was set as follows: maximum process removal with the range of hybrid copolymer dosage and within the pH range (5–9). By assay of 39 results of starting points in the optimization of RSM, the best optimum removal efficiency for turbidity and COD removal was 96.4% and 83.5% respectively. This optimum removal was acquired at the desirability function of 0.978 with the design variables as follows: hybrid copolymer dosage of 137 mg/L at wastewater pH 6.68.

Using an optimization technique from the Matlab optimization toolbox, GA was applied to the quadratic equations for turbidity and COD removal models to optimize the variables and responses. The removal optimization can be stated as follows:

Find: *(Dose and pH)*

Maximize removal process = *f (Dose and pH)*

Subjected to the constraint: *removal process ≤ 100%*
Parameter ranges: $-1 \leq Dose \leq +1$ and $-1 \leq pH \leq +1$

where -1 is the low level of the factor *Dose* or factor *pH* in the quadratic model equations [Equations (3) and (4)]; $+1$ is the high level of the factor *Dose* or factor *pH* in the quadratic model equations (Equations 3 and 4).

Figure 7a shows the results of the optimum coded factors $X_1 = 1$ and $X_2 = -0.154$ in GA optimization for turbidity removal. Based on Table 1, the real value of the coded factor X_1 was 150 mg/L while the real value of the coded factor X_2 was 6.69. The results of the GA optimization show that the turbidity removal efficiency was 96.56%. Meanwhile, the optimum results of the coded factors in GA optimization for COD removal were $X_1 = 0.6$ and $X_2 = -0.128$ (Figure 7b). Thus, the real value of the coded factor X_1 was 130 mg/L and the real value of the coded factor X_2 was 6.74. The optimized COD removal efficiency that depends on these optimized factors was 83.54%.

Figure 7. Plot of fitness value *vs.* generation for the variables in GA optimization for (**a**) turbidity removal; and (**b**) COD removal.

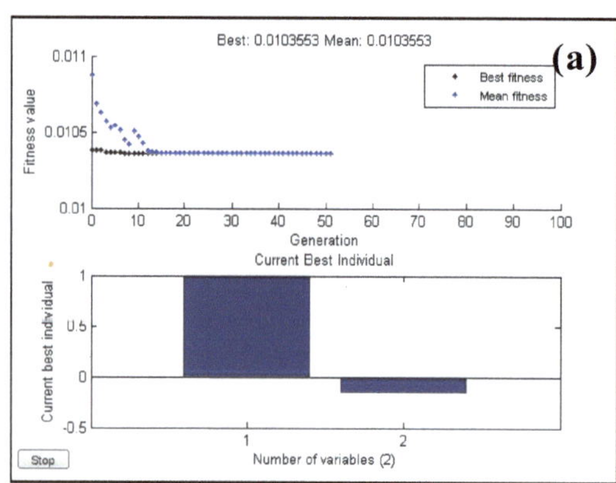

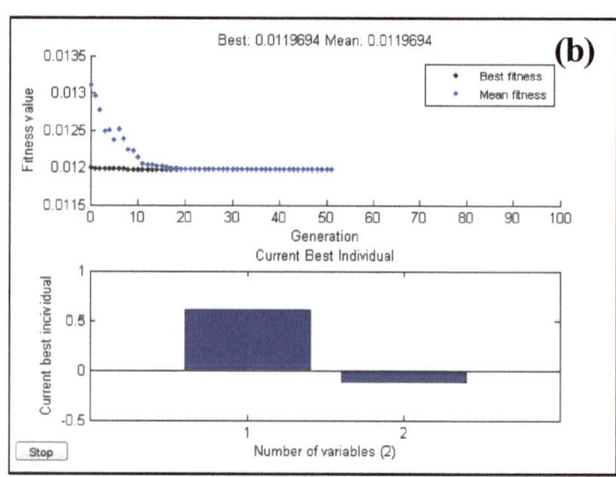

To select the best optimum removal efficiency with the lowest cost, a comparison of optimization between the desirable function in RSM and the GA results in terms of variables and the optimum removal efficiency is shown in Table 5. It is clear that the final best predicted values were almost the same in both optimization techniques but the optimum dosage of the hybrid copolymer required for the best optimized turbidity removal in GA is more than that in the RSM optimized method. While the optimum dosage of the hybrid copolymer required for achieving the best optimum COD removal according to the GA technique was lower than that required in RSM optimization by the desirable function.

The optimized hybrid copolymer dosage of (137 mg/L) for the best predicted turbidity removal can be contributed to a predication of 96.4% for the turbidity removal according to the desirable function in RSM optimization. Whereas, the optimized hybrid copolymer dosage of (130 mg/L) for best predicted COD removal can be contributed to a predication of 83.54% for COD removal based on GA optimization.

Table 5. Comparison between the desirable function and GA optimization techniques for optimum variables and predication for pollutants removal.

Model	Optimized Technique	Optimal Dose (mg/L)	Optimal pH	Best predicted Removal (%)
Turbidity removal	Desirable function	137	6.68	96.40
	Genetic algorithm	150	6.69	96.56
COD removal	Desirable function	137	6.68	83.50
	Genetic algorithm	130	6.74	83.54

Finally, three extra experiments were conducted under the optimum condition to confirm the validity of the statistical experimental strategies. The obtained removal results of these three experiments were close to those estimated by using response surface methodology. These validation experiments proved that the developed models could be considered to be accurate and reliable.

4. Conclusions

Physical-chemical methods are fast wastewater treatment processes. One such physical-chemical method is flocculation in which many types of commercial and conventional flocculants can be used. In this study, a new hybrid copolymer was synthesized, characterized and employed in wastewater treatment. The novel hybrid copolymer was accomplished by focusing on the influence of two important operating variables: hybrid copolymer dosage and wastewater pH. The experiments of the flocculation process were utilized by RSM. The results were arrived at by applying RSM modeling that had been verified by conducting analysis of variance (ANOVA). The effects of both hybrid copolymer dosage and wastewater pH on the optimal operational conditions are discussed according to the desirable function and GA optimization techniques. Under these optimized conditions, the removal efficiencies according to RSM optimization using the desirable function were 96.4% and 83.5% for turbidity and COD removal models respectively. The optimized desirability function was 0.978. GA optimization established the best prediction of 96.56% for turbidity removal and 83.54% for COD removal.

Acknowledgments

The financial support for this work from China University of Geosciences (Wuhan, China) is gratefully acknowledged.

References

1. Wang, J.P.; Chen, Y.Z.; Ge, X.W.; Yu, H.Q. Optimization of coagulation–flocculation process for a paper-recycling wastewater treatment using response surface methodology. *Colloid. Surf. A* **2007**, *302*, 204–210.
2. Zheng, H.; Zhu, G.; Jiang, S.; Tshukudu, T.; Xiang, X.; Zhang, P.; He, Q. Investigations of coagulation–flocculation process by performance optimization, model prediction and fractal structure of flocs. *Desalination* **2011**, *269*, 148–156.
3. Devi, R.; Dahiya, R. COD and BOD removal from domestic wastewater generated in decentralised sectors. *Bioresour. Technol.* **2008**, *99*, 344–349.

4. Duan, J.; Gregory, J. Coagulation by hydrolysing metal salts. *Adv. Colloid Interface Sci.* **2003**, *100*, 475–502.

5. Lee, K.E.; Teng, T.T.; Morad, N.; Poh, B.T.; Mahalingam, M. Flocculation activity of novel ferric chloride–polyacrylamide (FeCl$_3$-PAM) hybrid polymer. *Desalination* **2011**, *266*, 108–113.

6. Yang, W.; Qian, J.; Shen, Z. A novel flocculant of Al(OH)$_3$-polyacrylamide ionic hybrid. *J. Colloid Interface Sci.* **2004**, *273*, 400–405.

7. Tang, H.; Shi, B. The characteristics of composite flocculants synthesized with inorganic polyaluminium and organic polymers. In *Chemical Water and Wastewater Treatment VII*, Proceedings of the 10th Gothenburg symposium 2002, Gothenburg, Sweden, June 17–19, 2002; IWA Publishing: London, UK, 2002; pp. 17–28.

8. Kasiri, M.; Khataee, A. Photooxidative decolorization of two organic dyes with different chemical structures by UV/H$_2$O$_2$ process: Experimental design. *Desalination* **2011**, *270*, 151–159.

9. Omar, F.M.; Rahman, N.N.N.A.; Ahmad, A. COD reduction in semiconductor wastewater by natural and commercialized coagulants using response surface methodology. *Water Air Soil Poll.* **2008**, *195*, 345–352.

10. Woon, S.; Querin, O.; Steven, G. Structural application of a shape optimization method based on a genetic algorithm. *Struct. Multidiscip. Opitm.* **2001**, *22*, 57–64.

11. Mason, R.L.; Gunst, R.F.; Hess, J.L. *Statistical Design and Analysis of Experiments: With Applications to Engineering and Science*; Wiley-Interscience: Hoboken, NJ, USA, 2003; Volume 356752.

12. Yang, Y.; Li, Y.; Zhang, Y.; Liang, D. Applying hybrid coagulants and polyacrylamide flocculants in the treatment of high-phosphorus hematite flotation wastewater (HHFW): Optimization through response surface methodology. *Sep. Purif. Technol.* **2010**, *76*, 72–78.

13. Pretsch, E.; Bühlmann, P.; Badertscher, M. *Structure Determination of Organic Compounds: Tables of Spectral Data*; Springer: Berlin, Germany, 2009.

14. Ghafari, S.; Aziz, H.A.; Isa, M.H.; Zinatizadeh, A.A. Application of response surface methodology (RSM) to optimize coagulation–flocculation treatment of leachate using poly-aluminum chloride (PAC) and alum. *J. Hazard. Mater.* **2009**, *163*, 650–656.

15. Sharma, P.; Singh, L.; Dilbaghi, N. Optimization of process variables for decolorization of Disperse Yellow 211 by *Bacillus subtilis* using Box–Behnken design. *J. Hazard. Mater.* **2009**, *164*, 1024–1029.

16. Wantala, K.; Khongkasem, E.; Khlongkarnpanich, N.; Sthiannopkao, S.; Kim, K.W. Optimization of As(V) adsorption on Fe-RH-MCM-41-immobilized GAC using Box-Behnken design: Effects of pH, loadings, and initial concentrations. *Appl. Geochem.* **2011**, *5*, 1027–1034.

17. Tripathi, P.; Srivastava, V.C.; Kumar, A. Optimization of an azo dye batch adsorption parameters using Box–Behnken design. *Desalination* **2009**, *249*, 1273–1279.

18. Ahmadi, M.; Vahabzadeh, F.; Bonakdarpour, B.; Mofarrah, E.; Mehranian, M. Application of the central composite design and response surface methodology to the advanced treatment of olive oil processing wastewater using Fenton's peroxidation. *J. Hazard. Mater.* **2005**, *123*, 187–195.

19. Trinh, T.K.; Kang, L.S. Response surface methodological approach to optimize the coagulation–flocculation process in drinking water treatment. *Chem. Eng. Res. Design* **2011**, *89*, 1126–1135.

20. Stephenson, R.J.; Duff, S.J.B. Coagulation and precipitation of a mechanical pulping effluent—I. Removal of carbon, colour and turbidity. *Water Res.* **1996**, *30*, 781–792.

21. Tan, B.H.; Teng, T.T.; Omar, A. Removal of dyes and industrial dye wastes by magnesium chloride. *Water Res.* **2000**, *34*, 597–601.

Water **2012**, *4*, 770-784; doi:10.3390/w4040770

OPEN ACCESS

water

ISSN 2073-4441
www.mdpi.com/journal/water

Article

Fungal Waste-Biomasses as Potential Low-Cost Biosorbents for Decolorization of Textile Wastewaters

Valeria Prigione, Irene Grosso, Valeria Tigini, Antonella Anastasi and Giovanna Cristina Varese *

Department of Life Science and Systems Biology, University of Turin, Viale Mattioli, 25,
Turin 10125, Italy; E-Mails: valeria.prigione@unito.it (V.P.); irene.grosso84@gmail.com (I.G.);
valeria.tigini@unito.it (V.T.); antonella.anastasi@unito.it (A.A.)

* Author to whom correspondence should be addressed; E-Mail: cristina.varese@unito.it;
Tel.: +39-011-670-5984; Fax: +39-011-670-5962.

Received: 28 July 2012; in revised form: 18 September 2012 / Accepted: 24 September 2012 / Published: 12 October 2012

Abstract: The biosorption potential of three fungal waste-biomasses (*Acremonium strictum*, *Acremonium* sp. and *Penicillium* sp.) from pharmaceutical companies was compared with that of a selected biomass (*Cunninghamella elegans*), already proven to be very effective in dye biosorption. Among the waste-biomasses, *A. strictum* was the most efficient (decolorization percentage up to 90% within 30 min) with regard to three simulated dye baths; nevertheless it was less active than *C. elegans* which was able to produce a quick and substantial decolorization of all the simulated dye baths (up to 97% within 30 min). The biomasses of *A. strictum* and *C. elegans* were then tested for the treatment of nine real exhausted dye baths. *A. strictum* was effective at acidic or neutral pH, whereas *C. elegans* confirmed its high efficiency and versatility towards exhausted dye baths characterised by different classes of dyes (acid, disperse, vat, reactive) and variation in pH and ionic strength. Finally, the effect of pH on the biosorption process was evaluated to provide a realistic estimation of the validity of the laboratory results in an industrial setting. The *C. elegans* biomass was highly effective from pH 3 to pH 11 (for amounts of adsorbed dye up to 1054 and 667 mg of dye g^{-1} biomass dry weight, respectively); thus, this biomass can be considered an excellent and exceptionally versatile biosorbent material.

Keywords: biosorption; *Cunninghamella elegans*; fungi; textile industry wastewater; waste-biomass

1. Introduction

Control of pollution is one of the prime concerns of society today, since in both developing and industrialized nations a growing number of contaminants enter water supplies from human activity [1]. Actually, many industries, such as textile, paper, plastics and dyestuffs, consume substantial volumes of water, using chemicals during manufacturing and dyes to colour their products. As a result, a considerable amount of polluted wastewater is generated, which is a major source of aquatic pollution and can cause considerable damage to the receiving waters if the discharge is not adequately treated [2]. Indeed, many synthetic dyes are toxic, mutagenic, carcinogenic and represent a potential health hazard to all forms of life [3]. Prior to their release, therefore, coloured wastewaters should be treated to bring their dye concentrations down to nationally permitted levels.

Virtually all known physico-chemical techniques (coagulation, adsorption, filtration, membrane separation, *etc.*), including advanced oxidation processes (AOPs), photolysis by UV irradiation and sonolysis by means of ultrasound exposition have been explored to remove these toxic compounds from wastewater, but all of them present some drawbacks: excessive chemical usage, expensive plant requirements, high operational costs, lack of effective colour removal and sensitivity to variable wastewater input [4]. A single, universally applicable end-of-pipe solution appears to be unrealistic, and the combination of traditional and innovative techniques is deemed imperative to devise technically and economically feasible options.

Among the numerous techniques of dye removal, adsorption through activated carbons or organic resins has been so far one of the procedures of choice but the very high costs have resulted in the necessity to find alternative, cheaper adsorbent materials [2,4]. Biosorption is becoming an attractive technique thanks to its advantages over other techniques: high efficiency, cost effectiveness, and good removal performance [5]. In particular, fungi have a positive potential for the development of cost-effective biosorbents since they can be grown using unsophisticated fermentation techniques and inexpensive growth media, while producing high yields of biomass. Furthermore, many species are extensively used in a variety of large scale industrial fermentation processes where, after enzyme extraction and biochemical transformations, the biomass cannot be re-used and constitutes a waste material that is generally poorly valorised [6]. Hence, the use of waste-biomasses in biosorption application could be helpful not only to the environment, in solving the solid waste disposal problem, but also to the economy [2,7]. However, until now the potential use of fungal waste-biomasses for the removal of pollutants remains largely untapped and has been almost exclusively limited to heavy metals [6,8–11].

In the recent past, biosorption studies involving different kinds of selected organisms either dead or alive have dominated the literature. However, despite a large number of lab-scale studies on the decolorization of mono-component synthetic dye solutions, there is a need to generate relative performance data on real industrial effluents, which so far have been considered very rarely in biosorption experiments. Dye removal from real effluents should be included in studies on biosorption as this process is strongly dependent on pH, ionic strength and temperature, which are generally very variable in actual wastewaters; besides, the massive presence of salts, surfactants and other additives may hinder the dye biosorption performance [12].

In the present study, the biosorption potentials of three fungal waste-biomasses from the pharmaceutical industry (towards three simulated exhausted dye baths) were compared with that of a selected fungal biomass (*Cunninghamella elegans* Lendner) which in the light of previous experiments had already proven to be very effective in synthetic dyes and chromium removal with both mono and multi-component dye solutions, for simulated exhausted dye baths and a real tanning effluent [13–16]. Afterwards, in order to assess the value of the biosorption process under real conditions, the two most promising biomasses were tested against nine real exhausted dye baths representative of different dye types and dyeing processes. Finally, the effect of pH on the biosorption process was evaluated to provide a realistic estimation of the validity of the laboratory results in an industrial setting.

2. Results and Discussion

2.1. Decolorization

The results obtained by the biosorption tests with simulated dye baths (Figure 1) indicate that the waste-biomass of *A. strictum* provided excellent yields of decolorization of SABW (Simulated Acid Bath for Wool, 85%) and SDBC (Simulated Direct Bath for Cotton, 73%) and fair results towards SRBC (Simulated Reactive Bath for Cotton, 40%), displaying higher biosorptive capacities than *Acremonium* sp. and *Penicillium* sp. biomasses. Nevertheless, the good results obtained by *A. strictum* were lower than those obtained by *C. elegans*, which was able to effect a quick and substantial decolorization of all the three exhausted dye baths (up to 99%). Noteworthy is the fastness of the biosorption process with all the tested biomasses: in most cases the maximum yield of decolorization was obtained within the first 30 min of incubation. Only in the case of SDBC, for *C. elegans* and *A. strictum* was observed a significant increase between 30 min and 24 h.

Since different chemical groups of the fungal cell wall, such as carboxyl, amine, imidazol, phosphate, sulphydryl, sulphate, hydroxyl and the lipid fraction, have been suggested as potential binding sites [7], these different biosorption yields could be due to the different composition of the cell wall. Actually, there are marked differences in the structure and composition of the cell wall of fungi belonging to different classes such as Zygomycetes *C. elegans* and Ascomycetes *A. strictum* [17]. In particular, the main difference seems to be the extremely abundant presence of chitosan in the cell wall of Zygomycetes which is not present in Ascomycetes as *A. strictum*, and which is known to play a key role in the biosorption process [2]. Moreover, it has been reported that the culture medium (amount and type of C and N sources) and conditions (*i.e.*, fermentation process, static or agitated conditions, *etc.*), may affect the quali-quantitative composition of the cell wall [18]. In particular, Tigini [19] has recently highlighted by FT-IR analysis that *C. elegans* grown on different culture media can have great variation in the composition of its cell wall and the same *C. elegans* biomass used in the present study showed a high chitin and chitosan content.

The q_e values obtained from the tested biomasses (Figure 1), particularly those of *C. elegans* and *A. strictum*, are in line with the best results reported in the literature [5,12], but an added bonus stems from the fact that these exhausted dye baths are prepared mixing several commercially important industrial dyes which contain high concentrations of salts at different pH values, introducing real parameters that often bar the attainment of good biosorption yields [5]. It must be borne in mind that until now most of the data concerning the exploitation of fungal biomasses in dye biosorption have been obtained on single molecules at low concentrations and only a few studies with multicomponent dye solutions and high salt concentration have been carried out [13,14].

Figure 1. (**a–c**) Decolorization percentage of the simulated exhausted dye baths (SABW, SRBC, SDBC) after 30 min, 2 h, 6 h and 24 h incubation by the biomasses of *Acremonium* sp., *Acremonium strictum, Penicillium* sp. and *Cunninghamella elegans*; (**d**) amount of adsorbed dye (q_e), letters indicate significant differences among q_e of different biomasses for the same dye bath.

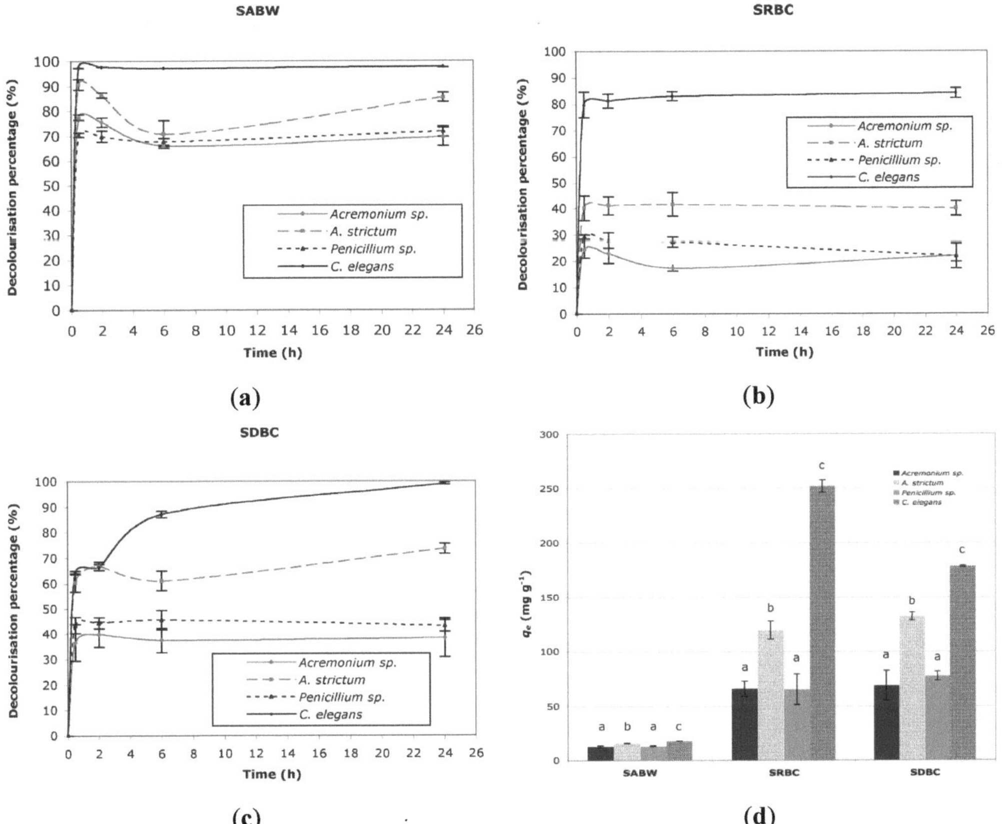

In order to compare the sorption performance of *C. elegans* and *A. strictum* biomasses, the Langmuir and Freundlich isotherms were calculated for SABW, the exhausted dye bath for which the highest *DP* values were obtained (Table 1).

The comparison of the R^2 values showed that, in both cases, the Langmuir model fits better with the experimental data than the Freundlich one. For both the biomasses, the isotherms were positive, regular and concave to the concentration axis, indicating an increase of dye uptake with an increase in the equilibrium dye concentration (Figure 2). See Table 2 for abbreviations used in the following Figures and Tables.

Table 1. Langmuir and Freundlich isotherm constants for the biosorption of SABW by *Cunninghamella elegans* and *Acremonium strictum* biomasses.

Species	Langmuir			Freundlich		
	q_{max} (mg g^{-1})	K_L (L mg^{-1})	R^2	K_F [mg$^{(n-1)/n}$ L$^{1/n}$ g^{-1}]	n	R^2
Cunninghamella elegans	594.4 ± 11.82	0.0048 ± 0.0005	0.993	54.29 ± 2.14	3.13	0.946
Acremonium strictum	289.5 ± 5.35	0.0114 ± 0.0012	0.982	48.29 ± 9.32	3.99	0.913

Figure 2. Comparison of the experimental equilibrium data with the estimated Langmuir isotherms of SABW obtained by *Cunninghamella elegans* and *Acremonium strictum* biomasses.

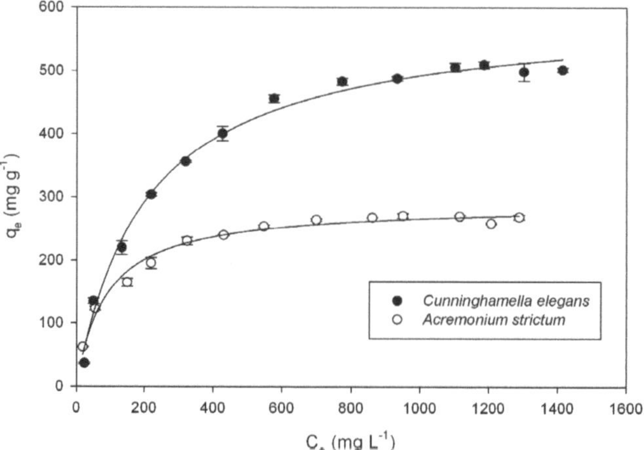

From a theoretical point of view, *C. elegans* has proven to be able to absorb a quantity of dye twice that of *A. strictum* (q_{max} 594 *vs.* 289 mg g^{-1}). Nevertheless, the high yields obtained with the *A. strictum* biomass are particularly relevant considering that to date there are very few references for the use of fungal waste-biomasses for the treatment of colored wastewaters. This biomass was more effective than the *Trichoderma harzianum* mycelia used to remove Rhodamine 6G from aqueous solution [20] or the industrial biomass of *Corynebacterium glutamicum*, tested for the treatment of a solution containing the dye Reactive Black 5 [21]. Certainly, the main advantage of the industrial waste-biomasses is the fact that they are cheap; however, the impact on costs of the multiple washings necessary to eliminate the residues of the industrial processes on the biomass that could negatively affect the decolorization yields should not be underestimated.

In order to validate a future real application in the textile industry of the biosorption process, *C. elegans* and *A. strictum* biomasses were tested for the treatment of nine real exhausted dye baths for natural and synthetic fibers (Figure 3).

Figure 3. Decolorization percentage of the real exhausted dye baths after 1 h, 2 h, 6 h and 24 h incubation by the biomasses of (**a**) *Cunninghamella elegans* and (**b**) *Acremonium strictum*.

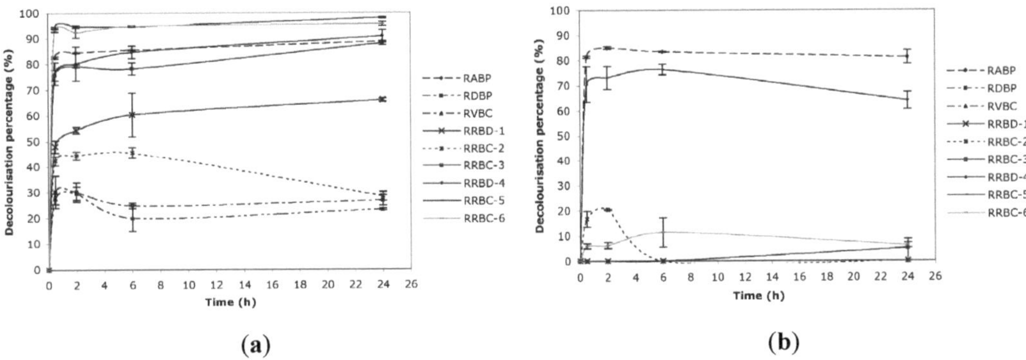

(**a**)   (**b**)

Although in the tests with simulated dye baths the results obtained by the biomass of *C. elegans* were comparable to those obtained by the biomass of *A. strictum*, in the case of real dye baths the difference between the two biomasses became substantial. The *A. strictum* biomass was only effective in the treatment of RABP and RRCB-5 (DP 81% and 64%, respectively), the two exhausted dye baths characterised by the lowest pH values. On the contrary, *C. elegans* biomass showed good yields of decolorization towards RABP and most of the reactive baths (DP up to 98%, complying to government standards), with the only exception being RRBC-2 (DP 28%); in this last case the low yield of decolorization could be due to the very high pH (11.3), the high concentration of salts (90 g L^{-1}), which may compete with the dye molecules for the same binding sites, and, probably, to the presence of a dye with low affinity for the biomass itself.

It is known from the literature that among the factors affecting the efficiency of biosorption processes, is undoubtedly the initial dye concentration: initial concentration provides an important driving force to overcome all mass transfer resistance of the dye between the aqueous and solid phases. Hence, a higher initial concentration of dye may enhance the adsorption process [5]. Nevertheless, the *C. elegans* biomass was efficient also for exhausted dye baths characterised by very low dye concentrations (e.g., RRBC-3, RRBC-4, RRBC-5 and RRBC-6).

The same biomass proved to be moderately effective against the exhausted dye baths containing disperse (RDBP) and vat (RVBC) dyes (DP 27% and 23%, respectively). These types of dye baths, however, are difficult to treat by biosorption as already stressed by other authors [22,23]. This result is probably due to the chemical characteristics of these insoluble dyes, since hydrophobic attractions, dye-dye aggregation mechanisms and dye-surfactants interactions can act simultaneously, reducing the biosorption effectiveness [2]. Moreover, in the case of disperse dyes, the presence of auxiliaries (*i.e.*, carriers), used to fix dyes to fibers, can also obstruct the dye biosorption onto the fungal biomass through competitive mechanisms [24].

The absence in the literature of works on fungal biosorption which take into account such a large number of real dye baths, differing in terms of dye types and dyeing processes, hampers a sound comparison between our data and those obtained with other fungal biomasses under similar conditions; however based on these results, the *C. elegans* biomass confirms its high efficiency and versatility for exhausted dye baths characterized by different classes of dyes, pH and ionic strength. On the other hand, the *A. strictum* biomass proved to be effective only against exhausted dye baths at acidic or neutral pH, but it should be noted that the *A. strictum* biosorption yields are quite comparable with those of other biomasses reported in the literature [11–19]. Once again, the *C. elegans* biomass can be considered a truly exceptional biosorbent in terms of performance and versatility, however also the *A. strictum* biomass could find future application for the treatment of acid exhausted dye baths. It should also be considered that the great variability of real textile wastewaters is certainly a limiting factor for many purification technologies now available, such as biological treatment with activated sludge or adsorption with activated carbons. In the first case the sudden change in operating conditions may cause an obvious decrease in degradative activity, in the latter case, pH is the main factor limiting the yields of adsorption, especially under conditions of high alkalinity.

2.2. Effect of Initial pH

The literature indicates pH as being one of the most important abiotic parameters in regulating the biosorption yields, regardless of the sorbent material used. Actually, pH can influence the interaction between adsorbent and solute in aqueous medium in two main ways: (i) by changing the ionization potential of the dye molecules; (ii) by changing the net charge of active sites of the adsorbent surface [25]. Most of the studies on dye biosorption have reported the necessity of strong acidic conditions for optimum biosorption [26]; however, actual textile wastewaters are generally basic and adjusting and maintaining extreme acidic conditions usually increases the overall process cost [21].

The q_e values of *C. elegans* and *A. strictum* biomasses towards SABW at different initial pH values (from pH 3 to pH 11) are shown in Figure 4. In all cases the *C. elegans* biomass displayed significantly higher q_e values than that of *A. strictum*. Both the biomasses showed the highest q_e values at pH 3 (103 and 63 mg g^{-1} for *C. elegans* and *A. strictum*, respectively). From pH 5 to pH 9 *C. elegans* biomass displayed q_e values still very high (up to 89 mg g^{-1}), although significantly lower than at pH 3; whereas in the case of *A. strictum* the q_e value fell by 50% from pH 3 to pH 5. At pH 11 a significant reduction of q_e was registered with both the biomasses, which was particularly evident for *A. strictum*.

In order to evaluate the effect of initial pH of the exhausted dye bath on the sorption potentials of *C. elegans* biomass until its complete saturation, biosorption experiments were conducted in batch mode in subsequent cycles. The biomass saturation was reached at pH 3 after 12 cycles at 300 ppm and 7 cycles at 900 ppm, at pH 5 after 20 cycles at 300 ppm and 9 cycles at 900 ppm, at pH 7 after 28 cycles at 300 ppm and 9 cycles at 900 ppm, at pH 9 after 28 cycles at 300 ppm and 11 cycles at 900 ppm, and finally at pH 11 after 10 cycles at 300 ppm and 9 cycles at 900 ppm.

The total amounts of adsorbed dye (q_{tot}), obtained by summing the values of q_e of each cycle, are reported in Figure 5. The highest q_{tot} values were observed at pH 3 (1035 mg g^{-1}), pH 7 (930 mg g^{-1}) and pH 9 (945 mg g^{-1}), followed by pH 5 (824 mg g^{-1}) and, finally, pH 11 (667 mg g^{-1}).

Figure 4. Amount of adsorbed dye at the equilibrium (q_e) of *Cunninghamella elegans* and *Acremonium strictum* biomasses towards SABW at different initial pH. Capital letters indicate significant differences for *C. elegans* biomass at different pH; small letters indicate significant differences for *A. strictum* biomass at different pH.

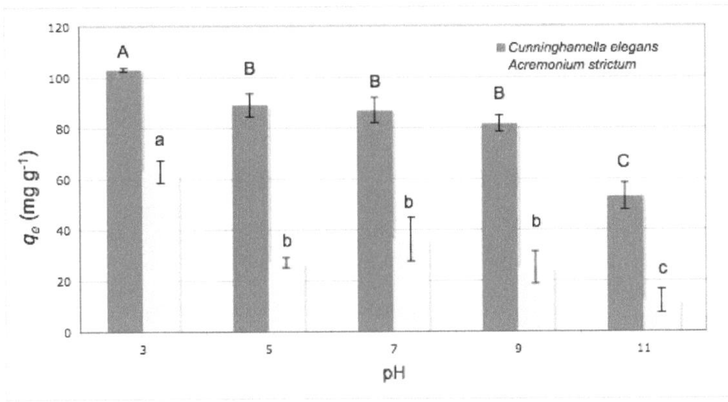

Figure 5. Total amount of adsorbed dye (q_{tot}) of *Cunninghamella elegans* biomass towards SABW at different initial pH. Letters indicate significant differences.

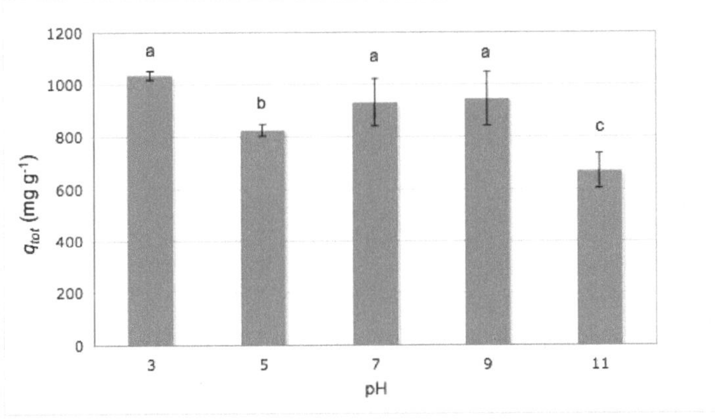

C. elegans biomass is, hence, highly effective over a wide pH range (3–11). This finding is particularly relevant in view of practical application for the treatment of real industrial wastewaters, which are often characterized by very high pH values (up to 12–13) which are of course always fluctuating. The relevance of our results is even more evident when they are compared with the literature data regarding different sorbent materials. Thus, for example, the biomass of bacterium *C. glutamicum* displayed the highest biosorption yield towards the dye Reactive Black 5 at pH 1, with an evident decrease at higher pH values [21]; similarly, the biosorptive potential of activated sludges towards Direct Black 38 drastically decreased by increasing the pH from 1 to 11 [27]. Among the fungal biomasses tested so far, *C. elegans* proved to be the most versatile with respect to this parameter: Kiran and collaborators [28], treating the dye Acid Red 57 with the biomass of *Cephalosporium aphidicola*, noting that q_e decreased proportionally to 0 with the increase of initial pH between 1 and 6. Aksu and Çağatay [29], studying the *Rhizopus arrhizus* biomass for the removal of the dye Germano Turquise Blue-G, observed the maximum q_e at pH 2 and total ineffectiveness at pH 4. Likewise, Khambhaty and collaborators [30] using *Aspergillus wentii* to remove Brilliant Blue G observed an evident and constant q_e decrease when the solution was increased from pH 2 to 10.

Moreover, the obtained results show that the biomass can be subjected to numerous biosorption cycles until complete saturation, reducing in this way the amount of waste generated by the fungal treatment. The exhausted biomass could then be treated by incineration.

3. Experimental Section

3.1. Industrial Waste-Biomasses and Selected Test Organisms

Three waste-biomasses from industrial pharmaceutical productions were kindly provided by Antibioticos S.p.a. (*Acremonium* sp. and *Penicillium* sp.) and ACS Dobfar S.p.a. (*Acremonium strictum*, synonym *Cephalosporium acremonium*). The waste-biomasses were obtained from the pharmaceutical factories as inactivated slurries (autoclaved at 121 °C for 30 min), as required by safety procedures.

C. elegans (MUT 2861) was obtained from the *Mycotheca Universitatis Taurinensis* Collection (MUT, University of Turin, Department of Life Science and Systems Biology). It was patented for dye biosorption [31] and deposited at the Deutsche Sammlung von Mikroorganismen und Zellkulturen (DSMZ, Braunschweig, Germany). Starting cultures of this fungus were lyophilised until use. They were revitalised on MEA and mature conidia for the inocula and biomass production were obtained from cultures grown on the same medium in the dark at 24 °C for one week.

3.2. Fungal Biomass Preparation

The waste-biomasses were rinsed in distilled water by centrifugation (9 cycles at 8000 rpm for 5 min) in order to eliminate the metabolites produced during fermentation and other colored impurities. The *C. elegans* biomass was produced according to Prigione *et al.* [15]. All the biomasses were lyophilised (Lyophiliser LIO 10P, Cinquepascal, Trezzano s/n, Italy) and powdered to particles of uniform size (300 μm < Ø < 600 μm).

3.3. Simulated and Real Exhausted Dye Baths

The composition and the characteristics of the exhausted dye baths used in this study are listed in Table 2. The three simulated exhausted dye baths (SABW, SRBC and SDBC), designed to mimic wastewater produced during wool or cotton textile dyeing processes, were prepared using mixing of industrial dyes at high concentrations. These simulated exhausted dye baths, previously developed by the industrial partners of the EC FP6 Project SOPHIED (NMP2-CT-2004-505899), were used with the permission of the SOPHIED Consortium. The industrial dyes used in these experiments were selected as being representative of different structural dye types, commercially important and with a wide range of applications across the textile industries. They are commercial products purchased from Town End plc (Leeds, UK), containing in addition to dye molecules which constitute the 30%–90% of the total weight also other organic molecules such as additives. These simulated exhausted dye baths mimic the industrial ones also with respect to the presence of different salts, often in high concentrations, and for the pH values.

Table 2. Exhausted dye bath name, acronym, composition and pH.

Exhausted dye bath	Acronym	Dyes	Dye concentration	Salt concentration	Auxiliaries	pH
Simulated Acid Bath for Wool	SABW	Mix of 3 dyes (Abu62, AY49, AR266)	300 mg L^{-1}	5 g L^{-1}	n.i.	5.0
Simulated Reactive Bath for Cotton	SRBC	Mix of 4 dyes (Rbu222, RR195, RY145, Rbk5)	5000 mg L^{-1}	70 g L^{-1}	n.i.	10.0
Simulated Direct Bath for Cotton	SDBC	Mix of 3 dyes (DrBu71, DrR80, DrY106)	3000 mg L^{-1}	2 g L^{-1}	n.i.	9.0
Real Acid Bath for Polyamide	RABP	Mix of 3 dyes	433 mg L^{-1}	n.i.	Surfactants, weak acid, fixatives	5.0
Real Disperse Bath for Polyester	RDBP	1 dye	658 mg L^{-1}	n.i.	Dispersants, weak acid, strong base	9.3
Real Vat Bath for Cotton	RVBC	Mix of 2 dyes	1424 mg L^{-1}	20 g L^{-1}	Weak acid, strong base, glucose	12.7
Real Reactive Bath for Cotton 1	RRBC-1	Mix of 3 dyes	542 mg L^{-1}	77 g L^{-1}	Weak acid, strong base	10.9
Real Reactive Bath for Cotton 2	RRBC-2	1 dye	343 mg L^{-1}	90 g L^{-1}	Ca and Mg sequestering, oil, weak acid, strong base	11.3
Real reactive Bath for Cotton—continuous dyeing	RRBC-3	Mix of 3 dyes	43 mg L^{-1}	200 g L^{-1}	Sodium carbonate	10.3
Real reactive Bath for Cotton—washing with cold water	RRCB-4	Mix of 3 dyes	34 mg L^{-1}	200 g L^{-1}	n.i.	10.0
Real reactive Bath for Cotton—neutralization and washing at 40 °C	RRCB-5	Mix of 3 dyes	98 mg L^{-1}	200 g L^{-1}	Acetic acid	5.9
Real reactive Bath for Cotton—boiling soaping	RRCB-6	Mix of 3 dyes	60 mg L^{-1}	200 g L^{-1}	Surfactants	7.8

Note: n.i. means not indicated in the dye bath formulation.

The nine real industrial exhausted dye baths used in this study were kindly provided by textile industry members of the BIOTEX Project (Call MD 2007 to promote excellence in the Lombardy Region meta-districts) and are representative of different dye types (acid, disperse, reactive, vat) and dyeing processes (batch and continuous). The exhausted dye baths RABP, RDBP, RVBC, RRBC-1 and RRBC-2 were taken from batch dyeing plants, after the dyeing process, while RRC-3, RRC-4, RRC-5 and RRC-6 all coming from a continuous dyeing plant, were collected at four stages of the process (*i.e.*, after dyeing, washing with cold water, neutralization/washing at 40 °C and boiling and soaping). The dye concentration was obtained using a sample of the dye bath at known concentration, taken before the dyeing process: the absorbance spectrum area of this sample was compared with that of the same bath taken after the dyeing process, in this way it was possible to calculate, indirectly, the residual dye concentration in the exhausted baths.

3.4. Sorption Experiments

Each biomass was weighed and 0.5 g dry weight was placed in a 50 mL Erlenmayer flask containing 30 mL of simulated or real exhausted dye bath. The flasks were incubated at 30 °C under agitated conditions at 130 rpm. Each trial was performed in triplicate. Exhausted dye baths without biomass were used as abiotic controls to assess decolorization other than that due to biosorption (e.g., photobleaching or complexation).

After 30 min (or 1 h), 2, 6 and 24 h, 200 μL were taken from each sample of the exhausted dye baths, centrifuged at 14,000 rpm for 5 min to remove disturbing mycelial fragments, and examined with a spectrophotometer (TECAN Infinite M200, Grödig, Austria) to obtain the complete absorbance spectra. Since a linear relationship existed between the area of absorbance spectrum and dye concentration, the percentage of removed dye (*DP*, decolorization percentage) was calculated as the extent of decrease of the spectrum area from 360 nm to 790 nm, with respect to that of the abiotic control. At the end of the experiment the amount of adsorbed dye (q_e), that is mg of adsorbed due g^{-1} of biomass dry weight, was determined by using the following equation, taking into account the dye concentration difference in the exhausted dye bath at the beginning and at equilibrium:

$$q_e = (C_i - C_e) \cdot V/m \tag{1}$$

where C_i and C_e are the initial and the equilibrium dye concentrations (mg L^{-1}); V is the volume of the solution (L); and *m* is the amount of the biosorbent used (g).

The significance of differences ($p \leq 0.05$) among the *DP* values at 30 min (or 1 h), 2, 6 and 24 h and among q_e values was calculated with the Mann-Whitney test (SYSTAT 10 for windows [32]).

3.5. Adsorption Isotherms

Adsorption isotherm experiments were carried out by bringing into contact a fixed amount of biomass with a suitable volume of SABW at an appropriate concentration. The equilibrium data were obtained by investigating a wide range of concentrations, representative of the application in wastewater remediation treatment.

Adsorption isotherms were obtained by correlating the amount of adsorbed solute (q_e) with the residual concentration of dye in solution at equilibrium (C_e). The two following equilibrium isotherm models were used to fit the experimental data.

Langmuir model:

$$q_e = q_{max} \cdot [(K_L \cdot C_e) / (1 + K_L \cdot C_e)] \tag{2}$$

Freundlich model:

$$q_e = K_F \cdot C_e^{1/n} \tag{3}$$

where q_{max} is the maximum solid phase concentration of adsorbate (forming a complete monolayer coverage on the sorbent surface); K_L is the Langmuir constant related to the solute affinity for the sorbent-binding sites; K_F and n are Freundlich constants; K_F and slope $1/n$ are defined as a sorption coefficient representing the amount of dye molecules for a unit equilibrium concentration and as a measure of the sorption intensity or surface heterogeneity, respectively; a value of $1/n = 1$ shows that the partition between two phases does not depend on the concentration; a value of $1/n < 1$ corresponds to a normal Langmuir isotherm; while $1/n > 1$ indicates a cooperative sorption involving strong interactions between the molecules of adsorbate [33].

3.6. Effect of Initial pH

Eighty-four mg of lyophilised biomass (corresponding to 0.5 g of biomass fresh weight) was placed in a 50 mL Falcon tube containing 30 mL of SABW at different pH values (3, 5, 7, 9 and 11). The flasks were incubated as previously described. Once the adsorption equilibrium was reached, *i.e.*, after three consecutive measurements resulting in an equal *DP* value, the samples were centrifuged at 8000 rpm for 5 minutes, and the treated dye bath was replaced by 30 mL of the untreated one. The test ended when the biomass was no longer able to adsorb dye. Finally, the total amount of adsorbed dye (q_{tot}) was calculated by adding up the q_e values obtained at the end of each cycle.

4. Conclusions

The results obtained in this study enable the following conclusions to be drawn:

(1) *A. strictum* biomass was more efficient than the other two industrial waste-biomasses, being able to substantially decolorize the simulated dye baths; however, it was effective only towards the real ones characterized by acidic pH. Hence, at the moment, industrial waste-biomasses such as *A. strictum* can be considered competitive and potentially useful for the treatment of specific types of wastewater (e.g., acid exhausted dye baths) only;

(2) *C. elegans* biomass is endowed with a high ability to remove dyes belonging to different chemical classes, not only from simulated dye baths but also from many real ones;

(3) The high applicative potentialities of *C. elegans* biomass for the decolorization of textile wastewater was demonstrated by the very good biosorption yields even under extreme conditions of pH (3–11); for this reason, this biomass can be considered an excellent and exceptionally versatile biosorbent material.

Hence, biosorption by means of *C. elegans* biomass could be considered a valid alternative to other techniques for wastewater treatment, being applicable to real industrial wastewaters representative of different dye types (acid, disperse, vat, reactive) and dyeing processes (batch and continuous) in a timely fashion. Future investigations are obviously needed to validate the data obtained at pilot scale and to verify under field conditions the possibility to couple the biosorption process with other physical and biological/chemical treatmen, aiming at complete decolorization of effluents and water reuse. Since biosorption is very effective towards wastewater with high concentrations of dyes, this technique seems to be particulary suitable for application as a primary treatment. Partially purified wastewater could then be sent for subsequent conventional treatment such as bioxidation by activated sludge.

Acknowledgments

The authors thank ACS Dobfar S.p.a. and Antibioticos S.p.a. for providing the waste-biomasses, the BIOTEX Project (Call MD 2007 to promote excellence in the Lombardy Region meta-districts) for providing the real dye baths and the SOPHIED Consortium (EC FP6 Project SOPHIED, NMP2-CT-2004-505899) for simulated dye baths composition. This work was supported by the project "Sviluppo di procedure di biorisanamento di reflui industriali" (BIOFORM, Compagnia di San Paolo, Turin, Italy) and by Marcopolo Engineering S.p.a. (Borgo San Dalmazzo, Italy).

References

1. Shannon, M.A.; Bohn, P.W.; Elimelech, M.; Georgiadis, J.G.; Marinas, B.J.; Mayes, A.M. Science and technology for water purification in the coming decades. *Nature* **2008**, *452*, 301–310.
2. Crini, G.; Badot, P.M. Application of chitosan, a natural aminopolysaccharide, for dye removal from aqueous solution by adsorption process using batch studies: A review of recent literature. *Prog. Polym. Sci.* **2008**, *33*, 399–447.
3. Anjaneya, A.; Santoshkumar, M.; Anand, S.N.; Karegoudar, T.B. Biosorption of acid violet dye from aqueous solutions using native biomass of a new isolate of *Penicillium* sp. *Int. Biodeterior. Biodegrad.* **2009**, *63*, 782–787.
4. Hai, F.I.; Yamamoto, K.; Fukushi, K. Hybrid treatment systems for dye wastewater. *Environ. Sci. Technol.* **2007**, *37*, 315–377.
5. Aksu, Z. Application of biosorption for the removal of organic pollutants: A review. *Process Biochem.* **2005**, *40*, 997–1026.
6. Svecova, L.; Spanelova, M.; Kubal, M.; Guibal, E. Cadmium, lead and mercury biosorption on waste fungal biomass issued from fermentation industry. I. Equilibrium studies. *Sep. Purif. Technol.* **2006**, *52*, 142–153.
7. Wang, J.; Chen, C. Biosorbents for heavy metals removal and their future. *Biotechnol. Adv.* **2009**, *27*, 195–226.
8. Gulati, R.; Saxena, R.K.; Gupta, R. Fermentation waste of *Aspergillus terreus*: A potential copper biosorbent. *World J. Microbiol. Biotechnol.* **2002**, *18*, 397–401.
9. Gadd, G.M. Biosorption: Critical review of scientific rationale, environmental importance and significance for pollution treatment. *J. Chem. Technol. Biotechnol.* **2009**, *1*, 13–28.
10. Gochev, V.K.; Velkova, Z.I.; Stoytcheva, M.S. Hexavalent chromium removal by waste mycelium of *Aspergillus awamori*. *J. Serb. Chem. Soc.* **2010**, *75*, 551–564.
11. Zhang, D.; He, H.J.; Li, W.; Gao, T.Y.; Ma, P. Biosorption of cadmium(II) and lead(II) from aqueous solutions by fruiting body waste of fungus *Flammulina velutipes*. *Desalin. Water Treat.* **2010**, *20*, 160–167.

12. Kaushik, P.; Malik, A. Fungal dye decolourisation: Recent advances and future potential. *Environ. Int.* **2009**, *35*, 127–141.

13. Prigione, V.; Tigini, V.; Pezzella, C.; Anastasi, A.; Sannia, G.; Varese, G.C. Decolourisation and detoxification of textile effluents by fungal biosorption. *Water Res.* **2008**, *42*, 2911–2920.

14. Prigione, V.; Varese, G.C.; Casieri, L.; Filippello Marchisio, V. Biosorption of simulated dyed effluents by inactivated fungal biomasses. *Bioresour. Technol.* **2008**, *99*, 3559–3567.

15. Prigione, V.; Zerlottin, M.; Refosco, D.; Tigini, V.; Anastasi, A.; Varese, G.C. Chromium removal from a real tanning effluent by autochthonous and allochthonous fungi. *Bioresour. Technol.* **2009**, *100*, 2770–2776.

16. Tigini, V.; Prigione, V.; Donelli, I.; Anastasi, A.; Freddi, G.; Giansanti, P.; Mangiavillano, A.; Varese, G.C. *Cunninghamella elegans* biomass optimisation for textile wastewater biosorption treatment: an analytical and ecotoxicological approach. *Appl. Microbiol. Biotechnol.* **2011**, *90*, 343–352.

17. Deacon, J. *Fungal Biology*, 4th ed.; Blackwell Publishing Ltd: Oxford, UK, 2006.

18. Anastasi, A.; Prigione, V.; Casieri, L.; Varese, G.C. Decolourisation of model and industrial dyes by mitosporic fungi in different culture conditions. *World J. Microbiol. Biotechnol.* **2009**, *25*, 1363–1374.

19. Tigini, V.; Prigione, V.; Donelli, I.; Freddi, G.; Varese, G.C. Influence of culture medium on fungal biomass composition and biosorption effectiveness. *Curr. microbiol.* **2012**, *64*, 50–59.

20. Sadhasivam, S.; Savitha, S.; Swaminathan, K. Exploitation of *Trichoderma harzianum* mycelial waste for the removal of rhodamine 6G from aqueous solution. *J. Environ. Manag.* **2007**, *85*, 155–161.

21. Vijayaraghavan, K.; Yun, Y.S. Utilization of fermentation waste (*Corynebacterium glutamicum*) for biosorption of Reactive Black 5 from aqueous solution. *J. Hazard. Mater.* **2007**, *141*, 45–52.

22. Fu, Y.; Viraraghavan, T. Dye biosorption sites in *Aspergillus niger*. *Bioresour. Technol.* **2002**, *82*, 139–145.

23. Ozyurt, M.; Ozer, A.; Atacag, H. Decolourisation of Setopers Black RD-ECO by *Aspergillus oryzae*. *Fresenius Environ. Bull.* **2005**, *14*, 531–535.

24. Tigini, V. Fungal Biosorption in Wastewater Treatment: Decolourisation and Detoxification of Textile and Tannery Effluents. Ph.D. Thesis, University of Turin, Turin, Italy, 2010.

25. Aksu, Z.; Tatli, A.İ.; Tunç, Ö. A comparative adsorption/biosorption study of Acid Blue 161: Effect of temperature on equilibrium and kinetic parameters. *Chem. Eng. J.* **2008**, *142*, 23–39.

26. Vijayaraghavan, K.; Lee, M.W.; Yun, Y.S. A new approach to study the decolorization of complex reactive dye bath effluent by biosorption technique. *Bioresour. Technol.* **2008**, *99*, 5778–5785.

27. Wang, X.; Xia, S.; Zhao, J. Biosorption of Direct Black 38 by dried anaerobic granular sludge. *Front. Environ. Sci. Engin. China* **2008**, *2*, 198–202.

28. Kiran, I.; Akar, T.; Ozcan, A.S.; Ozcan, A.; Tunali, S. Biosorption kinetics and isotherm studies of Acid Red 57 by dried *Cephalosporium aphidicola* cells from aqueous solution. *Biochem. Eng. J.* **2006**, *31*, 197–203.

29. Aksu, Z.; Çağatay, S.S. Investigation of biosorption of Gemazol Turquise Blue-G reactive dye by dried *Rhizopus arrhizus* in batch and continuous systems. *Sep. Purif. Technol.* **2006**, *48*, 24–35.

30. Khambhaty, Y.; Mody, K.; Basha, S. Efficient removal of Brilliant Blue G (BBG) from aqueous solutions by marine *Aspergillus wentii*: Kinetics, equilibrium and process design. *Ecol. Eng.* **2012**, *41*, 74–83.

31. Varese, G.C.; Prigione, V.P.; Casieri, L.; Voyron, S.; Bertolotto, A.; Filipello Marchisio, V. Use of *Cunninghamella elegans* Lendner in Methods for Treating Industrial Wastewaters Containing Dyes. Eur. Pat. Appl. EP07118877.5, 2007.

32. *SYSTAT*, version 10; SPSS Inc.: Chicago, CA, USA, 2000.

33. Ilhan, S.; Iscen, C.F.; Caner, N.; Kiran, I. Biosorption potential of dried *Penicillium restrictum* for Reactive Orange 122: Isotherm, kinetic and thermodynamic studies. *J. Chem. Technol. Biotechnol.* **2008**, *83*, 569–575.

Water **2012**, *4*, 759-769; doi:10.3390/w4040759

OPEN ACCESS

water

ISSN 2073-4441
www.mdpi.com/journal/water

Article

The CO$_2$ Emission Factor of Water in Japan

Yasutoshi Shimizu *, Satoshi Dejima and Kanako Toyosada

ESG Promotion Department, TOTO LTD, 2-1-1, Nakashima, Kokurakita-ku, Kitakyushu 802-8601, Japan; E-Mails: satoshi.dejima@jp.toto.com (S.D.); kanako.toyosada@jp.toto.com (K.T.)

* Author to whom correspondence should be addressed; E-Mail: yasutoshi.shimizu@jp.toto.com; Tel.: +81-93-952-3315; Fax: +81-93-952-3468.

Received: 9 August 2012; in revised form: 23 September 2012 / Accepted: 24 September 2012 / Published: 28 September 2012

Abstract: From the viewpoint of combating global warming in Japan, measures to reduce emissions from the activities involved in daily life have been accelerated in concurrence with the efforts made in the industrial sector to save energy. As one such measure, the reduction of energy consumption in waterworks and sewer systems by reducing the volume of water used in the housing sector is gaining attention; measures for the conversion of water saving into CO$_2$ reduction credit in the domestic credit system are also being examined. To address the credit development for CO$_2$ reduction by water saving, it was necessary to determine the CO$_2$ emission factor for water. Hence, we calculated the CO$_2$ emission factor of water use in Japan and determined the value to be 0.376 kg CO$_2$/m^3 which applied the generating end electricity value. In addition, since electricity contributes to 90% of the energy consumption of the waterworks and sewer systems of Japan and since the emission factor for electricity changes with the power source composition ratio, the CO$_2$ emission factor for water also needs to be updated to match the emission factor for electricity. We therefore developed a calculation equation for updating this emission factor.

Keywords: global warming; CO$_2$ reduction; water; saving-water; Japan

---

1. Introduction

With the first commitment period of the Kyoto Protocol (2012) coming to an end, the United Nations Framework Convention on Climate Change will enter a new stage. Japan has accepted the responsibility to reduce greenhouse gases (GHGs [1], hereafter referred to as CO_2 to represent all the gases) by 6% compared to the value in 1990 during the first commitment period of the Kyoto Protocol. However, Japan was compelled to change its energy policy in the aftermath of the earthquake disaster and the subsequent accident at the nuclear power plant. Japan then declared its inability to participate in the second commitment period of the Kyoto Protocol. As regards progress in the plan to address the objectives of the Kyoto Protocol, the Ministry of the Environment reported that although the targets had been met up to 2010, emission in the housing sector increased dramatically while the emission in the industrial sector decreased compared to the 1990 value, which is the base year for the Kyoto Protocol. The emission increased by 34.8% in the housing sector in the fiscal year (FY) 2010, and the efforts to reduce CO_2 emission from households may determine the progress that Japan can achieve with regard to the plan against the background of the resumed operation of thermal power plants after the earthquake disaster.

The breakdown of CO_2 emission from households in Japan is shown in Figure 1. The main sources of emission are automobiles, household electric appliances, and residential plumbing equipment. The Eco-Car Tax Reduction was started in FY 2009, and a subsidy called the Eco-Point system to help retrofit the appliances with more efficient them in FY 2010; both policies aim to promote the replacement of less efficient products with the latest high-energy-efficiency them and thus realize a reduction in CO_2 emission from automobiles and household electric appliances.

Figure 1. Breakdown of CO_2 emission from Japanese houses.

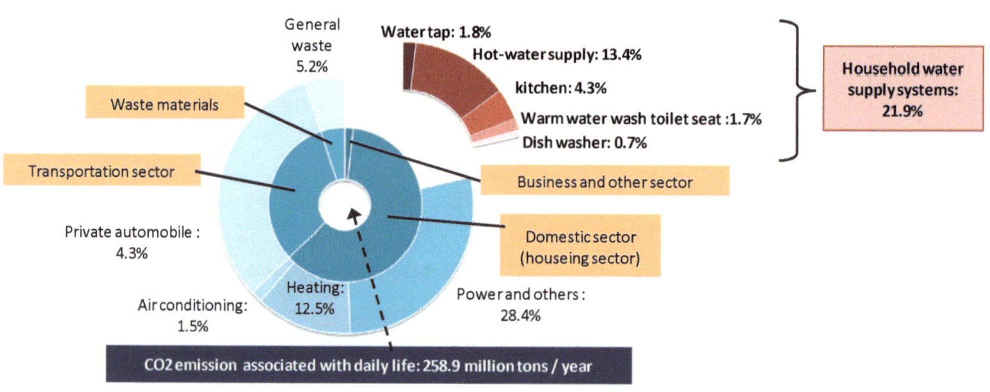

Notes: CO_2 emissions from the warm water wash toilet seat and the dish washer reported in the "electricity use by equipment at home" for FY 2005 were added to CO_2 emissions from water supply systems and reconstructed as CO_2 emissions from Japanese houses.

In recent years, researches associated water use and CO_2 emissions have been performed globally [2–4]. In Japan, research relating water-saving performance of bathroom fixtures such as toilets and showers with CO_2 reduction has also progressed, and the fact that the widespread use of water-saving fixtures can be effective in CO_2 reduction has been recognized [5,6]. Housing-Eco-Point subsidies have since been introduced to promote the replacement of traditional toilets with water-saving ones since January 2011.

CO_2 reduction by the widespread use of energy-saving household electric appliances and water-saving equipment is also being considered for inclusion in the Domestic Clean Development Mechanism

(CDM) and Bilateral Offset Credit Mechanism, which some of the methods adopted to realize the Government of Japan's objective of CO_2 emission reduction [7,8].

Carbon credits are calculated by measuring the reduction in the energy consumption or the amount of water saved by replacing conventional equipment with energy-saving or water-saving products, and multiplying this value with the CO_2 emission factor to convert the values into the amount of CO_2. The latest value of the CO_2 emission factor for electricity is announced every year by the Federation of Electric Power Companies Japan, and this value is utilized widely for calculating carbon credit conversion factors, CO_2 emission reports from the industries in the Keidanren Voluntary Action Plan on the Environment, *etc*.

The only known value of the CO_2 emission factor for water is 0.59 kg CO_2/m^3, which was mentioned in the Environment Agency of Japan Household Keeping Book in 1996 [9]; it was derived from the energy consumption of waterworks and sewer systems. It was later updated as a value only for the waterworks, and has not been updated since. Considering that electricity contributes over 90% of the energy required for the operating of waterworks and sewer systems and that the CO_2 emission factor for electricity changes annually depending on the composition ratio of the type of power-generation processes, such as nuclear and thermal power generations, the CO_2 emission factor derived from energy consumption should also be reexamined every year [10].

We therefore examined the CO_2 emission factor for water—in order to include it in the method for realizing carbon credits—by converting the volume of water saved into a reduction in CO_2 emission.

2. Analysis

The operation data for each treatment facility with boundary areas, as shown in Figure 2, are declared annually for the waterworks and sewer systems in Japan as Waterworks Statistics [11] and Sewerage Statistics [12]. Using these statistical values, we analyzed the energy consumption per cube meter treatment (described as energy consumption rate below) of water treatment in each waterworks and sewer system from FY 1990 to 2008.

Figure 2. Calculation boundary for CO_2 emission factor of water.

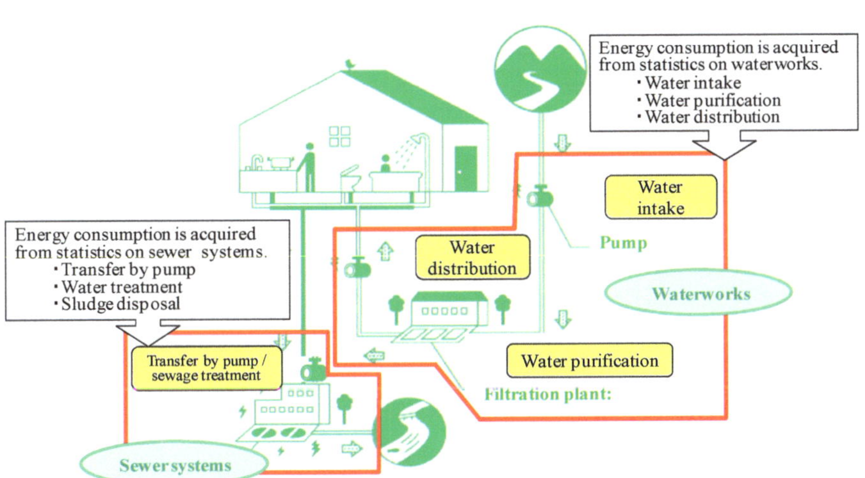

For electricity, the energy consumption can be converted into CO_2 emission by using the CO_2 emission factor per unit energy consumption. Along similar lines, we examined whether it was possible to introduce a CO_2 emission factor for unit volume of water used. That is, we examined whether energy consumption and water treatment volume were proportional in waterworks and sewer systems.

3. Results and Discussion

3.1. Waterworks System

The rate of increase in water supply coverage in Japan shows a monotonous increase by 2.8% from 94.7% in 1990 to 97.5% in 2008; its population increased by 3.4% from 123.53 million to 127.78 million in 2006, but stopped increasing thereafter. In spite of the factors that may cause an increase in water consumption, the amount of water supply (effective water volume) increased from 14.7 billion m^3/year in 1990 to quickly peak at 15.5 billion m^3/year in 1998, and thereafter. It has shown a decreasing trend owing to the widespread use of water-saving systems, *etc*. These changes in water supply are shown in Figure 3 along with the changes in the overall energy consumption of the entire waterworks systems in Japan.

Figure 3. Generated fresh water volumes and electricity consumptions for all water works in Japan.

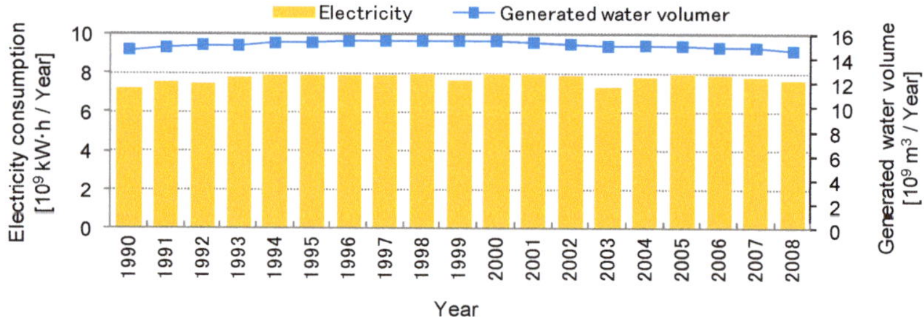

We analyzed these values to check whether the amount of fresh water generated and energy consumption were proportional. For energy consumption of whole waterworks, systems mainly consist of electricity (energy consumption ratio of electricity and fuel was 95:5). Fuel was thought to be used for drying and incineration of the sludge produced from water purification processes. Since the sludge treatment occurred with every purification method, the electricity and fuel use ratios in each processing method were regarded as the same. In addition, fuel consumption data of each facility was not reported. Therefore, analysis of the energy of the waterworks system in this research was only conducted as concerns electricity.

The breakdown of electricity consumption during the process is shown in Figure 4; Figure 5 shows the breakdown of the adopted water treatment processes by facility; and the breakdown of electricity consumption rate by facility and by treatment facility scale (treated water volume) is shown in Figure 6. Contrary to the expectation that the electricity consumption rate for treatment would vary dramatically among different water treatment methods, ranging from disinfection only to membrane filtration, we could not find any correlation between the electricity consumption rate of the facility and the treatment method, treated water volume, *etc.* in the same fiscal year data.

Figure 4. Breakdown of energy consumption for waterworks in Japan.

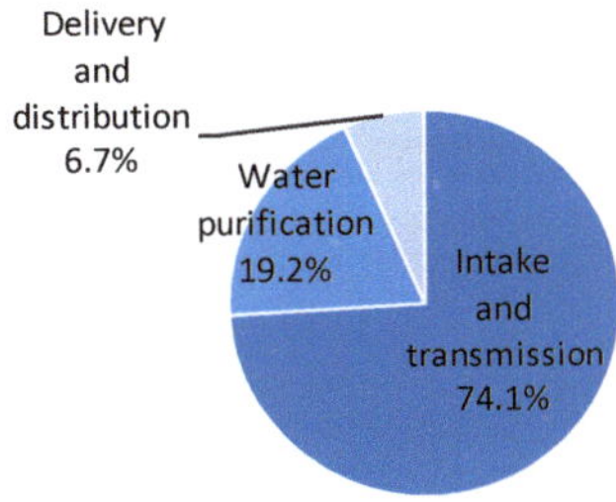

Figure 5. Adoption ratio of treatment processes for waterworks in Japan.

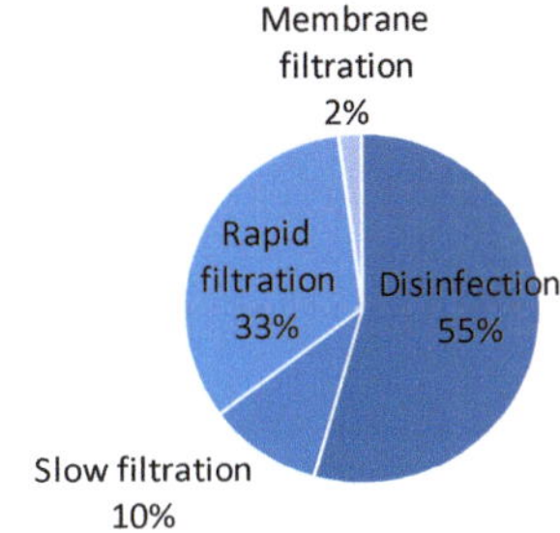

Figure 6. Relationship between energy consumption rates and treated water volumes of waterworks facilities in Japan.

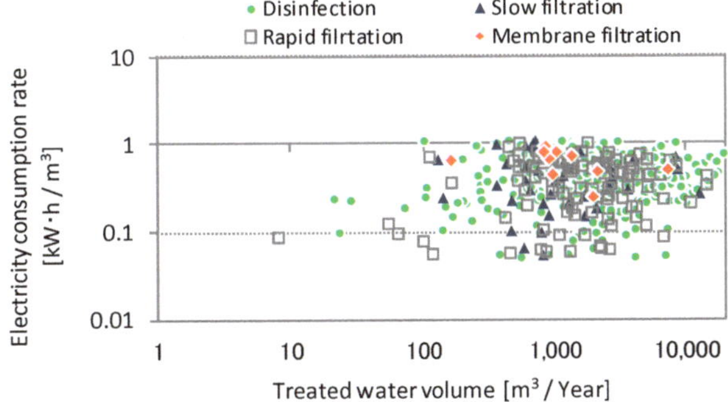

This was because of the fact that the overall electricity consumption at each facility was dominated by pumping and feeding processes, as described as intake, transmission delivery and distribution in Figure 4, as 81%, and that the electricity consumption of these pumping and feeding processes varied with the facility characteristics, such as water distribution. We thus studied the relationship between the average treatment consumption rate for all facilities in a fiscal year and the water treatment volume in order to compare the consumption rates for identical facility characteristics. The results are presented in Figure 7.

The results show that the water volume and the electricity consumption of the system are proportional to each other. As the pumping energy, which is attributed to 81% of the electricity consumption of the waterworks system, is proportional to the volume of water, by analogy the electricity consumed in the water purification processes can be considered to be proportional to the volume of water; thus, it is possible to determine a CO_2 emission factor for water in waterworks systems on the basis of the volume of water. The CO_2 emission factors for water are shown in Figure 8. For calculation, the emission factor of electricity was adopted as the average value for all power sources in Japan at the generating end and fuel consumption was taken into consideration.

Figure 7. Relationship between electricity consumption and treated water volume for waterworks.

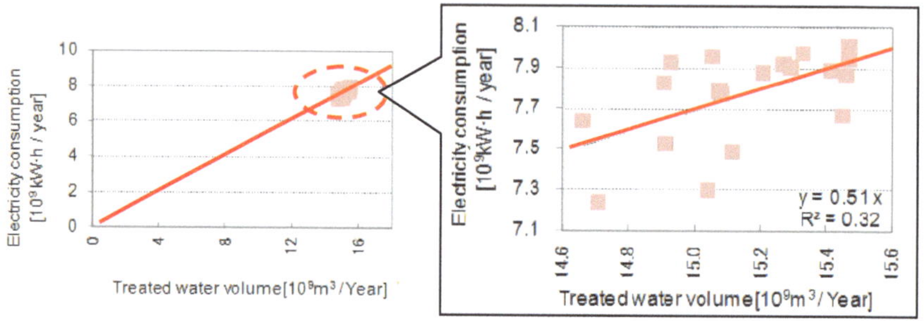

Figure 8. CO_2 emission factor of water in Japan incorporating the generating end electricity factor.

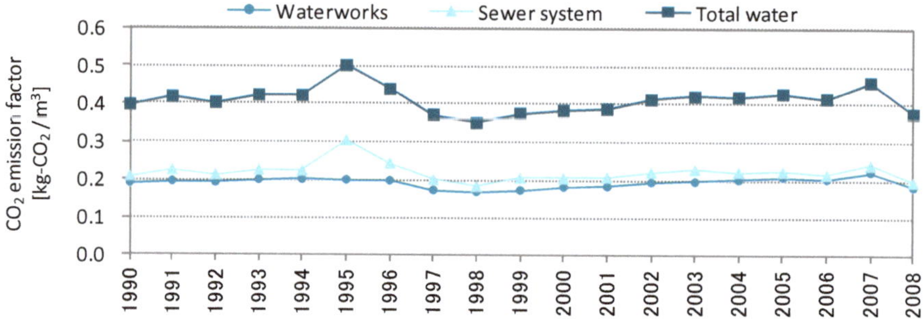

3.2. Sewer System

In the sewer system, fuel consumption occupied a bigger ratio than a waterworks, at 10%. The electricity and fuel consumption data for all facilities was released by the Sewerage Statistics. The energy consumption, which added electricity and fuel together, was then studied for the sewer system.

The breakdown of energy consumption of the sewer systems and that for the treatment processes adopted by the facility is shown in Figures 9 and 10, respectively. The processes can be largely classified into water feeding process, conversion of biochemical oxygen demand (BOD) into sludge and water purification, and the concentration and disposal of the converted sludge. Although the energy consumption during the water feeding process, which contributes to 21% of the total energy consumption, is assumed to be proportional to the volume of water, we expected that the energy consumption during purification and sludge treatment would depend on both the amount of BOD (pollution load) and the volume of water.

Figure 9. Energy consumption ratio of sewer systems in Japan.

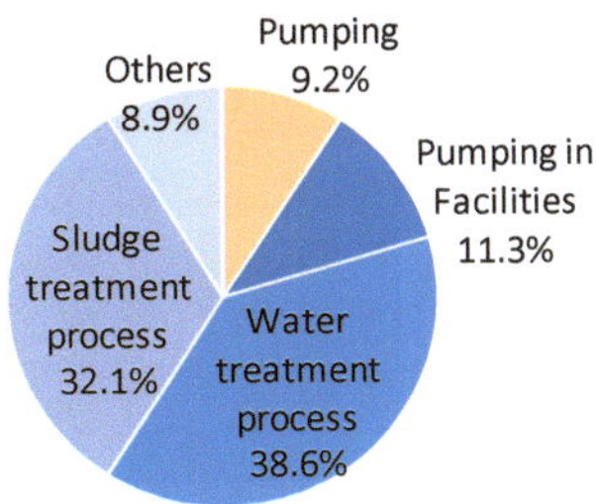

Figure 10. Adoption ratio of treatment processes in sewer systems in Japan.

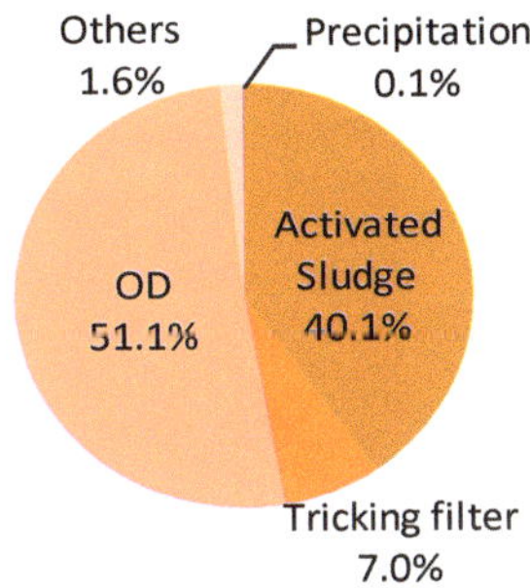

Since the amount of pollution load generated by households does not change even when water saving measures are adopted for residential systems, we used the operation data to analyze the factors that may affect energy consumption of the sewerage treatment systems when water-saving measures are popularized. The changes in the volume of water treated and in energy consumption of the sewer systems in Japan are presented in Figure 11. Owing to the increase in the coverage of the sewerage system, the water treatment volume increased by 40% since 1990. The relationship between energy consumption and water treatment volume, as observed from the data shown in Figure 11, is plotted in Figure 12. While energy consumption was proportional to the volume of water for all systems, the total pollution load also increased in proportion to the volume of water. We therefore analyzed this in detail.

We analyzed the effect of the treated water volume of the facilities on the energy consumption rate based on the data from FY 2008. The results are presented in Figure 13; the energy consumption rate clearly varies depending on the treated water volume in sewer systems. To homogenize the effects of the treated water volume, we extracted data on facilities with treatment capacities of 100 to 500 km^3/year and studied the relationship between the influent BOD concentration and energy consumption. While the influent BOD concentration into sewerage treatment facilities varied from about 90 to 380 kg/m^3 depending on the facility, there was no correlation between the influent BOD concentration and energy consumption.

Figure 11. Treated wastewater volumes and electricity consumptions for all sewer systems in Japan.

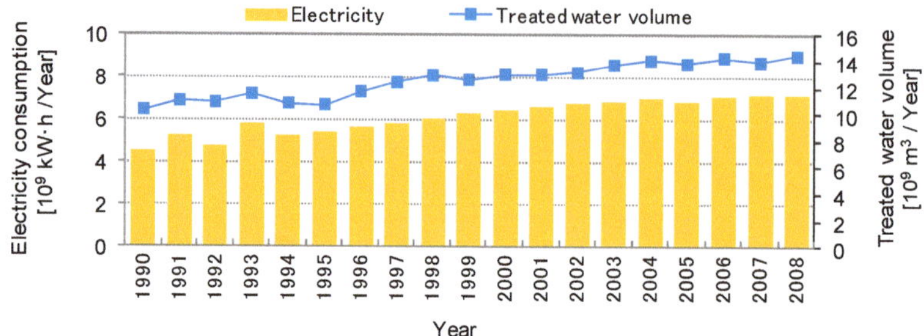

Figure 12. Relationship between energy consumption and treated wastewater volume for sewer systems.

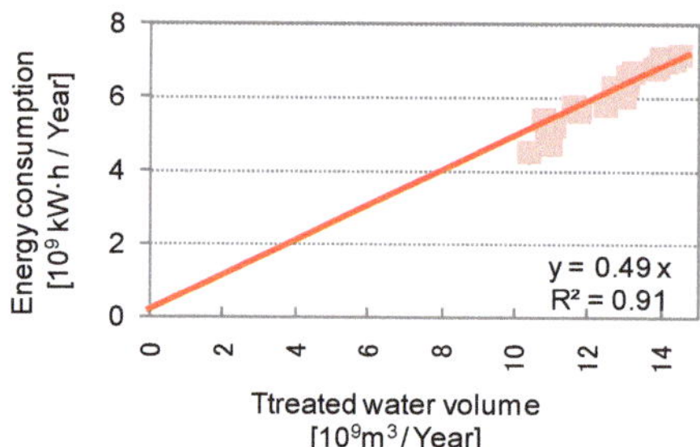

Figure 13. Relationship between energy consumption ratcs and trcatcd wastewater volumes for sewer systems in Japan.

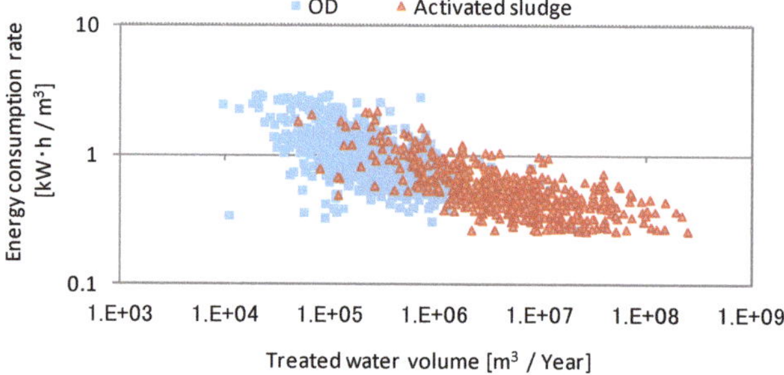

We found that energy consumption converged to a certain value (Figure 14). Thus, we concluded that the energy consumption of each facility can be considered to depend largely on the volume of water treated and that there is little effect of the changes in influent BOD concentration on the energy consumption.

On the basis of the above results, we surmised that the energy consumption of sewer systems is also proportional to the volume of water in the range of volume changes that can be expected to result from the adoption of water-saving measures, *etc.* in the future. The changes in the CO_2 emission factor for the sewer systems are shown in Figure 8. For this calculation, electricity and fuel consumption was taken into consideration.

The changes in influent sewage volume and BOD concentration fluctuate dramatically with time. On the other hand, the volume of processed water released from the system is controlled so that the effluent regulations including a BOD of less than 20 kg/m^3 are ensured. Thus, the sewer systems can be considered highly stable against external disturbances in water volume, pollution load, *etc*. Although we assume that the effects of an increase in pollution load resulting from water-saving measures are not noticeable because of the highly stable nature of the system, we await the results of future studies to identify the detailed mechanism explaining the stability of the system to changes in pollution load.

Figure 14. Relationship between energy consumptions and influent BOD concentrations for sewerage treatment facilities.

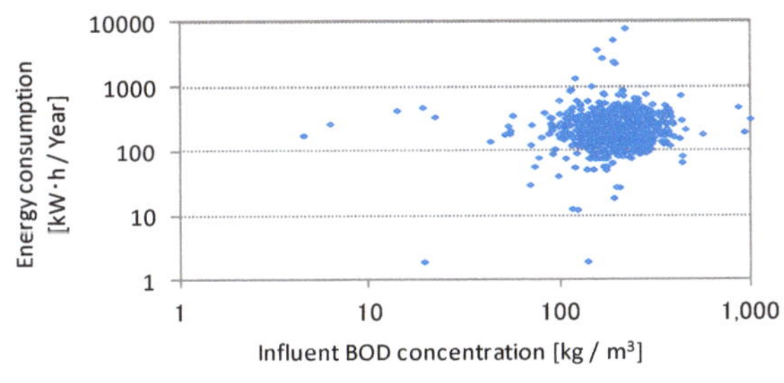

3.3. CO$_2$ Emission Factor for Water

We found that energy consumption was proportional to the volume of water treated in both waterworks and sewer systems; the latest CO$_2$ emission factors for water are shown in Table 1. For the calculation, the emission factor of electricity was adopted as the average value for all power sources in Japan at the both receiving and generating ends by assuming its application in CDM and Bilateral Offset Credit Mechanism. The calculated energy consumption rate value was about the same value as foreign countries values, such as Taiwan of 0.78 kWh/m^3 [13], UK of 0.98 kWh/m^3 [14] and China of 1.37 kWh/m^3 [15]. Detailed analysis will be conducted by future research.

Table 1. Energy consumption rate and CO$_2$ emission factor of water in Japan (FY 2008, kg CO$_2$/m^3).

Treatment process	Energy consumption rate (kWh/m^3)	CO$_2$ emission factor of water (kg CO$_2$/m^3)	
		Calculated with generating end electricity (0.335 kg CO$_2$/kW·h)	Calculated with receiving end electricity (0.373 kg CO$_2$/kW·h)
Waterworks system	0.499	0.181	0.201
Sewer system	0.512	0.195	0.214
Total	1.012	0.376	0.415

In the future, it will be possible to update the CO$_2$ emission factor for water using Equation (1) and the latest values from Waterworks Statistics, Sewerage Statistics, and CO$_2$ emission factor for electricity, which are updated annually.

$$CF_w = \left(\frac{\sum E_{wi}}{\sum Q_{wi}} + \frac{\sum E_{si}}{\sum Q_{si}} \right) * CF_e \qquad (1)$$

where CF_w is CO_2 emission factor for water (kg CO_2/m^3); CF_e is CO_2 emission factor for electricity (kg CO_2/kWh); E_{wi} is energy consumption by waterworks system (kW·h/year); E_{si} is energy consumption by sewer system (kW·h/year); Q_{wi} is volume of fresh water generation by waterworks (m^3/year); Q_{si} is volume of water treated by sewer system (m^3/year).

In structures such as apartment buildings and office buildings, water supplied from the waterworks is first accumulated in water tanks and then distributed to each residential unit or equipment by using pumps. Since the energy consumption for pumping in buildings could not be standardized, it is not included in the range of calculation in this study. Further study along these lines will be necessary in the future.

4. Conclusions

Conventionally in Japan, studies on future prediction of CO_2 reduction effect by widespread use of water-saving equipment was implemented on the basis of the CO_2 emission factor for water (0.59 kg CO_2/m^3), as announced by the Environment Agency of Japan. However, no reports on points such as the basis of calculation were published, making it difficult to verify the data, including whether there was a proportional relationship to the volume of water.

In this study, we confirmed that it was possible to determine a CO_2 emission factor for water as a value proportional to the volume of water. We successfully established a procedure to update the latest value, making possible the development of a methodology for the conversion the water saving volume directly into CO_2 reduction and thus, into carbon credit.

With regard to the developing countries in Asia, measures to conserve water resources and remedial measures for global warming are urgent issues because of rapid urbanization. The widespread use of water saving fittings is a solution for these issues. In addition, the realization of a carbon credit by water saving can become a means to tell society the co-benefit of it intelligibly. Therefore, there are high expectations on the realization of carbon credits through water saving measures, as evident from the joint implementation of a study by Japan and China to evaluate the potentials of the Bilateral Offset Credit System. This result of research expects to lead to realization of the carbon credits by water saving globally.

References

1. The Greenhouse Gas Inventory Office of Japan—The GHGs Emissions Data of Japan (1990–2010). Available online: http://www-gio.nies.go.jp/aboutghg/nir/nir-j.html (accessed on 28 June 2012).
2. Kenwey, S.J.; Priestly, A.; Cook, S.; Inman, M.; Gregory, A.; Hall, M. *Energy Use in the Provision and Consumption of Urban Water in Australia and New Zealand*; Water Services Association of Australia (WSAA): Sydney, Australia, 2008. Available online: http://www.csiro.au/files/files/pntk.pdf (accessed on 28 June 2012).

3. Hackett, M.J.; Gray, N.F. Carbon dioxide emission savings potential of household water use reduction in the UK. *J. Sustain. Dev.* **2009**, *2*, 36–43.

4. Walker, G. Water and energy use efficiency are increasingly linked. *Energy World* **2009**, *369*, 18–19.

5. Yasutoshi, S.; Kanako, T. The economic and environmental impact of remodeling in the case of a water-saving toilet bowl (in Japanese). *J. Soc. Heat. Air-Cond. Sanit. Eng. Jpn.* **2009**, *152*, 9–14.

6. Yasutoshi, S.; Kanako, T.; Kiyoshi, N. Prediction of CO_2 emission associated with residential plumbing equipment (in Japanese). *J. Soc. Heat. Air-Cond. Sanit. Eng. Jpn.* **2010**, *163*, 11–18.

7. Domestic Clean Development Mechanism (in Japanese). Available online: http://jcdm.jp/process/methodology.html (accessed on 24 September 2012).

8. New Mechanism Information Platform (in Japanese). Available online: http://www.mmechanisms.org/initiatives/index.html (accessed on 24 September 2012).

9. The Environment Agency of Japan. *Environmental Household keeping Book* (in Japanese); The Environment Agency of Japan: Tokyo, Japan, 1996.

10. Keidanren (Japan Business Federation). *Voluntary Action Plan on the Environment, the Electricity Emission Factor in Global Warming Countermeasure* (in Japanese); Japan Business Federation: Tokyo, Japan, 2012. Available online: http://www.keidanren.or.jp/japanese/policy/2011/113/honbun.pdf (accessed on 24 September 2012).

11. Japan Water Works Association. *Waterworks Statistics* 1990–2008 (in Japanese); Japan Water Works Association: Tokyo, Japan, 1990–2008.

12. Japan Sewage Works Association. *Sewerage Statistics* 1990–2008 (in Japanese); Japan Sewage Works Association: Tokyo, Japan, 1990–2008.

13. Cheng, C.L.; Liao, W.J.; Liu, Y.C.; Tseng, Y.C.; Chen, H.J. Evaluation of CO_2 emission for saving water strategies. In *Proceedings of 38th International Symposium of CIB W062 on Water Supply and Drainage for Buildings*, Edinburgh, UK, 27–30 August 2012.

14. Thames Water. Corporate Responsibility Report 2009/10. Available online: http://www.thameswater.co.uk/cr/ (accessed on 24 September 2012).

15. Toyosada, K.; Shimizu, Y.; Dejima, S.; Yoshitaka, M.; Sakaue, K. Evaluation of the potential of CO_2 emission reduction achieved by using water-efficient housing equipment in Dalian, China. In *Proceedings of 38th International Symposium of CIB W062 on Water Supply and Drainage for Buildings*, Edinburgh, UK, 27–30 August 2012.

Water **2012**, *4*, 650-669; doi:10.3390/w4030650

OPEN ACCESS

water

ISSN 2073-4441
www.mdpi.com/journal/water

Article

Upgrading of Wastewater Treatment Plants Through the Use of Unconventional Treatment Technologies: Removal of Lidocaine, Tramadol, Venlafaxine and Their Metabolites

Paola C. Rúa-Gómez [1,*]**, Arlen A. Guedez** [1]**, Conchi O. Ania** [2] **and Wilhelm Püttmann** [1]

[1]   Institute of Atmospheric and Environmental Sciences, Department of Environmental Analytical Chemistry, J.W. Goethe University Frankfurt am Main, Altenhöferallee 1, Frankfurt am Main 60438, Germany; E-Mails: arlenaguedez@hotmail.com (A.A.G.); puettmann@iau.uni-frankfurt.de (W.P.)

[2]   Instituto Nacional del Carbon (INCAR-CSIC), Apartado 73, Oviedo 33080, Spain; E-Mail: conchi.ania@incar.csic.es

*   Author to whom correspondence should be addressed; E-Mail: ruagomez@iau.uni-frankfurt.de; Tel.: +49-69-798-40230; Fax: +49-69-798-40240.

Received: 1 June 2012; in revised form: 15 August 2012 / Accepted: 3 September 2012 / Published: 11 September 2012

Abstract: The occurrence and removal efficiencies of the pharmaceuticals lidocaine (LDC), tramadol (TRA) and venlafaxine (VEN), and their major active metabolites monoethylglycinexylidide (MEGX), *O*-desmethyltramadol (ODT) and *O*-desmethylvenlafaxine (ODV) were studied at four wastewater treatment plants (WWTPs) equipped with activated sludge treatment technologies. In parallel to activated sludge treatment, the removal efficiency of the compounds in pilot- and full-scale projects installed at the WWTPs was investigated. Within these projects two different treatment methods were tested: adsorption onto powdered/granulated activated carbon (PAC/GAC) and ozonation. The metabolite MEGX was not detected in any sample. The concentrations of the target analytes in wastewater effluents resulting from activated sludge treatment ranged from 55 to 183 (LDC), 88 to 416 (TRA), 50 to 245 (ODT), 22 to 176 (VEN) and 77 to 520 ng L^{-1} (ODV). In the pilot project with subsequent treatment with PAC/GAC, the mean concentrations of the analytes were between <LOQs and 30 (LDC), 111 (TRA), 140 (ODT), 45 (VEN) and 270 ng L^{-1} (ODV). In the pilot project with subsequent ozonation of the effluent from the conventional treatment the mean concentrations were below the limit of quantification (LOQ) for all of the investigated compounds. The results showed limitations of activated sludge treatment technologies in removing the target compounds but highlighted both

PAC/GAC adsorption and ozonation technologies as effective post-treatment processes for the elimination of the target compounds from wastewater in WWTPs. Possible oxidation by-products formed during ozonation were not analyzed.

Keywords: lidocaine; tramadol; venlafaxine; metabolites; ozonation; activated carbon adsorption

---

1. Introduction

The occurrence of pharmaceuticals and personal care products in the aquatic environment has received increasing scientific and public attention in recent years. A large number of these compounds, unchanged or as active metabolites, are continuously transferred into the sewage water. Their removal by wastewater treatment plants (WWTPs) is a major subject of concern. In a typical European WWTP, conventional treatment including screening, grit removal, preliminary sedimentation, activated sludge treatment, chemical phosphate removal and final sedimentation is used. In this way, mechanical and biological degradation are the only elimination processes applied. Some pharmaceuticals such as ibuprofen and bezafibrate have been demonstrated to be effectively removed (removal rates >95%) by biological wastewater treatment [1,2]. However, several pharmaceuticals are only poorly removed/degraded by conventional wastewater treatment [3–7], causing their continuous discharge into recipient waters and their presence in different water matrices at concentrations ranging from nanograms to low micrograms per liter [8–10]. Due to their therapeutic and biological activity, pharmaceutical discharges pose a great risk to the aquatic environment affecting water and soil-dwelling organisms. Many studies report adverse effects on different aquatic organisms after their exposure to pharmaceutical compounds at environmentally relevant concentrations [11–13]. An actual challenge in wastewater treatment is to optimize existing treatment technologies and/or to upgrade existing treatment plants with new end-of-pipe technologies in order to improve removal efficiencies of several micropollutants including pharmaceuticals.

Many additional treatment technologies for wastewater have been discussed over the last decades. One of them is the chemical oxidation by ozone. Ozonation of wastewater is an end-of-pipe technology, which was traditionally used for disinfection purposes and just recently has been investigated for the removal of micropollutants. Results from both pilot- and full-scale plants using ozonation after the biological treatment reported removal efficiencies of about 95% for several micropollutants [14,15]. The major issues of concern arising from ozonation of wastewater are related to the formation of oxidation by-products from matrix components and transformation products from micropollutants [16].

Another technology discussed for the improvement of waste water treatment is the adsorption of micropollutants onto activated carbon (AC), which can either be implemented as an end-of-pipe technology or can be added to an existing technology in a WWTP, e.g., AC in a pumped bed-membrane bioreactor [17,18]. The most common applications of AC for wastewater treatment are known as granular activated carbon (GAC) and powdered activated carbon (PAC). Both GAC and PAC had been commonly used for sorption of organic micropollutants like pesticides and taste compounds [19,20]. In recent years, studies of AC adsorption in laboratory systems, pilot and full-scale drinking water treatment plants have been carried out reporting successful removal of some micropollutants including pharmaceuticals such as antibiotics and endocrine disrupting compounds [21–25]. There are many technologies available for the implementation of AC for wastewater treatment and each of them should be evaluated separately.

In a previous study the continuous WWTP discharge of the non-extensively studied pharmaceuticals lidocaine (LDC, anesthetic), tramadol (TRA, analgesic) and venlafaxine (VEN, antidepressant), and their major active metabolites desmethyltramadol (ODT) and desmethylvenlafaxine (ODV) has been reported [7,26], denoting the need improvement of available treatment units or application of alternative treatment technologies, which could mitigate the exposure of the aquatic organisms to such pharmaceutical compounds. Recent studies have demonstrated the presence of these compounds and their metabolites in some rivers and lakes in Europe and North America [6,8,27,28]. So far, studies concerning the removal of the compounds through unconventional technologies are scarce [14,29,30]. The main objectives of the present study were (a) to determine the efficiency of AC adsorption and ozonation for the removal of LDC, TRA, VEN and their major active metabolites; (b) to compare removal efficiencies of the analytes using alternative treatment technologies with the removal efficiencies using only biological treatment; and (c) to evaluate the influence of the investigated unconventional technologies on different performance parameters of a WWTP. Lab-scale experiments were carried out and wastewater samples from pilot- and full-scale projects at four different WWTPs in Germany were investigated.

2. Materials and Methods

2.1. Chemicals

LDC, TRA, VEN, and squalane (internal standard) were purchased from Sigma Aldrich (Steinheim, Germany). Monoethylglicinexylidide (MEGX) was kindly supplied by Astra Zeneca (Wedel, Germany). ODT, ODV and d6-TRA (internal standard) were obtained from Toronto Research Chemical Inc. (Toronto, Canada). The suppliers stated a chemical purity of 98% or greater for all reference compounds. Acetone was obtained from LS Labor Service (Griesheim, Germany) and was used as received. All other organic solvents were analytical grade (Carl Roth, Karlsruhe, Germany) and were distilled before use. Ultrapure water was generated using an Astacus ultrapure water purification system from MembraPure (Bodenheim, Germany). Individual stock solutions of each compound were prepared in methanol (1 µg µL^{-1}). Stock solution of the internal standard squalane (1 µg µL^{-1}) was prepared in hexane. Working standard solutions were obtained by appropriate dilution of stock solutions.

2.2. Characterization of ACs and Lab-Scale Adsorption Experiments

Textural characterization of the ACs Carbopal AP (Donau Carbon Corporation, Frankfurt am Main, Germany), Norit SAE Super (Norit Activated Carbon, Riesbürg, Germany), and Hydraffin XC30 (Donau Carbon Corporation, Frankfurt am Main, Germany) was carried out by measuring the N$_2$ adsorption isotherms at −196 °C. Before the experiments, the samples were outgassed under vacuum at 120 °C overnight. The isotherms were used to calculate the specific surface area, total pore volume, and micropore volume evaluated applying the Dubinin-Radushkevich method [31]. Preliminary adsorption tests of selected analytes (LDC and TRA) onto the ACs were carried out at room temperature in stirred batch systems. Different amounts of the ACs (ranging from 10 to 110 mg) were weighed and added to glass beakers containing 200 mL aqueous solutions of each compound. Due to the high adsorption capacities of the activated carbons, high initial concentrations of 100 mg L^{-1} of the pollutants were used in the batch experiments. The solutions were allowed to shake for 72 h at a constant temperature. The amount adsorbed was determined according to $Q_t = (C_0 - C_t)V/m$, where Q_t is the amount (mg g^{-1}) adsorbed at time t, C_0 is the initial concentration (mg L^{-1}), C_t is the concentration at time t (mg L^{-1}), V is the volume (L) of the adsorbate solution and m is the weight (g) of the activated carbon. All adsorption assays and the corresponding blank experiments were made in duplicate.

2.3. Projects at the WWTPs and Sample Collection

Pilot- and full-scale projects at four different German WWTPs located in Langen (WWTP-A), Kaarst (WWTP-B), Schwerte (WWTP-C) and Wuppertal (WWTP-D) were investigated in this study over the period from June to October 2011. The schematic diagrams and the various sampling points are shown in Figure 1, while details regarding population served, applied treatment technologies, characteristics of installed projects at the WWTPs and operational settings during sample collection are summarized in Table 1. The yearly treatment volumes of the WWTPs vary from *ca.* 3 to 47 million m^3 a^{-1}. The installed technologies at all of the investigated WWTPs include mechanical, chemical and biological treatment. WWTP-A, WWTP-C and WWTP-D use conventional activated sludge wastewater treatment, including a secondary clarifier after the aeration tank. After the secondary clarifier the WWTP-D has a flocculation system consisting of 28 filter chambers, in which sludge and residual solids formed by the addition of a soluble iron compound are filtered. The WWTP-B uses a membrane bioreactor (MBR) as biological step, substituting the secondary clarifier of a conventional WWTP by membranes.

The project at the WWTP-A consists of a pilot-scale plant, where effluent from the secondary clarifier is treated by two parallel technologies: PAC adsorption coupled to membrane filtration (PAC-MEM), and down flow fixed-bed columns filled with GAC (GAC-columns). From the second clarifier, treated wastewater is continuously pumped through a microfilter with a mesh size of *ca.* 0.3 mm and then deposited in a stirred collection tank. From this tank the wastewater is pumped to each investigated treatment system. The PAC-MEM system consists of a stirred contact tank (1 m^3), in which the PAC Carbopal AP is mixed with the biologically treated wastewater, followed by a hollow fiber ultrafiltration membrane (membrane pore size = 0.1 µm). The GAC-columns system operating in parallel at the WWTP-A, consists of two acrylic glass columns filled each one with 9.6 kg of GAC Hydraffin XC30. Another filter with a mesh size of *ca.* 0.1 mm is installed before the GAC columns. Each column has an internal diameter of 14.5 cm and an active length of 127 cm, resulting in a bed volume of 21 L.

In the project at the WWTP-B (PAC-in-MBR), the wastewater influent is pumped into a pilot-scale MBR (685 L). Experiments are conducted adding PAC Carbopal AP to the ultrafiltration membrane module in the MBR [membrane pore size = 0.05 µm (manufacturer's data)], which attempts to reproduce the operation conditions of the MBR at the WWTP-B [membrane pore size = 0.04 μm (manufacturer's data)].

> **Figure 1.** Schematic diagrams of the selected WWTPs (**a**) WWTP-A; (**b**) WWTP-B; (**c**) WWTP-C; (**d**) WWTP-D and the pilot-/full-scale projects. Sampling points are indicated by a cross.

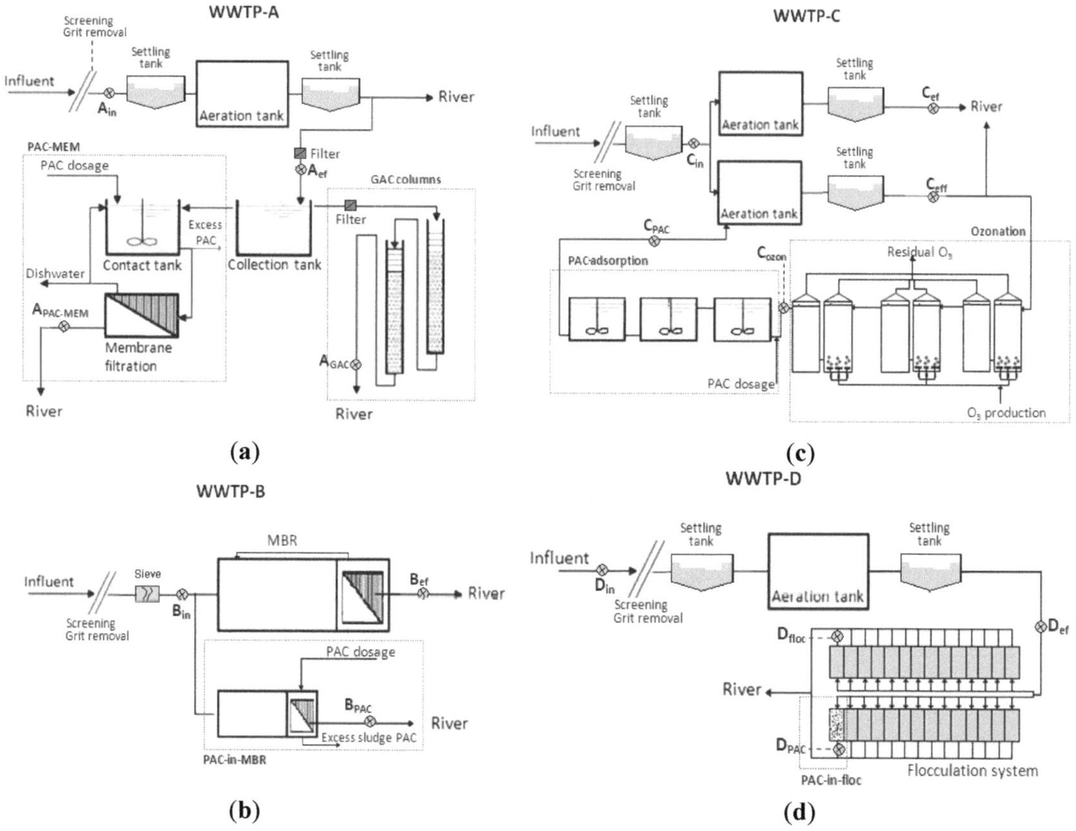

Experiments are performed at the full scale project at the WWTP-C. WWTP-C influent wastewater is evenly distributed between two conventional treatment systems with similar dimensions and operational settings. The effluent from one of these systems is discharged into the river. A fraction of the effluent of the other system (the difference between the maximum hydraulic loading of the secondary clarifier and the amount of wastewater by dry weather) is pumped to an ozonation plant consisting of six reactors (192 m^3) dosing O$_3$ per liter of wastewater. Thereafter, wastewater flows into a PAC-adsorption system consisting of three stirred tank reactors (450 m^3). The adsorbent is the PAC Norit SAE Super. After spending time in the stirred tank reactors, the wastewater/PAC mixture is pumped into the nitrification stage at the aeration tank.

The full-scale project at the WWTP-D (PAC-in-floc) consists of adding PAC Norit SAE Super to one filter chamber of the flocculation unit. In the filter chamber, the PAC is retained in the different heights of the filter bed and quantitatively removed every 24 h by the filter backwash.

Seven-day composite samples of the WWTP-A effluent (A_{ef}, $n = 4$), the permeate of the PAC-MEM system ($A_{PAC-MEM}$, $n = 4$) and effluent of the GAC columns (A_{GAC}, $n = 4$) were collected for analysis. A_{ef}, $A_{PAC-MEM}$ and A_{GAC} samples were initially adjusted to a pH < 2 in order to avoid possible further degradation in the flask collection due to the relative long collection period. Unfortunately, due to technical difficulties it was not possible to collect WWTP-A influent samples in the same way as the other samples at A_{ef}, $A_{PAC-MEM}$ and A_{GAC}. For this reason, twenty-four-hour composite samples of the WWTP-A influent (A_{in}, $n = 2$) and of A_{ef} ($n = 2$) were additionally collected in order to calculate the removal efficiencies of the target analytes by conventional wastewater treatment.

Table 1. Characteristics, operating conditions and description of the pilot- and full-scale projects at the investigated WWTPs during the sampling collection.

Characteristics	WWTPs			
	WWTP-A	**WWTP-B**	**WWTP-C**	**WWTP-D**
Treatment technology	Conventional activated sludge	Membrane bioreactor (MBR)	Conventional activated sludge	Conventional activated sludge
Sampling period	01/08/2011–06/09/2011	01/09/2011–06/10/2011	08/08/2011–18/08/2011	09/08/2011–15/08/2011
Population served (PE)	74,000	69,000	50,000	370,000
Average flow (m^3/a)	6,000,000	5,500,000	6,000,000	47,000,000
Wastewater type	98% R, 2% I	98% R, 2% I	92% R, 8% I	81% R, 19% I
Collection system	Combined	Combined	Combined	Combined
HRT (h)	18	28.5	38	51
SRT_{sludge} (d)	25–30	26	22	12
T (°C)	19.4	17.5	19.7	18.4

	Investigated projects at each WWTP					
Characteristics	1. PAC adsorption followed by membrane filtration (PAC-MEM) 2. GAC columns		MBR-PAC integrated system (PAC-in-MBR)	Ozonation followed by PAC-adsorption		PAC adsorption in flocculation system (PAC-in-floc)
Dimension	Pilot-scale		Pilot-scale	Full-scale		Full-scale
Project	**PAC-MEM**	**GAC-columns**	**PAC-in-MBR**	**Ozonation**	**PAC-ad.**	**PAC-in-floc**
AC type	Carbopal AP	Hydraffin XC30	Carbopal AP	-	Norit SAE Super	Norit SAE Super
AC dosage (mg AC/L of wastewater)	5	-	10	-	15	20
Transferred ozone dose (mg O_3/L of wastewater)	-	-	-	0.6	-	-
HRT (h)	0.9	-	24	0.4	0.9	0.4
SRT_{carbon} (day)	1	-	25[a]	-	22[a]	0.5
EBCT (h)	-	0.4	-	-	-	-

[a] SRT of the AC with the activated sludge; PE: population equivalent; R: residential; I: industrial/commercial; HRT: hydraulic retention time; SRT: solid retention time; EBCT: empty bed contact time.

Twenty-four-hour composite samples were collected at the other investigated WWTPs. At WWTP-B: influent (B_{in}, $n = 4$), effluent from the MBR (B_{ef}, $n = 4$) and permeate of the PAC-in-MBR (B_{PAC}, $n = 4$)

were collected. At WWTP-C samples collected were: influent (C_{in}, $n = 8$), effluent from the conventional treatment (C_{ef}, $n = 8$), effluent from ozonation unit (C_{ozon}, $n = 8$), effluent from the PAC-adsorption system (C_{PAC}, $n = 8$) and final effluent from the secondary clarifier containing wastewater treated by ozonation and PAC-adsorption units (C_{eff}, $n = 8$). At WWTP-D investigated samples were: influent (D_{in}, $n = 6$), effluent from the secondary clarifier (D_{ef}, $n = 6$), effluent from a filter chamber of the flocculation system with no addition of PAC (D_{floc}, $n = 6$) and effluent from the filter chamber, to which PAC was added (D_{PAC}, $n = 6$). Samples were collected simultaneously in the sample locations at each WWTP using automatic samplers (time-proportional) and then stored in brown glass bottles and cooled at 4 °C in the dark until processing in laboratory within 7 days after sampling.

2.4. Analytical Methods

Target analytes were extracted from the wastewater samples (untreated wastewater: 250 mL, wastewater treated by conventional technologies: 500 mL, wastewater treated by unconventional technologies: 1 L) by solid phase extraction (SPE) using Bond Elute PPL cartridges (100 mg/1 mL, Varian, Darmstadt, Germany). Samples from the sampling locations at the WWTP-A were neutralized to pH between 7.2 and 7.5 by addition of a NaOH solution before SPE. Water samples were filtered by pressure filtration using 1 μm borosilicate glass fiber filter (Type A/E, Pall, Dreieich, Germany) prior to SPE. Cartridges were eluted with methanol/acetone (1/1, *v*/*v*) and extracts were dried and dissolved in methanol. Squalane and d6-TRA were added to the extracts as internal quantification standards. Quantification of the analytes in extracts was performed using a Trace GC Ultra gas chromatograph (equipped with a TG-5MS capillary column) coupled to a DSQ II mass spectrometer (Thermo Scientific, Dreieich) operated in full scan mode (*m/z* 50–650) with electron impact ionization (70 eV). See Rúa-Gómez and Püttmann [7] for a detailed description of the applied analytical method. Briefly, ultra pure helium ($\geq$99.999%) was used as the carrier gas (1.1 mL min^{-1} flow), and the column oven temperature was increased from 80 to 300 °C at 4 °C min^{-1}, and maintained for 30 min at 300 °C. Sample aliquots of 1 μL were injected in the splitless mode (injector temperature 240 °C). Acquired data were processed using Xcalibur software Version 2.0.7 (Thermo Scientific, Dreieich, Germany).

Samples were collected in 1-L brown glass bottles. These were rinsed before use with ultrapure water and methanol and then heated to 110 °C for a minimum of 2 h. Before use glass fiber filters were washed with dichloromethane and then heated in an oven for 2 h at 400 °C. Blank samples, consisting of ultrapure water, were extracted and treated in the same way as field samples to test for sample contamination during transportation and preparation.

In accordance with DIN 32645 German Institute for Standardization [32] limit of detection (LOD) for LDC, TRA, VEN and the metabolites MEGX, ODT and ODV was calculated from measured calibration curves. The limit of quantification (LOQ) was estimated as three times the LOD and provided values of 16 (LDC), 50 (MEGX), 25 (TRA), 18 (VEN), 35 (ODT) and 23 ng L^{-1} (ODV). Recovery rates were calculated for the entire method by spiking the target analytes into 1-L groundwater samples ($n = 6$), 0.5-L treated wastewater samples ($n = 6$) and 0.2-L untreated wastewater samples ($n = 6$), at a spiking level of 200 ng L^{-1}. Mean recoveries in groundwater were: 81 ± 5% (LDC), 59 ± 3% (MEGX), 82 ± 7% (TRA), 94 ± 9% (VEN), 71 ± 9% (ODT) and 94 ± 4% (ODV); in treated wastewater: 66 ± 8% (LDC), 53 ± 8% (MEGX), 64 ± 5% (TRA), 82 ± 5% (VEN), 59 ± 4% (ODT) and

$86 \pm 4\%$ (ODV); and in untreated wastewater: $56 \pm 4\%$ (LDC), $49 \pm 9\%$ (MEGX), $61 \pm 6\%$ (TRA), $79 \pm 6\%$ (VEN), $51 \pm 9\%$ (ODT) and $71 \pm 11\%$ (ODV). No adjustments to concentrations in the samples were made in regard to the SPE recovery rates.

After equilibration, LDC and TRA concentrations in the solutions from the lab-scale adsorption tests were determined using a UV-VIS spectrophotometer UVPC2400 (Shimadzu, Duisburg, Germany) equipped with tungsten and deuterium lamps as light sources. After calibration of the instrument for each analyte, detection wavelength was set to 230 nm (LDC) and 270 nm (TRA).

Conventional physicochemical parameters of WWTP samples [Chemical Oxygen Demand (COD), total nitrogen (TN) as the sum of ammonium, nitrite and nitrate nitrogen, total phosphorus (TP) and dissolved organic carbon (DOC)] were determined according to standard methods indicated by the German Federal Ministry of Justice [33].

3. Results and Discussion

3.1. Occurrence and Removal of Target Analyses through Activated Sludge Treatment

Concentrations of the target analytes found in each sampling location at the investigated WWTPs are shown in Figure 2. The pharmaceuticals LDC, TRA and VEN and the metabolites ODT and ODV were detected above the LOQ in all of the influent and effluent samples from the activated sludge treatment, while MEGX, the metabolite of LDC, as expected in accordance with a previous study, was not detected in any sample. The concentrations of the investigated pharmaceuticals varied from 70 to 257 (LDC), 232 to 615 (TRA), 60 to 299 (ODT), 54 to 336 (VEN), and 235 to 723 ng L^{-1} (ODV) in the WWTP influents, whereas the concentrations in the effluent samples ranged from 55 to 183 (LDC), 88 to 416 (TRA), 50 to 245 (ODT), 22 to 176 (VEN) and 77 to 520 ng L^{-1} (ODV). Although a reduction in the concentrations of the target analytes was observed, the continuous discharge of the target analytes at these effluent concentrations into surface waters could cause adverse effects on the aquatic environment. A recent study has shown that concentrations of VEN at picogram per liter levels cause significant foot detachment from the substrate in freshwater snails, a sublethal effect that could have lethal consequences for these species [13]. Moreover, the potential toxicological effects caused by the interaction of different pharmaceuticals (among many other compounds present in wastewater effluents) cannot be discarded [34].

Due to their physiochemical characteristics (high water solubilities, low *n*-octanol/water partition coefficients and low Henry coefficients) all target analytes are expected to be found in the water phase rather than being volatilized or retained in the activated sludge [7]. Thus, elimination of the compounds achieved by activated sludge treatment corresponded to the difference between influent and effluent mass loads of the target analytes in the water phase. Removal efficiencies of the target analytes obtained by activated sludge treatment at the investigated WWTPs are presented in Table 2. The removal efficiencies of the target analytes obtained in each WWTP are difficult to compare, since the investigated WWTPs work with different operational settings and have different influent characteristics (see Table 1). The removal rates reported for the WWTP-A and WWTP-D indicate mechanical and biological treatment because the influent samples were collected before the first settling tank (at the WWTP-A) and before the screening step (at the WWTP-D). The WWTPs using conventional activated sludge treatment, WWTP-A, WWTP-C and WWTP-D, working with different operational settings, showed maximal removal rates of 35% (LDC), 56% (TRA), 27% (ODT), 56% (VEN) and 41% (ODV). These results were consistent with the findings from a previous study showing maximal removal efficiencies of 37% (LDC), 41% (TRA), 24% (ODT), 48% (VEN) and 29% (ODV) during conventional wastewater treatment [7], confirming that the investigated pharmaceuticals could only be partially removed using mechanical and biological treatment. Increasing aerobic solid retention times (SRT) can enhance the biological degradation of various pharmaceuticals such as bezafibrat and ibuprofen [35]. A similar effect for the target analytes has not been observed in the present study.

Figure 2. Average, maximum and minimum concentrations of target analytes along the treatment process at the investigated WWTPs: (**a**) WWTP-A; (**b**) WWTP-B; (**c**) WWTP-C; (**d**) WWTP-D. Number of samples are given in parenthesis.

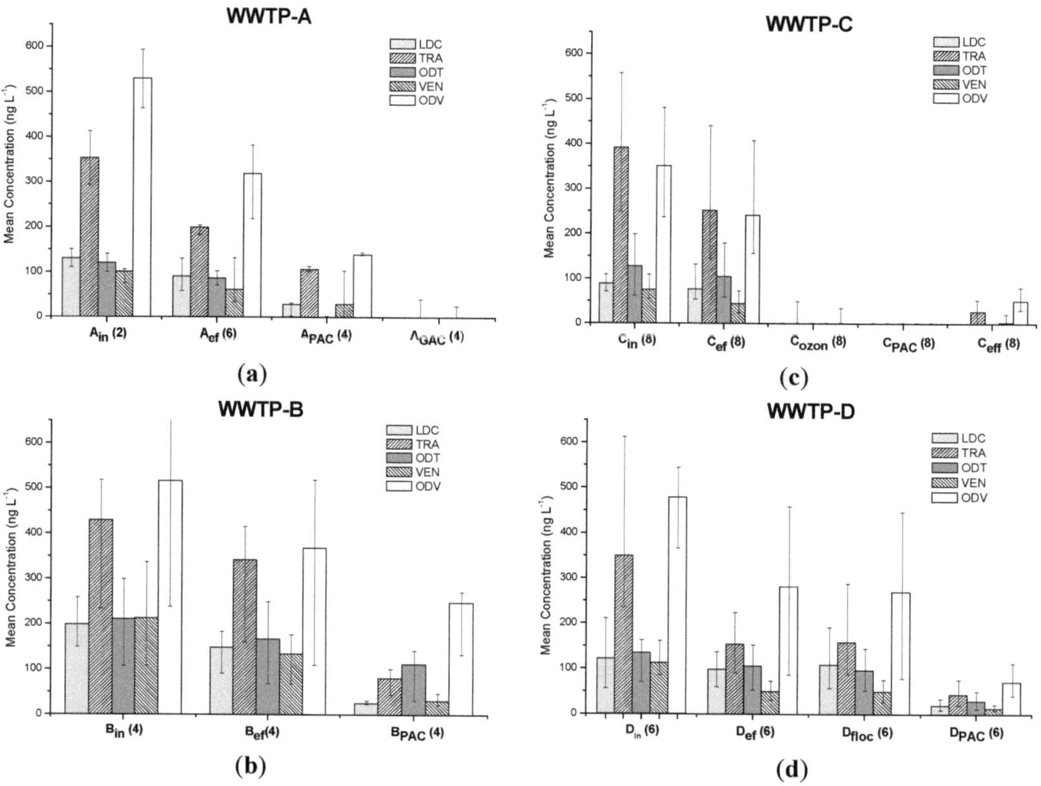

The WWTP-B uses a combination of a membrane process with a suspended growth bioreactor as the biological step. This system is actually being widely used for wastewater treatment, as it allows smaller sludge aeration basin volume, exceeds significantly the efficiency of conventional sand filtration and presents higher SRTs above the levels that can be obtained with secondary clarifiers [1,36]. However, removal efficiencies of the target compounds obtained by this treatment system were also insufficient (Table 2).

Table 2. Percentage of removal of target analytes during activated sludge treatment at the investigated WWTPs.

WWTP	% Removal				
	LDC	**TRA**	**ODT**	**VEN**	**ODV**
WWTP-A	29	43	27	40	39
WWTP-B	25	20	21	37	29
WWTP-C	14	36	17	40	31
WWTP-D	35	56	21	56	41

In activated sludge treatment, two mechanisms are considered for the removal of micropollutants from wastewater: adsorption on the sludge and degradation by microorganisms. Poor removal efficiencies of the target analytes achieved during activated sludge treatment (conventional and in a MBR) may be explained by the tendency of the compounds to remain in the water phase, which suppresses their sorption onto the sludge. Thus, the most plausible mechanism for the removal of the investigated compounds from wastewater seems to be the degradation by bacteria present in the sludge, which have been previously reported for other pharmaceuticals [37,38]. Furthermore, considering that many other compounds are also present in wastewater, some easier to degrade than others, a competition for the degradation of the compounds by the bacteria is expected to take place, thereby decreasing the removal efficiencies of the target analytes [39].

3.2. Removal of Target Analytes through Unconventional Technologies

3.2.1. Adsorption Experiments

Selected textural characteristics of the ACs used are summarized in Table 3. All three selected carbons showed similar porous features, with well-developed micro/mesoporosity as indicated by the type I/IV N_2 adsorption isotherms (data not shown). Some small differences concerning the pore volumes were observed; for instance, micropore volumes are rather close for the three activated carbons, whereas a slightly higher total pore volume was measured for PAC Norit SAE Super. It is well known that micropores are the active sites for the retention of micropollutants in both gas and liquid phase and that the transport pores (mesopores) and average particle size mainly affect the adsorption kinetics [40]. Bearing this in mind, it can be expected that these carbons would show similar adsorptive behaviors. Additionally, further characterization of the carbons confirmed that these adsorbents have a hydrophobic nature.

Table 3. Physical features of the used ACs.

Characteristics	Carbopal AP	Hydraffin XC30	Norit SAE super
Type	PAC	GAC	PAC
Avg. particle size diameter (μm)	33.6	1400	15
S_{BET} (m^2 g^{-1})	899	1036	965
V_T (cm^3 g^{-1})	0.524	0.619	0.69
W_o (cm^3 g^{-1})	0.40	0.44	0.40

S_{BET}: specific surface area evaluated from the BET equation; V_T: total pore volume evaluated at p/po 0.95; W_o: micropore volume evaluated applying the Dubinin-Radushkevich equation.

The lab-scale adsorption test data for the investigated ACs were fitted to both Freundlich and Langmuir models (Table 4), with higher correlation coefficients for the Langmuir equation. For all of the activated carbons higher adsorption capacities of LDC were obtained in comparison to TRA, which is in good agreement with the expected trend based on their chemical composition and size. No information about LDC and TRA adsorption on AC could be found in the literature for comparison. Nevertheless, more relevant to our study was the fact that the preliminary test data confirmed the adsorption of the selected micropollutants onto AC.

Table 4. Fitting parameters to the Langmuir and Freundlich models of the equilibrium adsorption isotherms (at 25 °C) of lidocaine (LDC) and tramadol (TRA) onto the activated carbons used in this study.

Activated carbon	Freundlich isotherms constants			Langmuir isotherm constants		
	K_F (mg g^{-1}/(mg L^{-1})$^{1/n}$)	$1/n$	r^2	q_m (mg g^{-1})	K_L (L mg^{-1})	r^2
			LDC			
Carbopal AP[a]	137	0.13	0.773	215	0.81	0.998
Hydraffin XC30-A[a]	56	0.31	0.918	196	0.17	0.997
Hydraffin XC30-B[b]	89	0.27	0.851	246	0.28	0.999
Norit SAE Super[a]	156	0.06	0.980	204	0.79	0.999
Activated carbon			**TRA**			
Carbopal AP[a]	46	0.15	0.964	84	0.36	0.995
Hydraffin XC30-A[a]	25	0.23	0.965	76	0.10	0.999
Hydraffin XC30-B[b]	29	0.23	0.948	85	0.15	0.999
Norit SAE Super[a]	61	0.08	0.987	87	0.39	0.999

[a] contact time of 72 h; [b] contact time of 120 h; Freundlich isotherm: $q_e = K_F (C_e)^{1/n}$; Langmuir isotherm: $q_e = (K_L \cdot q_m \cdot C_e)/(1 + K_L \cdot C_e)$; q_e: amount adsorbed per unit mass of adsorbent; C_e: equilibrium concentration of compound in liquid; q_m = maximum adsorption capacity K_F: Freundlich coefficient; K_L: Langmuir coefficient.

Despite the similarities in the chemical and porous features of the carbons (Table 3), slightly higher adsorption capacities were obtained for Carbopal AP and Norit SAE Super, compared to Hydraffin XC30. This is attributed to the different adsorption kinetics as a result of the particle size diameter of the carbons. Indeed, the lab-scale adsorption tests were carried out after 72 h of contact between the carbon/solution suspensions. Under these conditions, kinetic studies (data not shown) revealed the slow uptake of the GAC system compared to PAC, thereby resulting in slightly lower adsorption capacity (lower than the theoretical expected uptake at equilibrium conditions for this carbon). This was further confirmed by the increase in the uptake (q_m parameter from Langmuir model) when the contact time is risen up to 120 h, obtaining values in agreement with the porosity of the carbons.

3.2.2. Projects PAC-MEM and GAC-columns

In the final effluent of the system PAC-MEM at the WWTP-A, the target analytes were found at mean concentrations of 30 (LDC, VEN), 106 (TRA), <LOQ (ODT) and 139 ng L^{-1} (ODV), whereas the mean concentrations found in the effluent of the system consisting of fixed-bed columns filled with GAC were below the LOQ for all of the compounds (Figure 2). The decrease in the concentrations

compared with the concentrations in the WWTP-A effluent showed the adsorption of the target analytes both onto PAC and GAC.

Removal efficiencies of the target analytes observed in both investigated systems at the WWTP-A are listed in Table 5. Higher removal efficiencies were obtained during treatment with the GAC-columns than during treatment in the PAC-MEM system. These results are in good agreement with the expectation based on the porous features (Table 3) and the lack of restricted diffusion, thus confirming the suitability of the GAC in the fixed-bed columns.

Table 5. Percentage of removal of the target analytes obtained by the unconventional treatment systems at the investigated WWTPs.

Treatment system	LDC		TRA		ODT		VEN		ODV	
	$\%R_{UP}$	$\%R_{total}$	$\%R_{UP}$	$\%R_{total}$	$\%R_{UP}$	$\%R_{total}$	$\%R_{UP}$	$\%R_{total}$	$\%R_{UP}$	$\%R_{total}$
WWTP-A										
PAC-MEM	68	77	47	70	>80	>88	51	67	56	74
GAC columns	>72	>93	>90	>97	>80	>88	>70	>90	>92	>98
WWTP-B										
PAC-in-MBR	-	87	-	81	-	47	-	85	-	52
WWTP-C										
Ozonation	>89	>91	>95	>97	>84	>87	>80	>88	>95	>97
PAC-adsorption	nc	>91	nc	>97	nc	>87	nc	>88	nc	>97
Secondary clarifier	nc	>91	nc	93	nc	>87	nc	>88	nc	85
WWTP-D										
Flocculation system	4	37	2	55	5	28	2	56	5	44
PAC-in-floc	76	84	72	88	73	79	73	87	71	85

R_{UP}: Removal efficiency of the specific unit process; R_{total}: Removal efficiency after activated sludge treatment and the respective unit process; *nc*: not calculated because of values below the LOQ in the wastewater entering the unit process. For the data below the LOQ, $0.5 \times LOQ$ was used for the calculation of the removal efficiency.

The GAC-columns system located as a post-technology after activated sludge treatment increased significantly the removal efficiencies of the target analytes in the WWTP-A, achieving removal rates above 88% for all of the compounds. In the present work, the columns were filled with GAC about 1 month before the collection of the samples so its adsorption capacity was not yet exhausted. It can be expected that, with time, the adsorption capacity of the GAC-columns will decrease and be depleted while biological activity will be developed in the columns contributing to degradation of the compounds [41]. The use of filled GAC-columns has already been reported as an effective post-treatment technology for the removal of several pharmaceuticals. Nguyen *et al.* [42] reported removal efficiencies ≥98% for other hydrophilic pharmaceuticals such as carbamazepine and diclofenac, which during activated sludge treatment showed efficiencies below 40%.

3.2.3. Project PAC-in-MBR

With the addition of PAC into the MBR of the PAC-in-MBR system at the WWTP-B, the mean concentrations of the target analytes in the effluent were 26 (LDC), 80 (TRA), 111 (ODT), 31 (VEN) and 248 ng L^{-1} (ODV). The addition of PAC into the MBR achieved removal efficiencies of the target analytes (Table 5). A comparison of these values with the removal efficiencies obtained during activated sludge treatment at the WWTP-B (Table 2) has to be done carefully as the constant flow into the PAC-in-MBR system resulted in a flow-proportional sampling at B_{PAC}, whereas sampling at B_{in} was time-proportional and no monitoring of the daily variation of wastewater was carried out at this sampling location. The metabolites ODT and ODV appeared to be poorly adsorbed by the PAC within the MBR, since the removal efficiencies after addition of PAC were also deficient. Observing the performance of the PAC Carbopal AP in the PAC-in-MBR system and PAC-MEM system at the WWTP-A, it can be concluded that the PAC Carbopal AP is not suitable for an effective removal of the target analytes from wastewater. The lower adsorption capacities obtained for LDC and TRA in the lab-scale experiments (Table 4) also support this finding, and suggest that such poor performance of the AC might be related to hindered accessibility of these compounds in microporous ACs. This statement remains an assumption as no operating settings were tested to optimize the performance of both systems at WWTP-A and WWTP-B. Other authors reported satisfactory results in the removal process of the persistent hydrophilic pharmaceuticals carbamazepine and sulfamethoxazole by simultaneous PAC adsorption within a MBR [18]. However, little information about the chemical composition and the micro/mesoporosity of therein used PAC is provided. A study of the Ministry for Climate Protection, Environment, Agriculture, Nature Conservation and Consumer Protection of the German State of North Rhine-Westphalia [43] reported high removal efficiencies of sulfamethoxazole from wastewater using Carbopal AP as adsorbent. Thus, the PAC-in-MBR technology and the use of Carbopal AP at the WWTP-B are not discarded for an effective removal of the target analytes and argue for further investigation.

3.2.4. Project Ozonation Followed by PAC-Adsorption

The samples collected after the ozonation step at the WWTP-C showed mean concentrations below the LOQ for all of the target analytes (Figure 2). Thus, for a dosage of 0.6 mg O_3 per liter of wastewater ($\approx$0.1 mg O_3 mg^{-1} DOC during sample collection) and a contact time of 54 min, removal rates greater than 89% (LDC), 95% (TRA), 84% (ODT), 80% (VEN) and 95% (ODV) were calculated, presenting the ozonation as an effective step for removal of the selected compounds from wastewater after activated sludge treatment (Table 5). Removal efficiencies of LDC, VEN and ODV during ozonation were consistent with recently conducted studies. Hollender *et al.* [14] reported elimination rates of 98%, 99% and 96% for LDC, VEN and ODV, respectively, for a dosage of 0.6 mg O_3 mg^{-1} DOC and a contact time of about 9 min. An ozonation stage with a dosage of 0.5 mg O_3 mg^{-1} DOC and a contact time of 15 min showed removal rates of about 88% for both TRA and VEN [29]. The good removal efficiencies obtained by the ozonation at the WWTP-C with low ozone dosage relative to the DOC content was due to the long contact time (54 min).

Parallel to the beneficial effects of oxidation of the target compounds, the use of ozonation can cause the formation of undesired by-products through the reaction of ozone and OH radicals with different compounds present in wastewater. As biological treatment technologies have been reported as an effective tool for the removal of organic by-products [14,44], it is assumed that the recirculation of the effluent of the PAC-adsorption system at the WWTP-C to the aeration tank contribute to the degradation of eventually formed organic by-products.

No target compounds were detected in effluent samples from the PAC-adsorption system at the WWTP-C. Because the concentrations of the target analytes after ozonation were below the LOQs, removal efficiencies at the PAC-adsorption system could not be calculated. Although at the WWTP-C the implementation of a PAC-adsorption system after ozonation for the removal of the target analytes appears to be unnecessary, Reungoat *et al.* [29] highlighted the importance of a post-filtration with AC, in order to achieve total removal of different micropollutants and non-target compounds including transformation products. Additional to the removal of several compounds, the use of ozone in wastewater treatment provides disinfection, viral inactivation and sterilization of the final effluent [14].

The concentrations of the target analytes found in the effluent samples from the secondary clarifier containing wastewater treated by the ozonation and the PAC-adsorption systems were <LOQ (LDC, ODT and VEN), 27 (TRA) and 51 ng L^{-1} (ODV). The increment in the concentrations was due to the continuous entry of fresh wastewater influent to the aeration tank.

3.2.5. Project PAC-in-floc

The effluent samples from the flocculation filter chamber operated with PAC presented mean concentrations of 19 (LDC), 44 (TRA), <LOQ for ODT and VEN, and 71 ng L^{-1} (ODV), which are significantly lower than the concentrations found in the effluent samples from the filter chamber operated without PAC [108 (LDC), 157 (TRA), 97 (ODT), 50 (VEN) and 270 ng L^{-1} (ODV)] (Figure 2). Acceptable removal efficiencies were observed through the addition of PAC which increase the total removal of the target analytes at the WWTP-D. As expected due to the hydrophilic character of the target analytes, no significant elimination of these micropollutants was observed by the flocculation process at the WWTP-D (Table 5).

3.2.6. Mechanisms for Adsorption onto AC and Ozonation

The concentrations of the target analytes found in influent and effluent samples from the treatment systems at WWTP-A, WWTP-B and WWTP-D demonstrated the adsorption of the compounds onto the AC. Due to the moderately alkaline or alkaline character of the target analytes [pK_a values of 8.01 (LDC, [45]), 9.13 (TRA, [46]), 9.12 (ODT, [46]), 9.4 (VEN, [47]) and 14.46 (ODV, [48])], and the amino groups in their structures which under neutral conditions will be protonated forming cations [49], electrostatic interactions between the cations and the AC are expected to occur. However, due to the hydrophobic nature (basic character) of the ACs used, at neutral conditions their surface is expected to be neutral or slightly positive charged, for which electrostatic interactions (attractive) between the pollutants and the ACs are not expected to be the main driving force controlling the uptake on all three studied ACs [50]. Thus, the uptake strongly depends on the nature of the pollutant (*i.e.*, boiling point, molecular size, solubility).

The ozonation step investigated at the WWTP-C showed good removal efficiencies for all of the target compounds. Considering that the ozonation process is dependent on the functional groups and that ozone reacts relatively fast with compounds containing an activated aromatic moiety, double bonds or amino groups [16], the removal of the target analytes during ozonation at the WWTP-C is explained by the tertiary amino groups present in their chemical structures.

3.3. Evaluation of other Parameters Relevant to the Wastewater Treatment

Parallel to the removal of the target analytes, the implications of using AC and ozonation technologies on different performance parameters of the WWTP were analyzed. Concentrations of the parameters COD, TN, TP and DOC measured at the sampling locations at WWTP-B, WWTP-C and WWTP-D are represented in Figure 3. Due to technical difficulties at some sampling dates, statistical values of COD and TP could not be calculated at WWTP-B and WWTP-D, respectively. DOC concentrations were measured only at sampling locations at WWTP-D. Unfortunately, performance parameters could not be measured during the WWTP-A sampling. The mean COD concentrations found in samples at C_{eff} and D_{PAC} (10 and 7 mg L^{-1}) were lower than the mean COD concentrations measured at C_{ef} and D_{floc} (15 and 10 mg L^{-1}). This indicates that the addition of PAC to nitrification step of the aeration tank at the WWTP-C and to the flocculation filter chamber at WWTP-D enhance the COD removal process, and thus the overall removal of organic compounds at the WWTP. This affirmation can be confirmed by the mean DOC concentrations measured at D_{floc} and D_{PAC} ($[DOC]_{Dfloc}$ = 5 mg L^{-1} > $[DOC]_{DPAC}$ = 3 mg L^{-1}).

Mean TN concentrations measured in samples from B_{PAC} (5 mg L^{-1}) and C_{eff} (6 mg L^{-1}) were lower than the mean TN concentrations in samples from B_{ef} (7 mg L^{-1}) and C_{ef} (8 mg L^{-1}). The results showed that the addition of PAC to activated sludge processes increase the N removal at WWTP-B and WWTP-C. The nitrification enhancement is probably due to retention onto PAC of inhibitory compounds of the nitrification processes [51]. As expected due to the absence of growing bacteria in the flocculation chambers, no N removal was observed by addition of PAC to the flocculation chamber.

TP concentrations in the samples from B_{ef} and B_{PAC} varied so that it was not possible to establish any tendency at the PAC-in-MBR system at the WWTP-B. A slight decrease of the TP concentrations at the WWTP-C by addition of PAC to the activated sludge process was observed ($[TP]_{Cef}$=0.6 mg L^{-1} > $[TP]_{Ceff}$=0.4 mg L^{-1}). An increase in phosphorus removal efficiencies using AC in activated sludge systems were reported by Serrano *et al.* [52].

The concentrations of the performance parameters measured in the final effluents after the treatment projects at WWTP-B, WWTP-C and WWTP-D were far below the threshold values established by the German Federal Ministry of Justice [33] for discharges from WWTPs magnitude 4 (WWTP-B and WWTP-C; 90 (COD), 18 (TN) and 2 mg L^{-1} (TP)) and magnitude 5 [WWTP-D; 75 (COD), 13 (TN) and 1 mg L^{-1} (TP)].

Figure 3. Average concentrations and standard deviations (**a**) Chemical Oxygen Demand (COD); (**b**) total nitrogen (TN); (**c**) total phosphorus (TP); and (**d**) dissolved organic carbon (DOC) measured during sampling at WWTP-B, WWTP-C and WWTP-D. Threshold values of each parameter according the WWTP characteristics are indicated with a line [33].

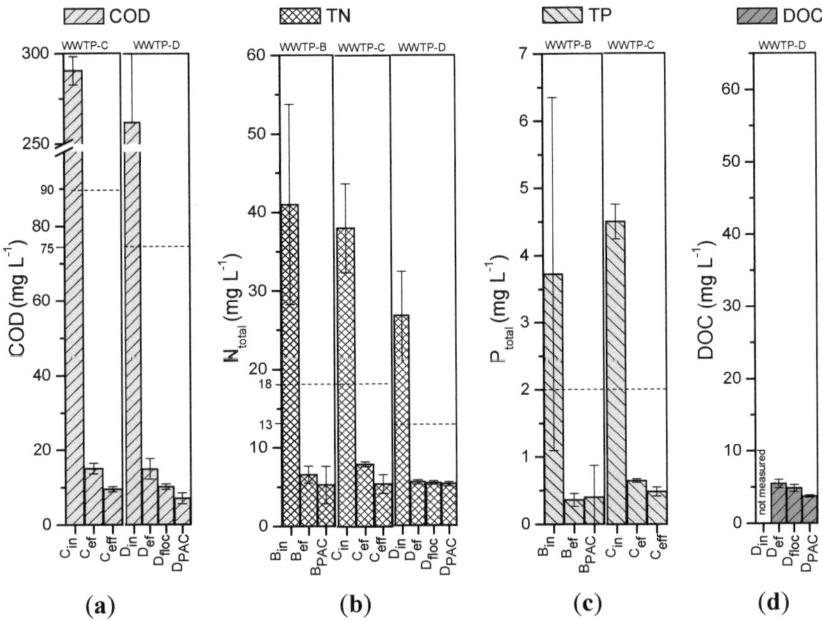

(a)   (b)   (c)   (d)

4. Conclusions

Different treatment systems including adsorption onto PAC/GAC and ozonation were demonstrated as viable post-treatment technologies in order to enhance the removal efficiencies of LDC, TRA, VEN and the metabolites ODT and ODV from biologically treated wastewater at a WWTP. Lab-scale adsorption tests on the selected carbon adsorbents showed high removal efficiencies for LDC as opposed to TRA, indicating that the overall uptake is governed by several factors: the affinity of the pollutant towards the aqueous solution, the structural shape of the pollutants, and the adsorbent particle size. The removal of the target analytes using AC was explained based on their chemical nature, and obtained data show their adequateness to be used in post-treatments for upgrading WWTPs. In the case of ozonation, the removal efficiency seems to be related to the presence of tertiary amino groups in the chemical structure of the micropollutants. Taking into account that the concentrations of the target analytes in the WWTP effluents are diluted to a great extent when they are discharged into surface waters, it can be expected, according to the removal efficiencies obtained in this study, that the concentrations of the pharmaceuticals and their metabolites will be below the LOQs at the discharge points of WWTPs upgraded with unconventional technologies. The addition of PAC to activated sludge processes appears to improve other water quality parameters such as TN and COD.

Parallel to the removal of micropollutants, different aspects such as waste production, electricity consumption and operation costs will have to be discussed for the implementation of AC and ozonation technologies at large-scale in WWTPs. Moreover, the formation of by-products has to be considered in case of ozonation.

Acknowledgments

The first author would like to thank the Catholic Academic Exchange Service (KAAD) for financial support. The support of colleagues and staff of the Institute of Atmospheric and Environmental Sciences at the Goethe University Frankfurt am Main and of the Instituto Nacional del Carbon in Oviedo is gratefully acknowledged. Projects in Schwerte and Wuppertal are financial supported by the Ministry for Climate Protection, Environment, Agriculture, Nature Conservation and Consumer Protection of the German State of North Rhine-Westphalia. Thanks are due to Gregor Lorenz and Dieter Thöle from the Ruhrverband, Jochen Herr from the RWTH Aachen University, Peter Cornel and Gregor Knopp from Institut IWAR at the Technische Universität Damstadt, Helmut Volpert and Stefan Hoffmann from the Abwasserverband Langen-Egelsbach-Erzhausen and Catrin Bornemann from the Wupperverband, for sampling and for their assistance in questions about technical characteristics of the investigated WWTPs. COA thanks MICINN for financial support (grant CTM2008-01956). Thanks go to German Academic Exchange Service (DAAD) for covering traveling costs within the project PPP-Spain.

References

1. Clara, M.; Strenn, M.; Gans, O.; Martinez, E.; Kreuzinger, N.; Kroiss, H. Removal of selected pharmaceuticals, fragrances and endocrine disrupting compounds in a membrane bioreactor and conventional wastewater treatment plants. *Water Res.* **2005**, *39*, 4797–4807.
2. Smook, T.M.; Zho, H.; Zytner, R.G. Removal of ibuprofen from wastewater: Comparing biodegradation in conventional, membrane bioreactor, and biological nutrient removal treatment systems. *Water Sci. Technol.* **2008**, *57*, 1–8.
3. Gros, M.; Petrović, M.; Barceló, D. Wastewater treatment plants as a pathway for aquatic contamination by pharmaceuticals in the Ebro River Basin (Northeast Spain). *Environ. Toxicol.Chem.* **2007**, *26*, 1553–1562.
4. Kasprzyk-Hordern, B.; Dinsdale, R.M.; Guwy, A.J. The removal of pharmaceuticals, personal care products, endocrine disruptors and illicit drugs during wastewater treatment and its impact on the quality of receiving waters. *Water Res.* **2009**, *43*, 363–380.
5. Palmer, P.M.; Wilson, L.R.; O'Keefe, P.; Sheridan, R.; King, T.; Chen, C.-Y. Sources of pharmaceutical pollution in the New York City watershed. *Sci. Total Environ.* **2008**, *394*, 90–102.
6. Metcalfe, C.D.; Chu, S.; Judt, C.; Li, H.; Oakes, K.D.; Servos, M.; Andrews, D.M. Antidepressants and their metabolites in municipal wastewater, and downstream exposure in an urban watershed. *Environ. Toxicol. Chem.* **2010**, *29*, 79–89.
7. Rúa-Gómez, P.; Püttmann, W. Occurrence and removal of lidocaine, tramadol, venlafaxine and their metabolites in German wastewater treatment plants. *Environ. Sci. Pollut. Res.* **2012**, *19*, 689–699.

8. Kasprzyk-Hordern, B.; Dinsdale, R.M.; Guwy, A.J. The ocurrence of pharmaceuticals, personal care products, endocrine disruptors and illicit drugs in surface water in South Wales. *Water Res.* **2008**, *42*, 3498–3518.

9. Peng, X.; Yu, Y.; Tang, C.; Tan, J.; Huang, Q.; Wang, Z. Occurrence of steroid estrogens, endocrine-disrupting phenols, and acid pharmaceutical residues in urban riverine water of the Pearl River Delta, South China. *Sci. Total Environ.* **2008**, *397*, 158–166.

10. Tixier, C.; Singer, H.P.; Oellers, S.; Müller, S. Occurrence and fate of carbamazepine, clofibric acid, diclofenac, ibuprofen, ketoprofen, and naproxen in surface waters. *Environ. Sci. Technol.* **2003**, *37*, 1061–1068.

11. Nentwig, G.; Oetken, O.; Oehlmann, J. Effects of pharmaceuticals on aquatic invertebrates—The example of carbamazepine and clofibric acid. In *Pharmaceuticals in the Environment, Sources, Fate, Effects and Risks*, 2nd ed.; Kümmerer, K., Ed.; Springer-Verlag: Berlin, Germany, 2004; pp. 195–207.

12. Schulte-Oehlmann, U.; Oetken, M.; Bachmann, J.; Oehlmann, J. Effects of ethinyloestradiol and methyltestosterone in prosobranch snails. In *Pharmaceuticals in the Environment, Sources, Fate, Effects and Risks*, 2nd ed.; Kümmerer, K., Ed.; Springer-Verlag: Berlin, Germany, 2004; pp. 233–247.

13. Fong, P.P.; Hoy, C.M. Antidepressants (venlafaxine and citalopram) cause foot detachment from the substrate in freshwater snails at environmentally relevant concentrations. *Mar. Freshw. Behav. Phy.* **2012**, *45*, 145–153.

14. Hollender, J.; Zimmermann, S.G.; Koepke, S.; Krauss, M.; Mcardell, C.S.; Ort, C.; Singer, H.; von Gunten, U.; Siegrist, H. Elimination of organic micropollutants in a municipal wastewater treatment plant upgraded with a full-scale post-ozonation followed by sand filtration. *Environ. Sci. Technol.* **2009**, *43*, 7862–7869.

15. Huber, M.M.; Göbel, A.; Joss, A.; Herrmann, N.; Löffler, D.; MCardell, C.S.; Ried, A.; Siegrist, H.; Ternes, T.A.; von Gunten, U. Oxidation of pharmaceuticals during ozonation of municipal wastewater effluents: A pilot study. *Environ. Sci. Technol.* **2005**, *39*, 4290–4299.

16. Von Gunten, U. Ozonation of drinking water: Part I. Oxidation kinetics and product formation. *Water Res.* **2003**, *37*, 1443–1467.

17. Dosoretz, C.G.; Böddeker, K. Removal of trace organics from water using a pumped bed-membrane bioreactor with powdered activated carbon. *J. Memb. Sci.* **2004**, *239*, 81–90.

18. Li, X.; Hai, F.I.; Nghiem, L.D. Simultaneous activated carbon adsorption within a membrane bioreactor for an enhanced micropollutant removal. *Bioresour. Technol.* **2011**, *102*, 5319–5324.

19. Ternes, T.A.; Joss, A. *Human Pharmaceuticals, Hormones and Fragrances: The Challenge of Micropollutants in Urban Water Management*; IWA Publishing: London, UK, 2006.

20. Bandosz, T.J. Activated carbon surfaces in environmental remediation. In *Interface Science and Technology*; Elservier: Amsterdam, The Netherlands, 2006; Volume 7.

21. Kim, S.H.; Shon, H.K.; Ngo, H.H. Adsorption characteristics of antibiotics trimethoprim on powdered and granular activated carbon. *J. Ind. Eng. Chem.* **2010**, *16*, 344–349.

22. Ternes, T.A.; Meisenheimer, M.; Mcdowell, D.; Sacher, F.; Brauch, H.J.; Haist-Gulde, B.; Preuss, G.; Wilme, U.; Zulei-Seibert, N. Removal of pharmaceuticals during drinking water treatment. *Environ. Sci. Technol.* **2002**, *36*, 3855–3863.

23. Westerhoff, P.; Yoon, Y.; Snyder, S.; Wert, E. Fate of endocrine-disruptor, pharmaceutical, and personal care product chemicals during simulated drinking water treatment processes. *Environ. Sci. Technol.* **2005**, *39*, 6649–6663.

24. Carvalho, A.P.; Mestre, A.S.; Haro, M.; Ania, C.O. Advanced Methods for the removal of acetaminophen from water. In *Acetaminophen, properties, clinical uses and adverse effects*; Javaherian, A., Latifpour, P., Eds.; Nova Science Publishers Inc.: Hauppauge, New York, NY, USA, 2012.

25. Ania, C.O.; Pelayo, J.G.; Bandosz, T.J. Reactive adsorption of penicillin in activated carbons. *Adsorption* **2011**, *17*, 421–429.

26. Rúa-Gómez, P.C.; Püttmann, W. Impact of wastewater treatment plant discharge of lidocaine, tramadol, venlafaxine and their metabolites on the quality of surface waters and groundwater. *J. Environ. Monitor.* **2012**, *14*, 1391–1399.

27. Lajaunesse, A.; Gagnon, C.; Sauvé, S. Determination of basic antidepressants and their *n*-desmethyl metabolites in raw sewage and wastewater using solid phase extraction and liquid chromatography-tandem mass spectrometry. *Anal. Chem.* **2008**, *80*, 5325–5333.

28. Schultz, M.M.; Furlong, E.T.; Kolpin, D.W.; Werner, S.L.; Schoenfuss, H.L.; Barber, L.B.; Blazer, V.S.; Norris, D.O.; Vadja, A.M. Antidepressant pharmaceuticals in two U.S. effluent-impacted streams: occurrence and fate in water and sediment, and selective uptake in fish neural tissue. *Environ. Sci. Technol.* **2010**, *44*, 1918–1925.

29. Reungoat, J.; Macova, M.; Escher, B.I.; Carswell, S.; Mueller, J.F.; Keller, J. Removal of micropollutants and reduction of biological activity in a full scale reclamation plant using ozonation and activated carbon filtration. *Water Res.* **2010**, *44*, 625–637.

30. Lee, C.O.; Howe, K.J.; Thomson, B.M. Ozone and biofiltration as an alternative to reverse osmosis for removing PPCPs and micropollutants from treated wastewater. *Water Res.* **2012**, *46*, 1005–1014.

31. Dubinin, M.M. Microporous structures of carbonaceous adsorbents. In *Characterization of Porous Solids*; Greeg, S.J., Sing, K.S.W., Stoecki, H.F., Eds.; Society of Chemical Industry: London, UK, 1979.

32. German Institute for Standardization. *Limits of Detection, Identification and Quantitation* (in German); DIN 32645; German Institute for Standardization: Berlin, Germany, 1994.

33. German Federal Ministry of Justice. *Regulation on Requirements for the Discharge of Wastewater into Surface Waters* (in German); German Federal Ministry of Justice: Berlin, Germany, 1997.

34. Cleuvers, M. Mixture toxicity of the anti-inflammatory drugs diclofenac, ibuprofen, naproxen, and acetylsalicylic acid. *Ecotoxicol. Environ. Saf.* **2004**, *59*, 309–315.

35. Siegrist, H.; Joss, A.; Alder, A.; Gobel, A.; Keller, E.; McArdell, C.; Ternes, T.A. Micropollutants—New requirements for wastewater treatment (in German). *EAWAG News* **2003**, *57*, 7–10.

36. Larsen, T.A.; Lienert, J.; Joss, A.; Siegrist, H. How to avoid pharmaceuticals in the aquatic environment. *J. Biotechnol.* **2004**, *113*, 295–304.

37. Jones, O.A.H.; Voulvoulis, N.; Lester, J.N. The occurrence and removal of selected pharmaceutical compounds in a sewage treatment works utilising activated sludge treatment. *Environ. Pollut.* **2007**, *145*, 738–744.

38. Gomez, M.J.; Bueno, M.J.M.; Lacorte, S.; Fernandez-Alba, A.R.; Aguera, A. Pilot survey monitoring pharmaceuticals and related compounds in a sewage treatment plant located on the Mediterranean coast. *Chemosphere* **2007**, *66*, 993–1002.

39. Wang, L.; Govind, R. Sorption of toxic organic compounds on wastewater solids: Mechanism and modeling. *Environ. Sci. Technol.* **1993**, *27*, 152–158.

40. Ania, C.O.; Bandosz, T.J. Importance of structural and chemical heterogeneity of activated carbon surfaces for adsorption of dibenzothiophene. *Langmuir* **2005**, *21*, 7752–7759.

41. Pipe-Martin, C.; Reungoat, J.; Keller, J. *Dissolved Organic Carbon Removal by Biological Treatment*; CRC for Water Quality Research Australia: Adelaide, Australia, 2010.

42. Nguyen, L.N.; Hai, F.I.; Kang, J.; Price, W.E.; Nghiem, L.D. Removal of trace organic contaminants by a membrane bioreactor-granular activated carbon (MBR-GAC) system. *Bioresour. Technol.* **2011**, *113*, 169–173.

43. Grünebaum, T. *Final Report Phase 1: Elimination of Drug Residues in Sewage Treatment Plants* (in German); Ministry for Climate Protection, Environment, Agriculture, Nature Conservation and Consumer Protection of the German State of North Rhine-Westphalia: Düsseldorf, Germany, 2011.

44. Schmidt, C.K.; Brauch, H.-J. N,N-dimethylsulfamide as precursor for N-nitrosodimethylamine (NDMA) formation upon ozonation and its fate during drinking water treatment. *Environ. Sci. Technol.* **2008**, *42*, 6340–6346.

45. SRC PhysProp Database. Available online: http://www.syrres.com/what-we-do/databaseforms.aspx?id=386 (accessed on 3 September 2012).

46. Tzvetkov, M.V.; Saadatmand, A.R.; Lötsch, J.; Tegeder, I.; Stingl, J.C.; Brockmöller, J. Genetically polymorphic OCT1: Another piece in the puzzle of the variable pharmacokinetics and pharmacodynamics of the opioidergic drug tramadol. *Clin. Pharmacol. Ther.* **2011**, *90*, 143–150.

47. Ellingrod, V.L.; Perry, P.J. Venlafaxine: A heterocyclic antidepressant. *Am. J. Hosp. Pharm.* **1994**, *51*, 3033–3046.

48. *Molconvert*, version 5.8.2; Molecule File Conversion with MolConverter; ChemAxon: Budapest, Hungary, 2012.

49. Babić, S.; Horvat, A.J.M.; Pavlović, D.M.; Kaštelan-Macan, M. Determination of pKa values of active pharmaceutical ingredients. *Trends Anal. Chem.* **2007**, *26*, 1043–1061.

50. Cabrita, I.; Ruiz, B.; Mestre, A.S.; Fonseca, I.M.; Carvalho, A.P.; Ania, C.O. Removal of an analgesic using activated carbons prepared from urban and industrial residues. *Chem. Eng. J.* **2010**, *163*, 249–255.

51. Ng, A.S.; Stenstrom, M.K. Nitrification in powdered activated carbon-activated sludge process. *J. Environ. Eng.* **1987**, *113*, 1285–1301.

52. Serrano, D.; Suárez, S.; Lema, J.M.; Omil, F. Removal of persistent pharmaceutical micropollutants from sewage by addition of PAC in a sequential membrane bioreactor. *Water Res.* **2011**, *45*, 5323–5333.

Water **2013**, *5*, 1101-1115; doi:10.3390/w5031101

OPEN ACCESS

water

ISSN 2073-4441
www.mdpi.com/journal/water

Review

Synergistic Water-Treatment Reactors Using a TiO$_2$-Modified Ti-Mesh Filter

Tsuyoshi Ochiai [1,2,*], **Ken Masuko** [2,3], **Shoko Tago** [1], **Ryuichi Nakano** [4], **Kazuya Nakata** [2], **Masayuki Hara** [1], **Yasuhiro Nojima** [5], **Tomonori Suzuki** [2,3], **Masahiko Ikekita** [2,3], **Yuko Morito** [2,6], and **Akira Fujishima** [1,2]

[1] Kanagawa Academy of Science and Technology, KSP East 407, 3-2-1 Sakado, Takatsu-ku, Kawasaki, Kanagawa 213-0012, Japan; E-Mails: pg-tago@newkast.or.jp (S.T.); pg-hara@newkast.or.jp (M.H.)

[2] Photocatalysis International Research Center, Tokyo University of Science, 2641 Yamazaki, Noda, Chiba 278-8510, Japan; E-Mails: nakata@rs.tus.ac.jp (K.N.); fujishima_akira@admin.tus.ac.jp (A.F.)

[3] Department of Applied Biological Science, Tokyo University of Science, Yamazaki 2641, Noda, Chiba 278-8510, Japan; E-Mails: j6412647@ed.tus.ac.jp (K.M.); chijun@rs.noda.tus.ac.jp (T.S.); masalab@rs.noda.tus.ac.jp (M.I.)

[4] Department of Microbiology and Immunology, Teikyo University School of Medicine, 2-11-1 Kaga, Itabashi-ku, Tokyo 173-8605, Japan; E-Mail: nakano@med.teikyo-u.ac.jp

[5] Kitasato Research Center for Environmental Science, 1-15-1, Kitasato, Minami-ku, Sagamihara, Kanagawa 252-0329, Japan; E-Mail: nojima@kitasato-e.or.jp

[6] U-VIX Corporation, 2-14-8 Midorigaoka, Meguro-ku, Tokyo 152-0034, Japan; E-Mail: y.morito@u-vix.com

* Author to whom correspondence should be addressed; E-Mail: pg-ochiai@newkast.or.jp; Tel.: +81-44-819-2040; Fax: +81-44-819-2070.

Received: 26 April 2013; in revised form: 5 June 2013 / Accepted: 11 July 2013 / Published: 22 July 2013

Abstract: The recent applications of a TiO$_2$-modified Ti-mesh filter (TMiP™) for water purification are summarized with newly collected data including biological assays as well as sewage water treatment. The water purification reactors consist of the combination of a TMiP, a UV lamp, an excimer VUV lamp, and an ozonation unit. The water purification abilities of the reactor were evaluated by decomposition of organic contaminants, inactivation of waterborne pathogens, and treatment efficiency for sewage water. The UV-C/TMiP/O$_3$ reactor disinfected *E. coli* in aqueous suspension in approximately 1 min completely, and also decreased the number of *E. coli* in sewage water in 15 min dramatically. The observed rate constants of 7.5 L/min and 1.3 L/min were calculated by pseudo-first-order kinetic analysis respectively. Although organic substances in sewage water were supposed to prevent the UV-C/TMiP/O_3 reactor from purifying water, the reactor reduced *E. coli* in sewage water continuously. On the other hand, although much higher efficiencies for decomposition of organic pollutants in water were achieved in the excimer/TMiP reactor, the disinfection activity of the reactor for waterborne pathogens was not as effective as the other reactors. The difference of efficiency between organic pollutants and waterborne pathogens in the excimer/TMiP reactor may be due to the size, the structure, and the decomposition mechanism of the organic pollutants and waterborne pathogens. These results show that a suitable system assisted by synergy of photocatalysts and other technologies such as ozonation has a huge potential as a practical wastewater purification system.

Keywords: photocatalysis; TiO_2-modified Ti-mesh filter; ozonation; excimer lamp; advanced oxidation processes; sewage water treatment

---

1. Introduction

Photocatalytic environmental purification, especially wastewater treatment, has received intensive consideration on cost and enduring stability [1–4]. However, popularly used photocatalyst and photocatalytic filters significantly limit its application because of relatively low purification efficiency [5,6] and difficulty in handling the powder [7–9]. Thus, although many researchers have been working on photocatalytic water purification, it could not be developed to the stage of effective real industrial technology because of the difficulty in fabricating a practical water purifier. Recently we have developed an easy-to-handle photocatalytic filter material, TiO_2 nanoparticles modified titanium mesh (Titanium mesh impregnated photocatalyst, TMiP™), and its applications for environmental purification [4,10–16]. Due to the highly-ordered three-dimensional structure modified with TiO_2 nanoparticles, TMiP provides excellent breathability for both air and water while maintaining a high level of surface contact. Its high flexibility and mechanical stability allow the design of any shape for the unit and for use in relatively severe situations such as inner plasma [11,15]. Based on these results, we summarize in this paper the applications of TMiP for water purification with newly collected data including biological assay and sewage water treatment. The purification abilities of the reactors were evaluated by decomposition of organic contaminants, inactivation of waterborne pathogens, and treatment of sewage water.

2. Materials and Methods

2.1. General Methods

In our experiments, all the reagents were analytical grade and used without further purification. All solutions were made from Milli-Q ultrapure water. The irradiation was provided by the 10 W BLB lamp (FL10BLB, Toshiba, λ_{max} = 365 nm) or the 18 W UV-C lamp (ZW18D15Y-Z356, Cnlight,

λ_{max} = 254 nm). UV intensity was measured with UV power meter C9536/H9535-254 for 254 nm (Hamamatsu Photonics, Hamamatsu, Japan), UV power meter C9536-01/H9958 for 310–380 nm (Hamamatsu Photonics), and UV RADIO METER UV-M03A with UV-SN31 sensor head (ORC Manufacturing, Tokyo, Japan). An ozone gas stream was generated from oxygen gas with a concentration of 0.5 to 10 mg/L by a corona-discharge ozone generator (ED-OG-AP1, Ecodesign Inc., Tokyo, Japan). The concentration of dissolved O_3 was measured with a Digital pack test for O_3 (Kyoritsu Chemical-Check. Lab., Corp, Tokyo, Japan). The initial concentration of dissolved methylene blue (MB) was controlled to 40 μM and the MB concentration as a function of treatment time was measured with a UV–visible spectrophotometer at 660 nm (UV-2450, Shimadzu, Kyoto, Japan). The concentration of phenol was measured by HPLC using the previously reported method [17]. Suspended solids (SS) and PtCo Color of sewage water samples were measured by spectrophotometer (DR5000, HACH Company, Loveland, CO, USA). All experiments were carried out at room temperature and atmospheric pressure.

Escherichia coli NBRC13965 (*E. coli*), *Legionella pneumophila* ATCC49249 (*L. pneumophila*), Qβ phage NBRC20012 (Qβ), and *feline calicivirus* F-9 ATCC VR-782 (FCV) were used as the main test waterborne pathogens to assess the biological purification efficiency of the units. *E. coli* and Qβ were obtained from the Biological Resource Center of the National Institute of Technology and Evaluation (Chiba, Japan). *L. pneumophila* and FCV were obtained from American Type Culture Collection (Manassas, VA, USA). *E. coli*, *L. pneumophila*, and Qβ were propagated and assayed by previously described methods [14,18–20]. FCV was propagated and assayed by the plaque technique on confluent layers of Crandell-Reese feline kidney cell cultures grown in 12-well culture plates as described elsewhere [21]. Standard Plate Count (SPC), Total Coliform (TC), and *E. coli* in sewage water sample were also assayed by previously described methods [17]. The SPC is usually reported as the number of all of the bacteria per milliliter of sample. There are no drinking water standards for SPC, but if more than 500 bacteria are counted in one milliliter of sample, further testing for TC or *E.coli* is suggested. The TC is reported as the number of a whole group of the coliform bacteria which can cause and indicate potential health problems. *E. coli* is considered to be the major species of coliform bacteria that is the best indicator of fecal pollution and the possible presence of pathogens.

2.2. Fabrication of TMiP

The detailed preparation procedure and characterization of TMiP were described in a previous report [10]. The procedure is briefly shown in Figure 1. The Ti-mesh, obtained by controlled chemical etching of 0.2 mm^t titanium foil, was anodized at a voltage of 70 V to give a violet color in acid solution. The resulting structure was controlled by mask pattern and etching time as shown in Figure 1 and the colors were able to be controlled by anodizing voltage and time [13]. Then the Ti-mesh was treated at 550 °C for 3 h to produce a TiO_2 layer on the Ti-mesh surface. The treated Ti-mesh was dip-coated with 25 wt % of TiO_2 anatase sol (TKD-701, TAYCA, Osaka, Japan) and was heated at 550 °C for 3 h. The structural and surface morphology of TMiP were examined and studied by using X-ray diffraction and Scanning Electron Microscopy.

Figure 1. Fabrication method of TiO$_2$ nanoparticles modified titanium mesh (TMiP™). Reproduced with permission from Ochiai *et al.* [10], Catalysis Science and Technology; published by RSC Publishing, 2011.

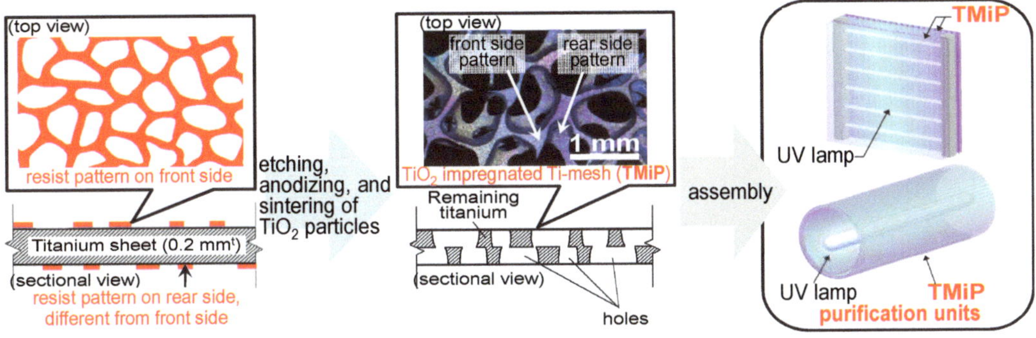

2.3. Fabrication of the Water Purification System Using TMiP

Figure 2a shows schematic illustrations of water-purification reactors. The reactors consist of an acrylic tube (49 mm i.d. × 400 mm length) with two ports, TMiP with UV lamps [BLB lamp (10 W, 6.0 mW cm^{-2} at 310–380 nm), UV-C lamp (10 W, 10 mW cm^{-2} at 254 nm), or excimer VUV lamp (7.2 mW cm^{-2} at 172 nm)], and a bubbling unit (air or ozone). For comparison of the effect of UV wavelength and air or O$_3$ bubbling, the combinations of the reactors were investigated [e.g., BLB/TMiP, BLB/TMiP/O$_3$, O$_3$ bubbling alone, and the excimer lamp unit wrapped with Ti-mesh without TiO$_2$ nanoparticles (excimer alone unit)]. O$_3$ dose was varied at 0.5–1.0 mg/L or 10 mg/L. The basic design and fabrication method of the excimer/TMiP reactor were described previously [12,16]. When the alternating current (AC) high voltage is applied to the electrodes, the dielectric barrier discharge occurs in the quartz tube which provides intense narrow band radiation at 172 nm from a xenon excimer (Xe$_2$*) [22]. UV intensity was measured by UV power meters as mentioned in Section 2.1. Figure 2b shows the schematic illustration of a water purification system which consists of the water-purification reactors shown in Figure 2a, a reservoir, a pump, and an O$_3$ production unit or an air pump. An aqueous solution containing organic contaminants or waterborne pathogens was circulated through the tube by the pump and was treated in the reactor.

2.4. Test Method for Evaluation of the Water Purification Ability of the System

Water purification ability of the units was evaluated by tests of MB decolorization, phenol decomposition, and inactivation of waterborne pathogens. In a typical run, one liter of an aqueous solution containing 40 µM MB was circulated through the reactor by the pump at a flow rate of 100 mL/min. The concentration of dissolved MB as a function of treatment time was measured with a UV–visible spectrophotometer. Similarly, 1 L of an aqueous suspension of *E. coli* was used as the biologically contaminated water sample and was circulated through the reactor by the pump at a flow rate of 1 L/min. The viability of waterborne pathogens in the artificially contaminated water samples was analyzed with previously reported methods [14,18–21].

Figure 2. Schematic illustrations of (**a**) water-purification reactor and (**b**) water-purification system. Reproduced with permission from Ochiai *et al.* [10,14,16], Catalysis Science and Technology; published by RSC Publishing, 2011.

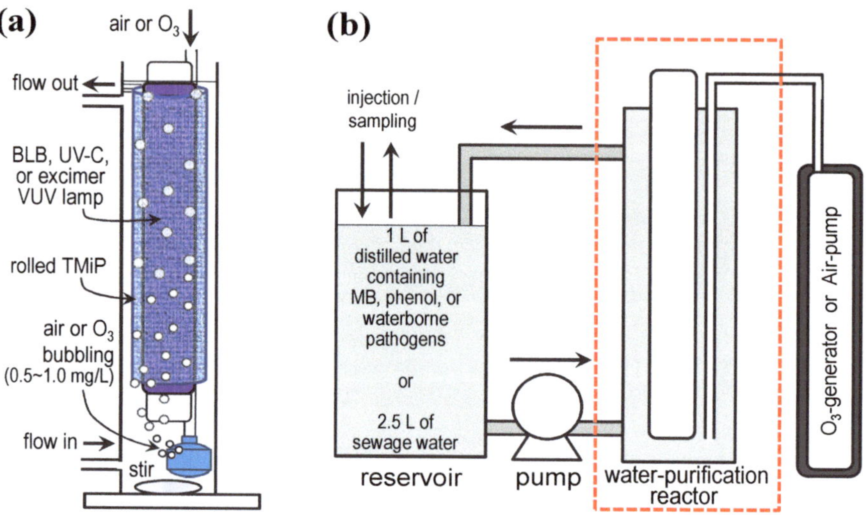

In this paper, a sewage water purification test was introduced as an experiment for practical use. The sewage water samples were collected from a sewage treatment plant which has a primary unit, two secondary units, and a tertiary treatment unit. The primary treatment unit consists of a mechanical screen, a grit removal tank, and a primary clarifier. The sewage treatment plant has two secondary treatment units in parallel; both units perform anoxic–oxic–anoxic–oxic-based biological nutrient removal. The main difference between them is the source of oxygen in the aeration tank: ambient air in unit I and pure oxygen in unit II. Units I and II are followed by ozonation as the tertiary treatment. To investigate the purification ability of the system for usage as an alternative for the tertiary treatment unit, sewage water samples were collected in 10 L plastic tank from the sewage treatment plant at the inlet of the tertiary treatment unit on 5 September 2011 and 18 July 2012. After addition of *E. coli* suspension to control initial viability of bacteria to approximately 10^6 Colony Forming Unit (CFU)/mL, the sewage water sample (2.5 L) was circulated through the reactor by the pump at a flow rate of 1 L/min. The viability of bacteria in the sewage water samples was analyzed with the above mentioned methods.

3. Results and Discussion

3.1. Water-Purification Ability of the Reactors Evaluated by Methylene Blue Decolorization

Figure 3 shows MB decolorization without any reactors (blank, crosses), by the BLB/TMiP reactor (open triangles), and by the BLB/TMiP/air reactor (solid diamonds). The MB concentrations were well fitted with a pseudo-first-order kinetics given by the following equation: $C = C_o \exp(-k_1 t)$. Where C_o is the initial MB concentration and k_1 is the observed rate constant. The values of k_1 were calculated by exponential fitting of Figure 3 to 0.60 and 0.82 h^{-1} for the BLB/TMiP and the BLB/TMiP/air reactors, respectively. Interestingly, the BLB/TMiP/air reactor showed a higher decolorization rate than the BLB/TMiP reactor. Tasbihi *et al.* reported that dissolved oxygen could improve the efficiency of the degradation of organics by enhancing the separation of photogenerated electron-hole pairs, thereby increasing ·OH concentration [23]. Thus, the excellent accessibility of the TMiP structure enhanced the air bubbling effect and resulted in higher decolorization. Evidence for this is shown in a previous report by comparison with TiO_2-modified commercial Ti-mesh (200 × 200 mm, 0.30 mmφ, 20 mesh, Nilaco) [10]. Interestingly, the SEM images of TiO_2-modified commercial Ti-mesh did not show the well dispersed spherical particles of TiO_2 on the surface. This may be caused by the difference between the good morphology of TMiP and the monotonous structure of commercial Ti-mesh. The BLB/TMiP/O_3 reactor was also evaluated by the MB decolorization test; however, the MB color suddenly disappeared due to extreme oxidation activity of the reactor. The water-purification ability of the BLB/TMiP/O_3 reactor is discussed in the next paragraph by comparison with the ability of O_3 bubbling alone.

Figure 3. Methylene blue (MB) decolorization by the water-purification reactors. Crosses: blank; open triangles: BLB/TMiP reactor; solid diamonds: BLB/TMiP/air reactor. Reproduced with permission from Ochiai *et al.* [10], Catalysis Science and Technology; published by RSC Publishing, 2011.

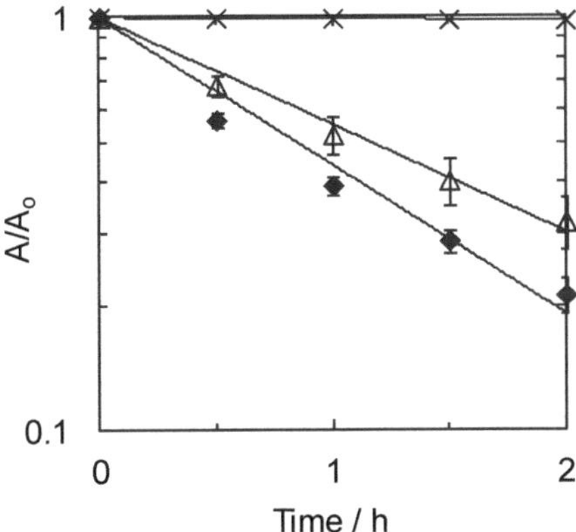

Figure 4 shows the phenol decomposition by BLB/TMiP (open triangles), O_3 bubbling alone (open squares), BLB/TMiP/O_3 (solid triangles), excimer alone (asterisks), and excimer/TMiP (open diamonds) reactors. The phenol concentrations were also well fitted with a pseudo-first-order kinetics in all reactors. The values of k_1 for the reactors were calculated and are indicated in Figure 4. It can be seen that phenol was effectively decomposed by the BLB/TMiP/O_3 reactor, compared with O_3 bubbling alone or the BLB/TMiP reactors. Interestingly, although without O_3 bubbling, the excimer/TMiP reactor shows the highest k_1 among reactors. We ascribed the results to reactive species and oxidative intermediates generated by VUV photolysis of water as mentioned by Oppenländer *et al.* [24–28]. Hydroxyl radicals and other reactive species are formed by VUV photolysis of water and can directly react with organics due to their strong oxidation potentials. As a result, the excimer/TMiP reactor is much more efficient than the other reactors.

Figure 4. Phenol decomposition with exponential curve fitting for the water-purification reactors. Open triangles: BLB/TMiP reactor; open squares: O_3 bubbling alone (0.5–1.0 mg/L); solid triangles: BLB/TMiP/O_3 reactor (0.5–1.0 mg/L); asterisks: excimer alone; open diamonds: excimer/TMiP reactor. Reproduced with permission from Ochiai *et al.* [16], Chemical Engineering Journal; published by Elsevier, 2013.

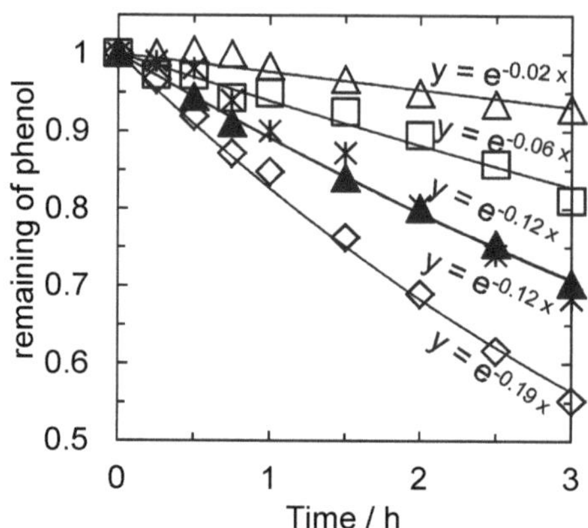

3.2. Water-Purification Ability of the Reactors Evaluated by the Inactivation of Waterborne Pathogens

Figure 5 shows the time courses of the log number of *E. coli* (a), *L. pneumophila* (b), Qβ (c), and FCV (d) in the water purification system with the O_3 bubbling alone (open squares), BLB/TMiP/O_3 (solid triangles), and excimer/TMiP (open diamonds) reactors. The disinfection activity of the BLB/TMiP/O_3 reactor is higher than the O_3 alone condition for *E. coli*, and *L. pneumophila*. On the other hand, the activity of the BLB/TMiP/O_3 reactor was not so effective on comparison with the BLB/TMiP reactor for Qβ and FCV. Moreover, disinfection activity of the excimer/TMiP was lower than the other reactors especially for *E. coli*. The difference in the activity of the reactors among the waterborne pathogens may be due to the size of molecules or bacteria, the surface composition of bacteria or viruses, the critical wavelength for disinfection, and the permeation ability of reactive species. Oppenländer *et al.* also reported VUV-induced oxidation of organics in homogeneous aqueous solution within a xenon-excimer flow-through photoreactor [24]. They found that the decomposition rate was strongly influenced by the size and the structure of the organics, e.g., the homologous series of saturated alcohols C_1–C_8 was decomposed in descending order of the TOC degraded after an irradiation time of 3 h. In this series 1-octanol was decomposed with the lowest efficiency because of the highest statistical possibility of formation of intermediate products. In addition, the path that 172 nm VUV light penetrates the water is very short due to the high absorption coefficients of water [29,30]. Therefore, no matter what oxidant was produced by VUV photolysis, it would not be consumed by the purification process due to the long distance (several millimeters) [31]. On the other hand, Cho *et al.* reported the difference of efficiency among waterborne pathogens by reactive oxygen species [32–37]. These studies indicate that bacteria could be inactivated by all forms of reactive oxygen species produced by photocatalysis, while viruses demonstrate notably higher resistance to photocatalytic inactivation. Since the structures of Qβ and FCV, compared to *E. coli* and *L. pneumophila*, are simpler, Qβ and FCV were susceptible to •OH and were not very susceptible to less reactive species such as $O_2^{\cdot-}$. In addition, the germicidal effect of UV-C is critical for disinfection [38,39]. On the contrary, VUV and UV-A are less important for it than the UV-C [40]. Our data clearly show this tendency.

Figure 5. Time courses of log number of (**a**) *E. coli*; (**b**) *L. pneumophila*; (**c**) Qβ; and (**d**) FCV in the water purification system. Open squares: O_3 bubbling alone (0.5–1.0 mg/L); solid triangles: BLB/TMiP/O_3 reactor (0.5–1.0 mg/L); open diamonds: excimer/TMiP reactor.

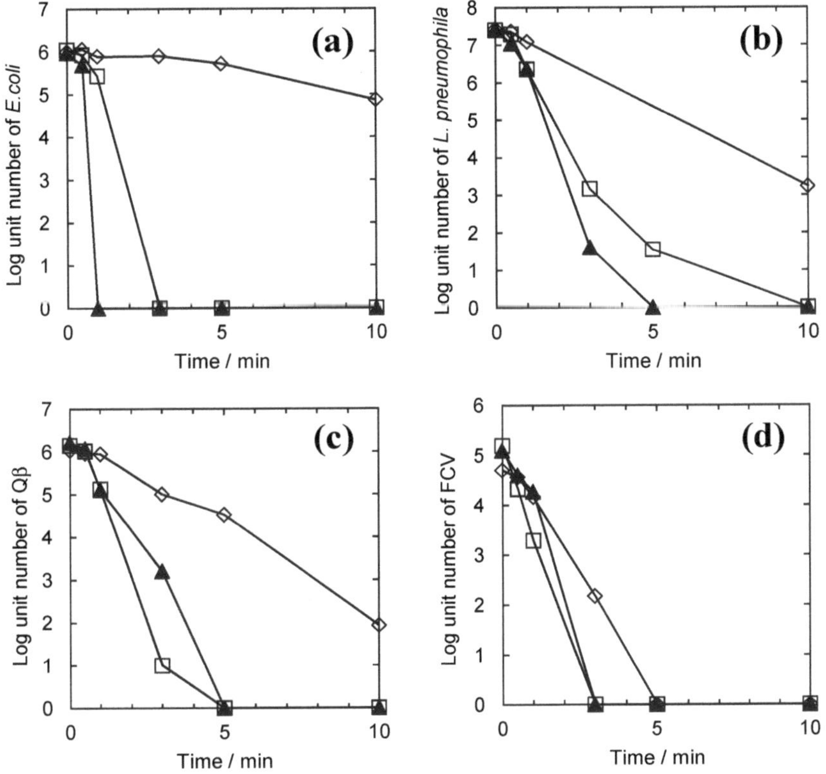

3.3. Water-Purification Ability of the Reactors Using the Higher Concentration of O_3 and UV-C

The result of the inactivation test by using the higher concentration of O_3 and UV-C shows the above-mentioned tendency more clearly (Figure 6). By using the UV-C/TMiP reactor, more than 5 min of treatment time is required for total inactivation of *E. coli* (Figure 6a, open circles). In contrast, 0.5–1.0 mg/L of O_3 bubbling (Figure 6a, open squares) takes a shorter treatment time (less than 3 min) for the inactivation and there is no difference between 10 mg/L of O_3 bubbling (Figure 6a, solid squares) and UV-C/TMiP/O_3 (10 mg/L) reactor (Figure 6a, solid circles, overlapped on solid squares). On the other hand, Qβ was not inactivated efficiently by the UV-C/TMiP reactor (Figure 6b, open circles) and 0.5–1.0 mg/L of O_3 bubbling (Figure 6b, open squares). However, with 10 mg/L of O_3 bubbling (Figure 6b, solid squares) the UV-C/TMiP/O_3 (10 mg/L) reactor (Figure 6b, solid circles) inactivated Qβ efficiently (less than 1 min).

Figure 6. Time course of log survival rate of (**a**) *E. coli*; and (**b**) Qβ in the water purification system. open circles: UV-C/TMiP reactor; open squares: O_3 alone (0.5–1.0 mg/L); solid squares: O_3 alone (10 mg/L); solid circles: UV-C/TMiP/O_3 reactor (10 mg/L). Reproduced with permission from Ochiai *et al.* [14], Catalysis Science and Technology; published by RSC Publishing, 2011.

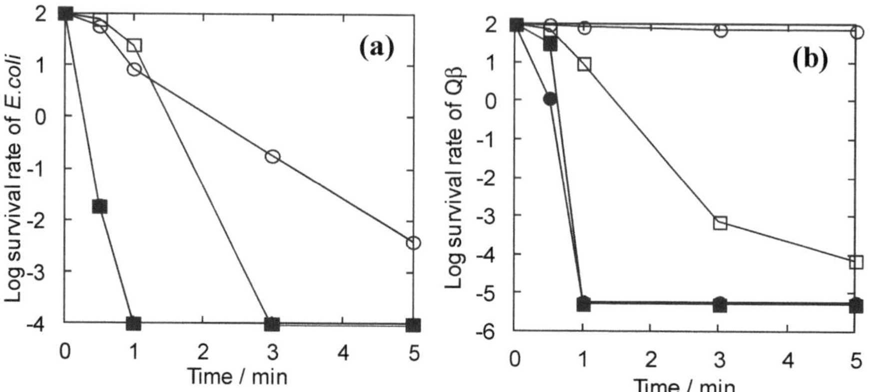

Similarly, the time course of phenol concentration in the water purification system with the UV-C/TMiP/O_3 (10 mg/L) reactor (Figure 7, solid circles) is larger than with the UV-C/TMiP reactor (Figure 7, open circles). This result indicates that the UV-C/TMiP/O_3 reactor was able to decompose phenol more efficiently than the UV-C/TMiP reactor. Here O_3 is a good acceptor of excited electrons the same as O_2. Thus, •OH production and photocatalytic oxidation are enhanced by the presence of O_3, which prevents carrier recombination in photocatalysis [41–44]. Moreover, O_3^- can oxidize organics with a relatively long lifetime [45,46]. On the other hand, there are many reports about photolytic/catalytic decomposition of O_3 [47–50]. Highly oxidative intermediates such as atomic oxygen could be produced from O_3 by UV-C irradiation and/or the presence of heterogeneous catalyst surfaces. In this case, UV-C irradiation and the large specific surface area of TMiP could decompose O_3 effectively [11].

Figure 7. Phenol decomposition by the water-purification system. open circles: UV-C/TMiP reactor; solid circles: UV-C/TMiP/O_3 reactor (10 mg/L). Reproduced with permission from Ochiai *et al.* [16], Chemical Engineering Journal; published by Elsevier, 2013.

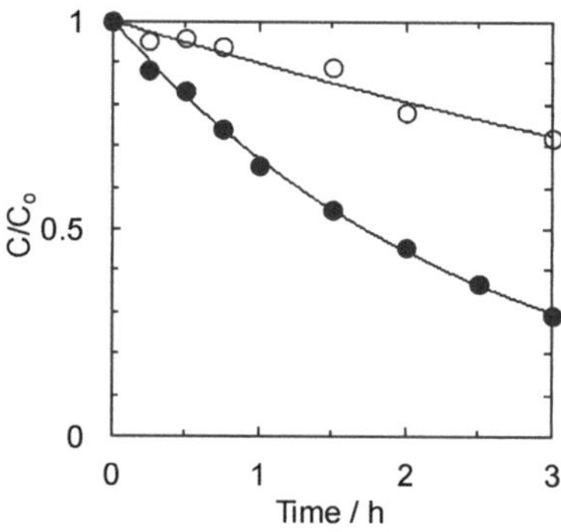

3.4. Treatment of Sewage Water Samples

Figure 8 shows the time course of SS and PtCo Color of the sewage water samples in the water-purification systems. There was almost no difference between the UV-C/TMiP/O$_3$ reactor (10 mg/L, solid circles) and the O$_3$ bubbling alone condition (10 mg/L, solid squares) for reduction of both SS (Figure 8a) and PtCo Color (Figure 8b). It is known that these factors ultimately decrease the photocatalytic efficiency of water treatment [3,51,52]. On the other hand, adding ozone to the water purification system results in an overall improvement in water quality due to more complete oxidation of color, organics, and suspended solids [53,54]. At the same time, ozone-induced microflocculation can reduce SS [55]. The present data well support these phenomena. However, interestingly, the time courses of survival rate of SPC (Figure 9a), TC (Figure 9b), and *E. coli* (Figure 9c) in the water purification systems show that the disinfection activity of the UV-C/TMiP/O$_3$ reactor is higher than the O$_3$ bubbling alone condition. These results were also ascribed to the synergistic effect of photocatalysis and ozonation as mentioned in Section 3.3. In addition, the germicidal effect of UV-C with the formation of pyrimidine dimers in the DNA is critical for disinfection [38,39]. On the contrary, VUV and UV-A are less important for sterilization than the UV-C [40]. Our data clearly show this tendency. However, the calculated rate constant, k_1, for pseudo-first-order kinetics of *E. coli* disinfection was 1.3 L/min (normalized by sample volume) in the treatment of a sewage water sample. This value is approximately 18% of the value for the kinetics of *E. coli* disinfection in aqueous suspension (7.5 L/min, calculated from Figure 6a). This result suggests that refractory organic substances such as humic substances or SS may affect the rate of disinfection [51,52].

Figure 8. Time course of (**a**) SS and (**b**) PtCo Color of the sewage water in the water-purification system. solid squares: O$_3$ bubbling (10 mg/L); solid circles: UV-C/TMiP/O$_3$ reactor (10 mg/L).

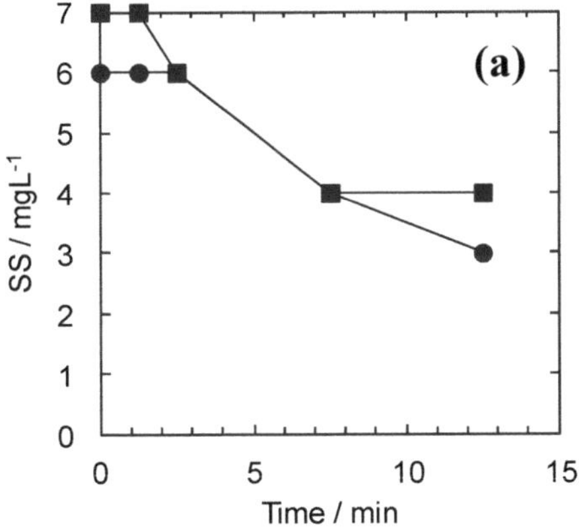
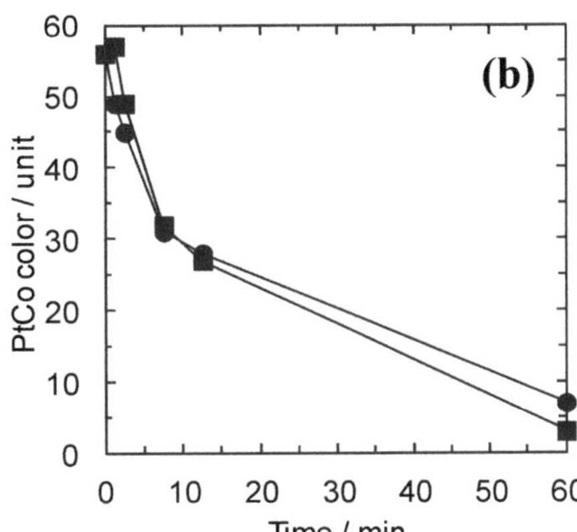

Figure 9. Time course of survival rate of (**a**) SPC; (**b**) TC; and (**c**) *E. coli* in the water purification system. Solid squares: O_3 bubbling (10 mg/L); solid circles: UV-C/TMiP/O_3 reactor (10 mg/L).

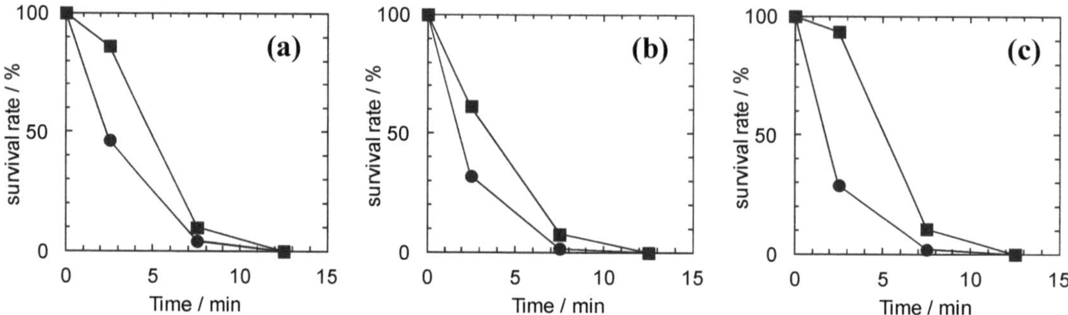

4. Conclusions

The high efficiency for decomposition of chemical and biological contaminants results from the highly-ordered three dimensional structure of TMiP and the synergistic effect of photocatalysis and other technologies such as ozonation or excimer VUV lamp. The excimer-lamp-assisted photocatalysis (excimer/TMiP unit) achieved high efficiency for decomposition of dissolved phenol in water compared with the photolysis (excimer alone unit), ozonation (O_3 alone unit), photocatalysis (BLB/TMiP and UV-C/TMiP units), and ozone-assisted photocatalysis (BLB/TMiP + O_3 unit). On the other hand, the disinfection activity of the excimer/TMiP unit for waterborne pathogens was not so effective compared with the other units. The reaction mechanism in this research can be formulated as follows. Photochemically generated •OH which is the dominant reactive species in the excimer/TMiP unit could decompose phenol and viruses efficiently. On the contrary, because of the difference of the size and the decomposition mechanism, it may be difficult to decompose the bacterial cell wall by an excimer/TMiP unit compared with phenol and viruses. Ozone could attack the cell wall and the surface components of bacteria and then disrupt its integrity by its extremely strong oxidation potential. Therefore, O_3 alone and BLB/TMiP + O_3 units could inactivate bacteria efficiently compared with an excimer/TMiP unit.

Acknowledgments

The work was supported in part by the Kurata Memorial Hitachi Science and Technology Foundation. The authors are grateful to Hayato Nanba, Toru Hoshi, Yoshihiro Koide (Kanagawa University), Touko Nakagawa (Tokyo University of Science), Etsuko T. Utagawa (National Institute of Infectious Diseases), Yasuji Niitsu, Go Kobayashi, Masahiro Kurano, Izumi Serizawa (ORC Manufacturing Co., Ltd.), Koji Horio, Koji Mori, and Hiro Tatejima (U-VIX Corporation) for the experiments, the helpful advice and discussions, and the supervisions.

Conflict of Interest

The authors declare no conflict of interest.

References

1. Fujishima, A.; Honda, K. Electrochemical photolysis of water at a semiconductor electrode. *Nature* **1972**, *238*, 37–38.
2. Fujishima, A.; Zhang, X.; Tryk, D.A. TiO$_2$ photocatalysis and related surface phenomena. *Surf. Sci. Rep.* **2008**, *63*, 515–582.
3. Chong, M.N.; Jin, B.; Chow, C.W.K.; Saint, C. Recent developments in photocatalytic water treatment technology: A review. *Water Res.* **2010**, *44*, 2997–3027.
4. Ochiai, T.; Fujishima, A. Photoelectrochemical properties of TiO$_2$ photocatalyst and its applications for environmental purification. *J. Photochem. Photobiol. C* **2012**, *13*, 247–262.
5. Ollis, D.F. Photocatalytic purification and remediation of contaminated air and water. *C. R. Acad. Sci. II C* **2000**, *3*, 405–411.
6. Ollis, D.F.; Pelizzetti, E.; Serpone, N. Photocatalyzed destruction of water contaminants. *Environ. Sci. Technol.* **1991**, *25*, 1522–1529.
7. Prieto-Rodriguez, L.; Miralles-Cuevas, S.; Oller, I.; Agüera, A.; Puma, G.L.; Malato, S. Treatment of emerging contaminants in wastewater treatment plants (WWTP) effluents by solar photocatalysis using low TiO$_2$ concentrations. *J. Hazard. Mater.* **2012**, *211–212*, 131–137.
8. Chen, M.; Chu, W. Degradation of antibiotic norfloxacin in aqueous solution by visible-light-mediated C-TiO$_2$ photocatalysis. *J. Hazard. Mater.* **2012**, *219–220*, 183–189.
9. Sánchez, L.; Peral, J.; Domènech, X. Aniline degradation by combined photocatalysis and ozonation. *Appl. Catal. B Environ.* **1998**, *19*, 59–65.
10. Ochiai, T.; Hoshi, T.; Slimen, H.; Nakata, K.; Murakami, T.; Tatejima, H.; Koide, Y.; Houas, A.; Horie, T.; Morito, Y.; *et al.* Fabrication of TiO$_2$ nanoparticles impregnated titanium mesh filter and its application for environmental purification unit. *Catal. Sci. Technol.* **2011**, *1*, 1324–1327.
11. Ochiai, T.; Nakata, K.; Murakami, T.; Morito, Y.; Hosokawa, S.; Fujishima, A. Development of an air-purification unit using a photocatalysis-plasma hybrid reactor. *Electrochemistry* **2011**, *79*, 838–841.
12. Ochiai, T.; Niitsu, Y.; Kobayashi, G.; Kurano, M.; Serizawa, I.; Horio, K.; Nakata, K.; Murakami, T.; Morito, Y.; Fujishima, A. Compact and effective photocatalytic air-purification unit by using of mercury-free excimer lamps with TiO$_2$ coated titanium mesh filter. *Catal. Sci. Technol.* **2011**, *1*, 1328–1330.
13. Ochiai, T.; Nakata, K.; Murakami, T.; Horie, T.; Morito, Y.; Fujishima, A. Anodizing effects of titanium-mesh surface for fabrication of photocatalytic air purification filter. *Nanosci. Nanotechnol. Lett.* **2012**, *4*, 544–547.
14. Ochiai, T.; Nanba, H.; Nakagawa, T.; Masuko, K.; Nakata, K.; Murakami, T.; Nakano, R.; Hara, M.; Koide, Y.; Suzuki, T.; *et al.* Development of an O$_3$-assisted photocatalytic water-purification unit by using a TiO$_2$ modified titanium mesh filter. *Catal. Sci. Technol.* **2012**, *2*, 76–78.
15. Ochiai, T.; Hayashi, Y.; Ito, M.; Nakata, K.; Murakami, T.; Morito, Y.; Fujishima, A. An effective method for a separation of smoking area by using novel photocatalysis-plasma synergistic air-cleaner. *Chem. Eng. J.* **2012**, *209*, 313–317.

16. Ochiai, T.; Masuko, K.; Tago, S.; Nakano, R.; Niitsu, Y.; Kobayashi, G.; Horio, K.; Nakata, K.; Murakami, T.; Hara, M.; *et al.* Development of a hybrid environmental purification unit by using of excimer VUV lamps with TiO$_2$ coated titanium mesh filter. *Chem. Eng. J.* **2013**, *218*, 327–332.

17. Ochiai, T.; Nakata, K.; Murakami, T.; Fujishima, A.; Yao, Y.Y.; Tryk, D.A.; Kubota, Y. Development of solar-driven electrochemical and photocatalytic water treatment system using a boron-doped diamond electrode and TiO$_2$ photocatalyst. *Water Res.* **2010**, *44*, 904–910.

18. Yao, Y.; Ochiai, T.; Ishiguro, H.; Nakano, R.; Kubota, Y. Antibacterial performance of a novel photocatalytic-coated cordierite foam for use in air cleaners. *Appl. Catal. B Environ.* **2011**, *106*, 592–599.

19. Yao, Y.; Kubota, Y.; Murakami, T.; Ochiai, T.; Ishiguro, H.; Nakata, K.; Fujishima, A. Electrochemical inactivation kinetics of boron-doped diamond electrode on waterborne pathogens. *J. Water Health* **2011**, *9*, 534–543.

20. Yamauchi, K.; Yao, Y.; Ochiai, T.; Sakai, M.; Kubota, Y.; Yamauchi, G. Antibacterial activity of hydrophobic composite materials containing a visible-light-sensitive photocatalyst. *J. Nanotechnol.* **2011**, *2011*, doi: 10.1155/2011/380979.

21. Kreutz, L.C.; Seal, B.S.; Mengeling, W.L. Early interaction of feline calicivirus with cells in culture. *Arch. Virol.* **1994**, *136*, 19–34.

22. Nohr, R.S.; MacDonald, J.G.; Kogelschatz, U.; Mark, G.; Schuchmann, H.P.; von Sonntag, C. Application of excimer incoherent-UV sources as a new tool in photochemistry: photodegradation of chlorinated dibenzodioxins in solution and adsorbed on aqueous pulp sludge. *J. Photochem. Photobiol. A* **1994**, *79*, 141–149.

23. Tasbihi, M.; Ngah, C.R.; Aziz, N.; Mansor, A.; Abdullah, A.Z.; Teong, L.K.; Mohamed, A.R. Lifetime and regeneration studies of various supported TiO$_2$ photocatalysts for the degradation of phenol under UV-C light in a batch reactor. *Ind. Eng. Chem. Res.* **2007**, *46*, 9006–9014.

24. Oppenländer, T.; Gliese, S. Mineralization of organic micropollutants (homologous alcohols and phenols) in water by vacuum-UV-oxidation (H$_2$O-VUV) with an incoherent xenon-excimer lamp at 172 nm. *Chemosphere* **2000**, *40*, 15–21.

25. Oppenländer, T.; Walddörfer, C.; Burgbacher, J.; Kiermeier, M.; Lachner, K.; Weinschrott, H. Improved vacuum-UV (VUV)-initiated photomineralization of organic compounds in water with a xenon excimer flow-through photoreactor (Xe* lamp, 172 nm) containing an axially centered ceramic oxygenator. *Chemosphere* **2005**, *60*, 302–309.

26. Oppenländer, T. Mercury-free sources of VUV/UV radiation: Application of modern excimer lamps (excilamps) for water and air treatment. *J. Environ. Eng. Sci.* **2007**, *6*, 253–264.

27. Afzal, A.; Oppenländer, T.; Bolton, J.R.; El-Din, M.G. Anatoxin—A degradation by advanced oxidation processes: Vacuum-UV at 172 nm, photolysis using medium pressure UV and UV/H$_2$. *Water Res.* **2010**, *44*, 278–286.

28. Wang, D.; Oppenländer, T.; El-Din, M.G.; Bolton, J.R. Comparison of the disinfection effects of Vacuum-UV (VUV) and UV light on *Bacillus subtilis* spores in aqueous suspensions at 172, 222 and 254 nm. *Photochem. Photobiol.* **2010**, *86*, 176–181.

29. Heit, G.; Neuner, A.; Saugy, P.-Y.; Braun, A.M. Vacuum-UV (172 nm) actinometry. The quantum yield of the photolysis of water. *J. Phys. Chem. A* **1998**, *102*, 5551–5561.

30. Weeks, J.L.; Meaburn, G.M.A.C.; Gordon, S. Absorption coefficients of liquid water and aqueous solutions in the far ultraviolet. *Radiat. Res.* **1963**, *19*, 559–567.

31. Han, W.; Zhang, P.; Zhu, W.; Yin, J.; Li, L. Photocatalysis of p-chlorobenzoic acid in aqueous solution under irradiation of 254 nm and 185 nm UV light. *Water Res.* **2004**, *38*, 4197–4203.

32. Cho, M.; Chung, H.; Yoon, J. Disinfection of water containing natural organic matter by using ozone-initiated radical reactions. *Appl. Environ. Microbiol.* **2003**, *69*, 2284–2291.

33. Cho, M.; Chung, H.; Choi, W.; Yoon, J. Linear correlation between inactivation of *E. coli* and OH radical concentration in TiO_2 photocatalytic disinfection. *Water Res.* **2004**, *38*, 1069–1077.

34. Cho, M.; Chung, H.; Choi, W.; Yoon, J. Different inactivation behaviors of MS-2 phage and *Escherichia coli* in TiO_2 photocatalytic disinfection. *Appl. Environ. Microbiol.* **2005**, *71*, 270–275.

35. Cho, M.; Yoon, J. Measurement of OH radical CT for inactivating cryptosporidium parvum using photo/ferrioxalate and photo/TiO_2 systems. *J. Appl. Microbiol.* **2008**, *104*, 759–766.

36. Cho, M.; Gandhi, V.; Hwang, T.M.; Lee, S.; Kim, J.H. Investigating synergism during sequential inactivation of MS-2 phage and *Bacillus subtilis* spores with UV/H_2O_2 followed by free chlorine. *Water Res.* **2011**, *45*, 1063–1070.

37. Cho, M.; Cates, E.L.; Kim, J.H. Inactivation and surface interactions of MS-2 bacteriophage in a TiO_2 photoelectrocatalytic reactor. *Water Res.* **2011**, *45*, 2104–2110.

38. Chatzisymeon, E.; Droumpali, A.; Mantzavinos, D.; Venieri, D. Disinfection of water and wastewater by UV-A and UV-C irradiation: Application of real-time PCR method. *Photochem. Photobiol. Sci.* **2011**, *10*, 389–395.

39. Oguma, K.; Katayama, H.; Ohgaki, S. Photoreactivation of legionella pneumophila after inactivation by low- or medium-pressure ultraviolet lamp. *Water Res.* **2004**, *38*, 2757–2763.

40. Halfmann, H.; Denis, B.; Bibinov, N.; Wunderlich, J.; Awakowicz, P. Identification of the most efficient VUV/UV radiation for plasma based inactivation of *Bacillus atrophaeus* spores. *J. Phys. D* **2007**, *40*, doi:10.1088/0022-3727/40/19/019.

41. Beltrán, F.J.; Rivas, F.J.; Gimeno, O.; Carbajo, M. Photocatalytic enhanced oxidation of fluorene in water with ozone. Comparison with other chemical oxidation methods. *Ind. Eng. Chem. Res.* **2005**, *44*, 3419–3425.

42. Nishimoto, S.; Mano, T.; Kameshima, Y.; Miyake, M. Photocatalytic water treatment over WO_3 under visible light irradiation combined with ozonation. *Chem. Phys. Lett.* **2010**, *500*, 86–89.

43. Nicolas, M.; Ndour, M.; Ka, O.; D'Anna, B.; George, C. Photochemistry of atmospheric dust: Ozone decomposition on illuminated titanium dioxide. *Environ. Sci. Technol.* **2009**, *43*, 7437–7442.

44. Rivas, F.J.; Beltrán, F.J.; Gimeno, O.; Carbajo, M. Fluorene oxidation by coupling of ozone, radiation, and semiconductors: A mathematical approach to the kinetics. *Ind. Eng. Chem. Res.* **2006**, *45*, 166–174.

45. Nakamura, R.; Sato, S. Oxygen species active for photooxidation of n-decane over TiO_2 Surfaces. *J. Phys. Chem. B* **2002**, *106*, 5893–5896.

46. Einaga, H.; Ogata, A.; Futamura, S.; Ibusuki, T. The stabilization of active oxygen species by Pt supported on TiO_2. *Chem. Phys. Lett.* **2001**, *338*, 303–307.

47. Huang, H.B.; Ye, D.Q.; Leung, D.Y.C. Removal of toluene using UV-irradiated and nonthermal plasma-driven photocatalyst system. *J. Environ. Eng.* **2010**, *136*, 1231–1236.

48. Einaga, H.; Futamura, S. Catalytic oxidation of benzene with ozone over alumina-supported manganese oxides. *J. Catal.* **2004**, *227*, 304–312.

49. Song, S.; Liu, Z.; He, Z.; Zhang, A.; Chen, J.; Yang, Y.; Xu, X. Impacts of morphology and crystallite phases of titanium oxide on the catalytic ozonation of phenol. *Environ. Sci. Technol.* **2010**, *44*, 3913–3918.

50. Guillard, C. Photocatalytic degradation of butanoic acid: Influence of its ionisation state on the degradation pathway: Comparison with O₃/UV process. *J. Photochem. Photobiol. A* **2000**, *135*, 65–75.

51. Phillips, S.L.; Olesik, S.V. Initial characterization of humic acids using liquid chromatography at the critical condition followed by size-exclusion chromatography and electrospray ionization mass spectrometry. *Anal. Chem.* **2003**, *75*, 5544–5553.

52. Jiang, J.; Kappler, A. Kinetics of microbial and chemical reduction of humic substances: Implications for electron shuttling. *Environ. Sci. Technol.* **2008**, *42*, 3563–3569.

53. Navarro, P.; Sarasa, J.; Sierra, D.; Esteban, S.; Ovelleiro, J.L. Degradation of wine industry wastewaters by photocatalytic advanced oxidation. *Water Sci. Technol.* **2005**, *51*, 113–120.

54. Miguel, N.; Ormad, M.P.; Mosteo, R.; Ovelleiro, J.L. Photocatalytic degradation of pesticides in natural water: Effect of hydrogen peroxide. *Int. J. Photoenergy* **2012**, *2012*, doi:10.1155/2012/371714.

55. Bullock, G.L.; Summerfelt, S.T.; Noble, A.C.; Weber, A.L.; Durant, M.D.; Hankins, J.A. Ozonation of a recirculating rainbow trout culture system I. Effects on bacterial gill disease and heterotrophic bacteria. *Aquaculture* **1997**, *158*, 43–55.

Water **2013**, *5*, 1346-1365; doi:10.3390/w5031346

OPEN ACCESS

water

ISSN 2073-4441
www.mdpi.com/journal/water

Review

Pharmaceuticals in the Built and Natural Water Environment of the United States

Randhir P. Deo [1] **and Rolf U. Halden** [2,3,4,*]

[1] Chemistry Program, College of Arts and Sciences, Grand Canyon University, Phoenix, AZ 85017, USA; E-Mail: randhir.deo@gcu.edu

[2] Center for Environmental Security, Biodesign Institute at Arizona State University, 781 E. Terrace Road, P.O. Box 875904, Tempe, AZ 85287-5904, USA

[3] Security and Defense Systems Initiative, Arizona State University, Tempe, AZ 85287-5904, USA

[4] Department of Environmental Health Sciences, Bloomberg School of Public Health, Johns Hopkins University, Baltimore, MD 21205, USA

* Author to whom correspondence should be addressed; E-Mail: halden@asu.edu;
Tel.: +1-480-727-0893; Fax: +1-480-965-6603.

Received: 22 June 2013; in revised form: 8 August 2013 / Accepted: 28 August 2013 / Published: 11 September 2013

Abstract: The known occurrence of pharmaceuticals in the built and natural water environment, including in drinking water supplies, continues to raise concerns over inadvertent exposures and associated potential health risks in humans and aquatic organisms. At the same time, the number and concentrations of new and existing pharmaceuticals in the water environment are destined to increase further in the future as a result of increased consumption of pharmaceuticals by a growing and aging population and ongoing measures to decrease per-capita water consumption. This review examines the occurrence and movement of pharmaceuticals in the built and natural water environment, with special emphasis on contamination of the drinking water supply, and opportunities for sustainable pollution control. We surveyed peer-reviewed publications dealing with quantitative measurements of pharmaceuticals in U.S. drinking water, surface water, groundwater, raw and treated wastewater as well as municipal biosolids. Pharmaceuticals have been observed to reenter the built water environment contained in raw drinking water, and they remain detectable in finished drinking water at concentrations in the ng/L to μg/L range. The greatest promises for minimizing pharmaceutical contamination include source control (for example, inputs from intentional flushing of medications for safe disposal, and sewer overflows), and improving efficiency of treatment facilities.

Keywords: drinking water; sewage sludge; pharmaceuticals; review

---

1. Introduction

Intended uses of pharmaceuticals in humans and animals are plentiful and include the prevention, diagnosis, and therapy of diseases as well as cosmetic and lifestyle purposes [1]. In recent years, however, their occurrence in the environment has raised concerns, both nationally and internationally, regarding implied risks posed to aquatic and terrestrial life forms, including humans [2–7]. In the United States (U.S.), pharmaceuticals have been found to occur throughout the water environment [8–13], including the drinking water supply [14–18]. Whereas the perceived and actual risks of trace levels of pharmaceuticals in drinking water is a topic of ongoing discussion, this review concentrates on the sources and pathways of water contamination in the U.S. to assess our understanding of the occurrence of pharmaceuticals in U.S. drinking water, and to identify opportunities for pollution control.

Environmental exposures of humans and aquatic organisms to pharmaceuticals have been reported [19–25] and the associated risks evaluated [17,26–28]. For example, human health impacts were assessed from exposure to pharmaceutically active compounds in drinking water [14–18] and edible fish [19–25]. Additionally, specific modes of action of pharmaceuticals have been evaluated in humans and mammals [29], including an analysis of metabolism and excretion by humans [30]. Overall, these studies conclude that, based on current knowledge, the presence of trace levels of pharmaceuticals poses negligible or only minor risks to humans. Exposure of aquatic organisms also is well established [31] and extends into coastal waters, as illustrated by reports on the antibacterial chemical triclosan, that was measured in blood plasma of wild bottlenose dolphins (*Tursiopstruncatus*) (0.025–0.11 ng/g wet weight), and in estuarine surface water samples (4.9–14 ng/L) [20].

Concerning the built (man-made) water environment, a significant volume of literature explored the performance of treatment plants for the removal of pharmaceuticals from raw wastewater (sewage) [32–36] and from raw drinking water [37–39]. Additional studies investigated various strategies for efficient removal and transformation of pharmaceuticals using advanced treatment employing processes of chemical [32,40–43], biological [44–47] and physical nature [48,49]. Since not all pharmaceuticals present in sewage are the result of intentional intake, metabolism and excretion, some researchers have investigated the composition of wastewaters from the pharmaceutical industry [50], and healthcare facilities [51], as well as the importance of disposal of unwanted or leftover pharmaceuticals into sanitary sewers [1,2,52–56].

The aforementioned reviews and articles carry valuable information and discussion on occurrences, treatment efficiencies, and risk assessments of pharmaceuticals in the environment. However, our focus in this review is to specifically examine the occurrences of pharmaceutical compounds in U.S. drinking water, in the context of the role of the different point sources of the built water environment, the interconnectivity of the built and natural water environments, and to identify opportunities for effectively controlling environmental contamination with pharmaceuticals in a sustainable fashion. A perspective on this issue is essential, however, for properly managing risks associated with the occurrence of pharmaceutical compounds in the built and natural water environment.

In this review, we first examine the interplay of the built water environment and the natural water environment to inform the management of pharmaceutical pollution; Second, we survey peer-reviewed publications for available data on the identity and concentration of pharmaceuticals present in various compartments of the water environment, exclusively in the U.S.; Third, we discuss the persistence and risk of pharmaceuticals in groundwater and drinking water. Fourth and finally, we propose strategies and criteria for minimizing occurrences of pharmaceuticals in drinking water, and in the preceding matrices of the natural and built water environment.

For the purpose of this review, we defined "pharmaceuticals" as prescription and non-prescription drugs that are either ingested or topically applied for prevention and/or cure of diseases and injuries. Thus, we included antimicrobial compounds that are used heavily in clinics and hospitals; however, we excluded other substances such as naturally occurring hormones, flavors and fragrances, cosmetics, and personal care products. The literature search was conducted using the Web of Science database, using the keyword "pharmaceutical(s)" in various aforementioned matrices of the built and natural water environments, including sewage sludge. Only maximum concentrations of the measured individual pharmaceuticals were included in this study. Also, we restricted our search to all the studies done exclusively in the U.S.

2. Built and Natural Water Environment

2.1. General Overview

The built water environment is a complex network of infrastructure comprising manmade lakes and reservoirs, canals, the sewerage and water distribution systems, drinking water treatment plants (DWTPs) and wastewater treatment plants (WWTPs), as well as rivers and aquifers reliant on WWTP effluent as the principal recharge mechanism. Significant additions to the built water environment resulted from the Clean Water Act, which was implemented in 1972 with the objective to "restore and maintain the chemical, physical, and biological integrity of the Nation's waters" [57]. One of the provisions of this regulation was to prohibit discharge of toxic pollutants from point sources, including domestic households and industrial facilities. The legislation also laid the foundation for the current practice of combining industrial and domestic wastewaters before treatment at the WWTP. In this context, it is important to note that pretreatment of high-strength industrial waters is widely practiced in the U.S. prior to their release into municipal sewer systems.

Despite these efforts, pharmaceuticals are known to occur in U.S. water resources, which behooves us to more closely study and better manage the fate and migration of pharmaceuticals through the water environment. Figure 1 shows a number of major components of the built water environment and the natural water environment. It identifies important compartments of the built water environment, such as sewage systems, WWTPs, DWTPs and the water distribution system, and shows how this manmade infrastructure is in communication with multiple components of the natural water environment, including surface water (rivers, lakes, oceans, as well as aquifers) and groundwater. Due to this connectivity, pharmaceuticals frequently straddle the interface of the built and natural water environments, thereby posing potential risks to humans as well as to aquatic and other terrestrial life forms that rely on water resources to survive and flourish.

Figure 1. Schematic showing inputs of pharmaceuticals to, and the interconnectivity of, the natural and built water environment.

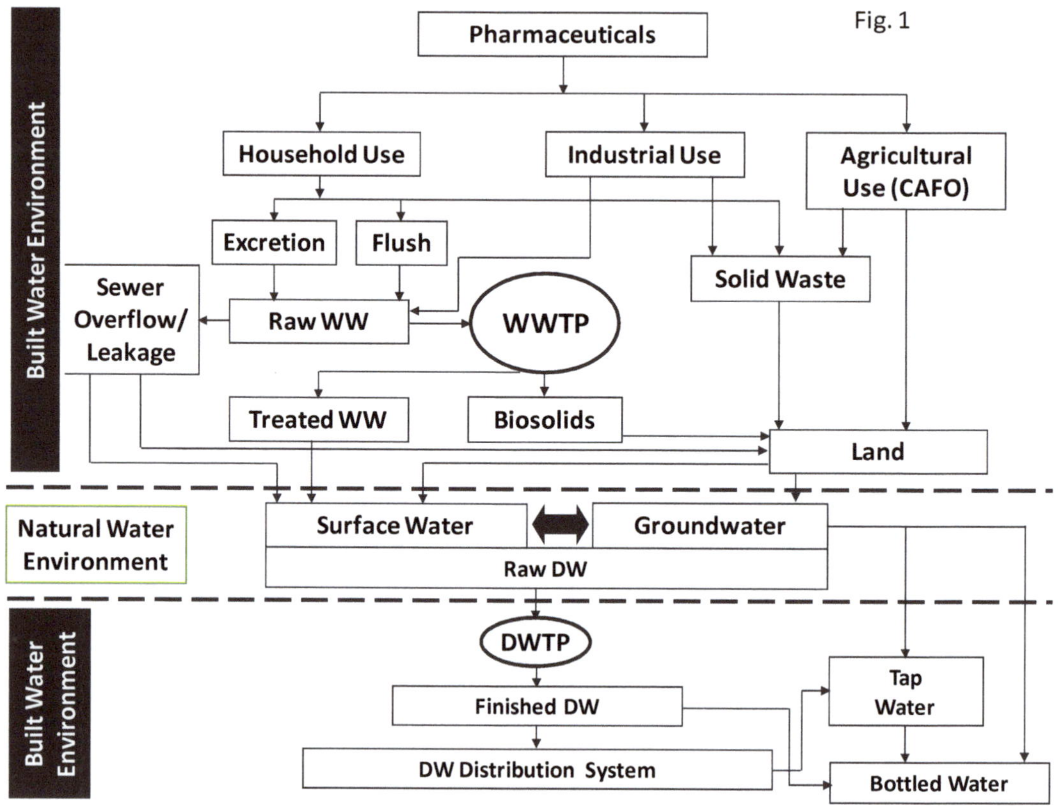

The major routes of pharmaceutical administration include enteral (e.g., oral), parenteral (e.g., injection), topical (e.g., skin surface), and inhalation. The ingested pharmaceuticals (mainly via enteral and parenteral administration) are excreted as un-metabolized or metabolized products, whereas the topically applied substances that do not enter the body by absorption also can be washed down the drain [33,52]. These pharmaceuticals are combined in the sewer system with black water (feces and urine) and gray water (domestic process waters from, e.g., washing, bathing, showering and kitchen use) to form raw wastewater or its synonym, sewage (Raw WW or RWW). Expired and unwanted (leftover) pharmaceuticals may be flushed down the drain, thereby leading to direct loading to wastewater [53]. Another source of pharmaceutical contamination in Raw WW is from the influx of waste from pharmaceutical manufacturing companies [50] and healthcare facilities [51].

Raw WW is conveyed to WWTPs whose primary goal is the removal of pathogens, turbidity, odor, color, Biological Oxygen Demand (BOD) and nutrients (primarily nitrogen and phosphorus) through a combination of physical, biological and chemical treatment [2]. However, their ability to also remove to a significant degree, the Pharmaceuticals and Personal Care Products (PPCPs), and other commodity chemicals is widely recognized and has been reviewed both from a mechanistic and quantitative perspective [58]. There are two process streams exiting the WWTPs: aqueous flow in the form of treated wastewater (Treated WW or TWW); and the municipal wastewater residuals or sewage sludge, which is an unwanted byproduct and that can be converted to so-called biosolids via additional treatment processes, including aerobic and/or anaerobic digestion, lime stabilization, and dewatering [59]. Treated WW is either reclaimed for land irrigation and farming or discharged into surface waters (streams, lakes, rivers, ponds, *etc.*) to close the water cycle. In coastal settings, biological sewage treatment is often omitted, and the effluent of primary treatment is directly discharged into the ocean. Pharmaceuticals discharged into surface waters may cause contamination of groundwater in aquifers, wells, springs and sumps either via direct leaching into the river bed or following application on land in irrigation water [8,11].

Pharmaceuticals may also enter surface water and groundwater through leaching of land-based pharmaceutical waste and solid waste contaminated with drugs. Major sources are unwanted or leftover drugs in domestic solid waste from residential households [52], from the pharmaceutical industry [50,60,61], from farm operations that land apply drug-tainted biosolids [62,63], and from animal waste that may contain excreted pharmaceuticals and partially metabolized drugs [40,64].

Another possible pathway for surface water and groundwater contamination is from leaking sewage distribution lines or overflows of combined sewer systems and, less prominent, overflows of sanitary sewer systems, both occurring under conditions of heavy rainfall or snowmelt. Although these overflows are seasonal/occasional, their impact of contamination can be significant, since the discharged sewer may contain untreated pharmaceuticals that are washed directly into surface water from where it may infiltrate into groundwater [65–67].

In the U.S., as in many other countries, groundwater can and is being used directly for consumption as drinking water without any treatment, particularly in rural and remote settings [68]. In contrast, surface water typically is subject to a multi-barrier treatment train to remove chemical and biological contaminants. Ocean water also may serve as a source of drinking water but it undergoes extreme treatment in the form of either distillation or reverse osmosis filtration. Raw drinking water (Raw DW or RDW) is processed in DWTPs to produce finished drinking water (Finished DW or FDW) that is ready for distribution as tap water or bottling, distribution and retail (Figure 1).

2.2. Occurrence and Distribution of Pharmaceuticals in the Built and Natural Water Environment

In the following section, we present the number of pharmaceuticals distributed according to their highest concentration reported in each of the concentration ranges in various compartments of the built and natural water environments (Figure 2). The highest concentration values and names of the individual pharmaceuticals occurring in Raw WW, Treated WW, Surface Water, Groundwater, Raw DW, and Finished DW are provided Tables S1 to S6, respectively, in the Supplementary Information (SI). For simplicity and to avoid redundancy, each individual pharmaceutical measured in each water matrix is represented in Figure 2 only once and only in the histogram representing the highest concentration range. No pharmaceutical is represented in more than one concentration range within each water matrix with the implicit understanding that its presence at sub-maximal concentrations constitutes the rule rather than the exception.

The numbers of pharmaceuticals detected in each of the down gradient matrices include 73 in RWW, 92 in TWW, 91 in SW, 33 in GW, 37 in RDW, and 23 in FDW. It is important not to draw potentially misleading conclusions from the data presented in Figure 2 regarding the treatment efficiency of infrastructure and the attenuation of drugs in the environment. This could be misleading because the maximum concentrations of pharmaceuticals included in this study are from different discrete geographical locations within U.S., and the measurement objective was not necessarily to investigate the same set of pharmaceuticals in the different water matrices. Nevertheless, occurrence of lesser number and lower concentration range of pharmaceuticals in drinking water is noted from a human health perspective, regardless of identity or geographical location.

Figure 2. Number of pharmaceuticals detected in various water matrices of the built water environment. Each pharmaceutical is represented only once and shown in the category reflecting its respective maximum concentration reported in a given aquatic compartment.

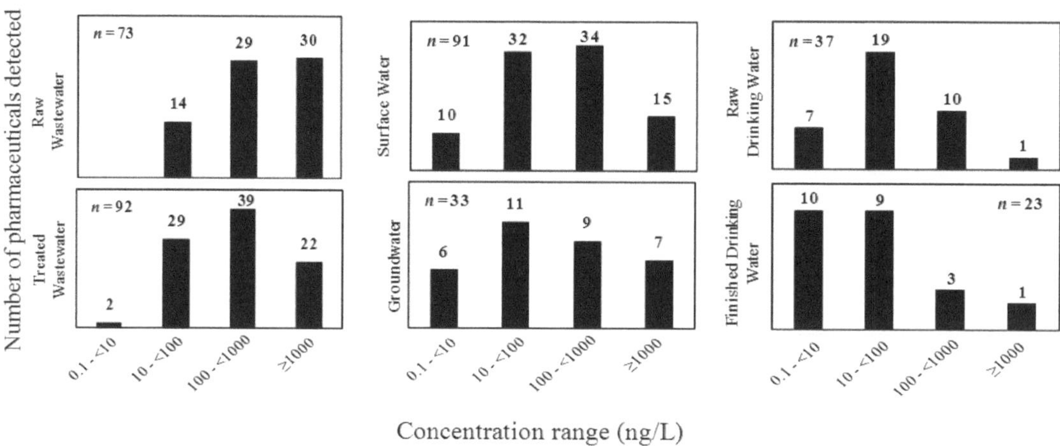

The apparent increase in the total number of pharmaceuticals from RWW (total of 73) to TWW (total of 92) is counter-intuitive. Only 62 pharmaceuticals detected were common between these two matrices, which may reflect that different sets of pharmaceuticals were measured in the individual studies examined here. Other potential explanations for this observation include (i) deconjugation of metabolites and release of the parent compounds during treatment; (ii) analytical difficulties that lead to higher detection limits in RWW when compared to TWW; (iii) a less comprehensive monitoring of RWW compared with TWW; and (iv) the common practice in the analytical laboratory of filtering RWW but not necessarily TWW prior to analysis. Since RWW represents a mixture of inputs from domestic, municipal and industrial sources entering the WWTP, it is possible that higher than reported concentrations of pharmaceuticals may be present in specific process waters prior to mixing and entry into WWTPs. Wastewaters discharged by the pharmaceutical industry could constitute a particularly strong source term, as revealed in a report by the United States Geological Survey (USGS) [50].

Use of filtration during standard sample processing can remove a significant fraction of the hydrophobic organic compound mass contained in the sample of interest. A recent study reported that up to 86% of the mass of tonalide can be sorbed to filterable material and thus be excluded from chemical analyses, due to the common practice of filtering aqueous samples in general and raw sewage in particular [69]. Thus, any differences from study to study in both the occurrence and concentration of drugs in a given water matrix may be real or only apparent. Biased data can result from both sample processing and analyte detection strategies utilized [69].

Thirty-three different pharmaceuticals have been reported in U.S. GW and are provided here in units of ng/L for maximum concentrations reported: acetaminophen (1890), caffeine (290), carbamazepine (420), ciprofloxacin (45), codeine (214), dehydronifedipine (22), diclofenac (46), dilantin (22), diltiazem (28), 1,7-dimethylxanthine (57), erythromycin (2380), 17-α-ethinylestradiol (230), fluoxetine (56), gemfibrozil (6860), ibuprofen (3110), lincomycin (1900), meprobamate (8.6), naproxen (0.7), oestriol (6.4), oestrone (1), oxybenzone (7.5), oxytetracycline (130), pentoxifylline (34), primidone (2.8), sulfadimethoxine (130), sulfamerazine (54), sulfamethazine (3600), sulfamethazole (170), sulfamethoxazole (1110), sulfathiazole (305), tetracycline (500), triclosan (53), and trimethoprim (18) (Figure 2, Table 1).

In order for a compound to become detectable in groundwater, it either must have passed through the wastewater treatment processes prior to injection into the subsurface for aquifer recharge or it must have resisted microbial transformation, and sorption to soils and sediments during the slow soil infiltration process following application of drug-laden biosolids [11]. Alternatively, compounds may enter shallow and deeper groundwater from urination, defecation, sewer overflows, leaking sewage distribution lines, and/or through leaching of pharmaceuticals contained in waste resulting from agriculture use, such as Concentrated Animal Feeding Operation (CAFO), as indicated in Figure 1.

Furthermore, twenty-three different pharmaceuticals have been reported in U.S. FDW. Their maximum concentrations are provided in units of ng/L in parentheses: acetaminophen (28), atenolol (26), caffeine (180.8), carbamazepine (258), codeine (30), cotinine (25), dehydronifedipine (4), diazepam (0.33), dilantin (32), erythromycin (1.3), fluoxetine (0.82), gemfibrozil (6.5), genistein (2.9), ibuprofen (1350), iopromide (31), lincomycin (4.4), meprobamate (43), naproxen (8), primidone (1.3), sulfamethoxazole (20), sulfathiazole (10), triclosan (734), and trimethoprim (1.7) (Figure 2, Table 1)).

The presence of pharmaceuticals in FDW may be related to multiple factors, including the pharmaceuticals' physical-chemical properties that allowed them to resist general biological, physical and chemical transformation processes, specific efficiency and/or overload of the treatment facilities (WWTP and DWTP) they passed through, and their respective initial mass loadings [66,70,71].

Whether long-term risks exist from chronic exposure to these compounds at low levels is a more difficult question to answer. Long-term, low-level exposures may involve toxicological mechanisms different from those observed in short-term, high-dose studies [71]. Furthermore, future demand for drinking water is expected to increase due to population growth and shortening of the water loop. Increased reliance on aggressive water reuse already is a key driver of research on contaminants of emerging concern (CECs). This notion is supported by the occurrence of pharmaceuticals in TWW (n = 92 drugs), SW (n = 91 drugs), GW (n = 33 drugs), and RDW (n = 37 drugs). Although generalizations are difficult to formulate, most immediate potential human health risks likely stem from elevated levels of pharmaceuticals in FDW, followed by drinking of untreated groundwater, which is more common in rural populations.

Table 1. Maximum concentrations of pharmaceuticals detected in Groundwater (GW) and in Finished Drinking Water (FDW) of the United States. Also shown are the Predicted No-Effect Concentration (PNEC) and the calculated Risk Quotient (RQ).

Pharmaceuticals	PNEC (ng/L)	GW (ng/L)	RQ (GW)	FDW (ng/L)	RQ (FDW)
Acetaminophen	1,000 [72]	1,890 [68]	1.890	28 [86]	0.028
Atenolol	3.1×10^5 [73]	NA	–	26 [37]	0.01
Caffeine	1.0×10^7 [64]	290 [68]	0.01	180.8 [86]	0.01
Carbamazepine	420 [74]	420 [68,79]	1.00	258 [87]	0.614
Ciprofloxacin	5 [75]	45 [8]	9.0	NA	–
Codeine	2,900 [a]	214 [68]	0.07	30 [88]	0.01
Cotinine	5,200 [a]	NA	–	25 [87]	0.01
Dehydronifedipine	15,000 [a]	22 [80]	<0.01	4 [87]	0.01
Diazepam	4,300 [76]	NA	–	0.33 [14]	0.01
Diclofenac	460 [77]	46 [79]	0.1	NA	–
Dilantin (Phenytoin)	1,800 [a]	22 [79]	–	32 [37]	0.018
Diltiazem	920 [a]	28 [80]	0.03	NA	–
1,7-dimethylxanthine	8,000 [a]	57 [80]	0.007	NA	–
Erythromycin	20 [77]	2,380 [81]	119	1.3 [89]	0.065
17-α-ethinylestradiol	1,800 [a]	230 [8]	0.128	NA	–
Fluoxetine	47 [77]	56 [80]	1.191	0.82 [14]	0.017
Gemfibrozil	780 [77]	6,860 [82]	–	6.5 [90]	0.008
Genistein	550 [a]	NA	–	2.9 [37]	0.005
Ibuprofen	1,000 [72]	3,110 [80]	3.11	1350 [91]	1.350
Iopromide	460,000 [a]	NA	–	31 [89]	0.01
Lincomycin	13,000 [a]	1,900 [83]	0.025	4.4 [86]	0.01
Meprobamate	110,000 [a]	8.6 [79]	–	43 [37]	0.01
Naproxen	640 [77]	0.7 [79]	–	8 [89]	0.013
Oestriol	14,000 [a]	6.4 [79]	<0.01	NA	–
Oestrone	4,800 [a]	1 [79]	<0.01	NA	–
Oxybenzone	3,500 [a]	7.5 [79]	<0.01	NA	–
Oxytetracycline	200 [76]	139 [84]	0.695	NA	–
Pentoxifylline	4,600 [a]	34 [79]	<0.01	NA	–
Primidone	4,300 [a]	2.8 [85]	–	1.3 [92]	0.01
Sulfadimethoxine	248,000 [a]	130 [83]	<0.01	NA	–
Sulfamerazine	116,000 [a]	54 [81]	<0.01	NA	–
Sulfamethazine	1.2×10^{6} [a]	3,600 [83]	0.001	NA	–
Sulfamethoxazole	27 [78]	1,110 [80]	41.11	20 [89]	0.741
Sulfathiazole	5,000 [a]	305 [81]	0.061	10 [88]	0.002
Tetracycline	90 [76]	500 [7]	5.6	NA	–
Triclosan	1,550 [72]	53 [8]	0.034	734 [91]	0.474
Trimethoprim	1,000 [64]	18 [68]	0.018	1.7 [86]	0.002

Notes: NA = Not Available; [a] Calculated = Chronic toxicity concentration (in ng/L) for fish (obtained from PBT Profiler [93])/100.

A comparison of the concentrations of pharmaceuticals occurring in GW and FDW against threshold concentrations of pharmaceuticals in drinking water is warranted when evaluating the magnitude of risks posed. Since quality standards are not yet available for pharmaceuticals in drinking water, we compared the highest concentrations of pharmaceuticals in GW and FDW against the predicted no-effect concentration (PNEC) value that is estimated from standard toxicity assays [72,76,94,95]. The result of this comparison is a calculated risk quotient (RQ), which is the ratio of the highest concentration of pharmaceuticals divided by the PNEC. We either used the lowest PNEC values available in the peer-reviewed literature, or estimated it by dividing the chronic (long-term) toxicity value for fish (obtained from PBT Profiler [93]) with an assessment factor of 100 [95] for extrapolating the test organism's chronic toxicity to the corresponding anticipated human no-effect concentration [72,94,96].

Table 1 shows PNEC and RQ values available for all the pharmaceuticals in GW and FDW. In GW, acetaminophen, carbamazepine, erythromycin, fluoxetine, ibuprofen, sulfamethoxazole, and tetracycline have RQ value > 1, and thus may pose a potential risk to humans, if the respective water is consumed without any (point-of-use) treatment. On the other hand, FDW has only ibuprofen with RQ value >1.

2.3. Pharmaceuticals in Municipal Sludge

An unwanted byproduct of wastewater treatment is sewage sludge that typically is treated to achieve stabilization and enable its application on land as biosolids according to federal and state guidelines. Municipal sludge used as fertilizer or soil conditioner is the subject of recent investigations as a source of organic pollutants in soils and adjacent aquatic environments [58,97,98]. The high organic carbon content of sewage sludge favors preferential sorption and enrichment of hydrophobic organic compounds during wastewater treatment process [58,99,100].

Figure 3 shows the number of pharmaceuticals distributed according to their maximum concentration (in units of µg/kg dry weight) reported in each of the concentration ranges in sewage sludge (Panel A), and the identity and maximum concentrations of pharmaceuticals occurring in >10,000 µg/kg dry weight concentration range (Panel B). The most abundant compounds are the two antimicrobial compounds, triclocarban (441,000) and triclosan (133,000). Other major contributors are the antibiotics ciprofloxacin (47,500), ofloxacin (58,100), and sulfanilamide (15,600). The antihistamine diphenhydramine (22,000) and the pain-reliever ibuprofen (11,900) contribute a lesser but still substantial mass fraction. Pharmaceuticals reported in biosolids at maximum concentrations of 1000 to <10,000 µg/kg include acetaminophen, anhydrotetracycline, azithromycin, caffeine, carbamazepine, chlortetracycline, cimetidine, clindamycin, 1,7-dimethylxanthine, doxycycline, 4-epianhydrotetracycline, 4-epitetracycline, fluoxetine, gemfibrozil, isochlortetracycline, metformin, miconazole, minocycline, naproxen, norfloxacin, ranitidine, sarafloxacin, and tetracycline. Maximum concentrations between 100 and <1000 µg/kg were reported for anhydrochlortetracycline, clofibric acid, codeine, cotinine, demeclocycline, diclofenac, diltiazem, 4-epichlortetracycline, erythromycin (total), erythromycin-H_2O, 17 alpha-ethinylestradiol, norfluoxetine, estriol, estrone, oxytetracycline, paroxetine, salicylic acid, sulfadiazine, sulfamerazine, sulfamethoxazole, thiabendazole, trimethoprim, and virginiamycin. Maximum levels in biosolids of 10 to <100 µg/kg were found for aspirin, albuterol, clarithromycin, dehydronifedipine, enrofloxacin, 4-epioxytetracycline, ketoprofen, lincomycin, lomefloxacin, oxolinic acid, roxithromycin, sulfachloropyridazine, sulfadimethoxine, sulfamethazine, sulfathiazole, and sulfisoxazole. At maximum concentrations of 0.1 to <10 µg/kg, only one drug, ormetoprim, was reported to occur in biosolids. The maximum concentration values and names of the individual pharmaceuticals occurring in Sewage Sludge are provided Table S7 in the Supplementary Information (SI).

Figure 3. (**A**) Number of pharmaceuticals detected in sewage sludge. Only maximum concentrations of the pharmaceuticals were included and categorized into different concentration ranges; (**B**) Identity and maximum concentrations of pharmaceuticals detected in sewage sludge at concentrations exceeding 10,000 µg/kg dry weight.

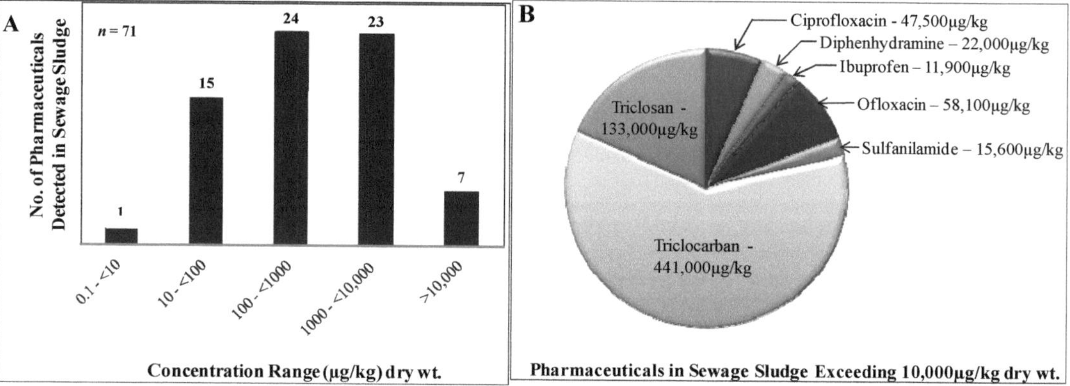

The accumulation of certain pharmaceuticals by sorption to sewage sludge can aid in the removal from the water environment but also can cause problems later on during disposal of these materials [8,97]. Today, approximately 50% of U.S. sewage sludge is applied on land as biosolids for inexpensive disposal and as a fertilizer or soil conditioner [101]. The application of biosolids laced with pharmaceuticals can pose secondary risks to water resources via leaching into groundwater and contamination of surface waters from runoff [63,102–105].

Additionally, pharmaceuticals contained in land-applied biosolids can directly pose risk to the environment and humans. Depending on their physical-chemical properties, pharmaceuticals can be strongly sorbed to soil and persist for a long time, or cause toxicity to soil bacteria and other microorganisms, thereby adversely impacting soil quality [59]. Furthermore, pharmaceuticals can be bioaccumulated into plants and crops that form part of the human diet. For example, bioaccumulation has been reported for sulfamethazine on lettuce, potato and corn [106], for chlortetracycline on corn, green onions and cabbage [107], and for carbamazepine, sulfamethoxazole, salbutamol, trimethoprim on Chinese cabbage (*Brassica campestris*) [108].

3. Sustainable Management of Pharmaceuticals

Reports of pharmaceuticals in the environment are projected to increase in the future due to increased monitoring efforts, a shortening of the water cycle, increased drug consumption, and the availability of advanced measurement technologies for trace analysis of pharmaceuticals in complex matrices. These projections demand sustainable management of pharmaceuticals in our water environment.

For example, progress has been made toward the use of advanced processing techniques that may minimize the occurrence of pharmaceuticals in drinking water. These advanced treatment techniques employ chemical [37], biological [36], and physical processes [34]. Chemical processes using advanced oxidation [109,110] and ozonation [39] have been shown to improve removal efficiency when used individually or in sequence [37,38,111,112]. For example, triclosan, ibuprofen, sulfamethoxazole, trimethoprim, naproxen, carbamazepine, phenytoin, diazepam, caffeine and fluoxetine removal efficiencies were improved by >70% through ozonation at a dose of 2.5 mg/L [37]. Additionally, progress has been made towards developing "green" chemistry approaches, for example, using Fe-TAML catalysts for purifying environmental waters [113].

The improved removal efficiency mentioned above, however, needs to be further investigated in order to assess whether the treatment technique will result in complete degradation of the pharmaceutical compounds (the safe and desired end result of any treatment process), or whether it transformed the pharmaceuticals into other products that may or may not be safe, when occurring both in isolation or in a mixture.

Besides improving the efficiency of DWTP, controlling important point sources of pharmaceuticals ought to be a priority in minimizing water and soil contamination. Source control of pharmaceutical contamination begins with proper disposal of pharmaceutical waste streams, as well as expired and leftover pharmaceuticals. Daughton and colleagues have thoroughly reviewed the reasons for accumulation of pharmaceuticals and repercussions of disposing leftover pharmaceuticals into sewage and solid waste [1,15,52,53,114]. The recommendation from these reviews is simple: do not throw unused or leftover pharmaceuticals in the toilet or flush them down the drain. Proper disposal methods should be practiced, for example, dropping off pharmaceuticals at local facilities that collect them.

Environmental occurrence of pharmaceuticals is not really a new problem but one that had been concealed for decades due to the lack of both suitable analytical techniques and financial resources to enable monitoring using available methods. Detection methods have improved substantially in the past decade and will have to be revised continuously in the future to include new pharmaceuticals entering the market. Additionally, pharmaceuticals should be routinely monitored in environmental compartments and process flows that are in communication with drinking water resources (Figure 1) for early detection of potential hazards.

In addition to continued efforts to improve the removal efficiencies of WWTPs and DWTPs, pharmaceutical industries and academic research institutions may harness new technologies to design "green pharmaceuticals" [115]. Next generation "green pharmaceuticals" ideally will be designed to undergo removal and destruction in sewage and drinking water treatment works, while maintaining the therapeutic qualities of their contemporary counterparts *in vivo* in the target organism (*i.e.*, humans and animals).

Furthermore, municipalities and academic institutions may engage in public education for proper use and disposal of pharmaceuticals in partnership with government agencies. Regulating agencies may take a more active role in implementing policies to curtail excessive application of pharmaceuticals in agribusiness, require risk assessment of new pharmaceuticals before market launch, and promote education on managing chemicals in a sustainable manner.

4. Summary and Conclusions

Pharmaceuticals are subject to repeated transfer between the built water environment and the natural water environment. Typical cycles for polluting drugs include excretion or disposal into wastewater, incomplete removal during sewage treatment and entry of attenuated drug quantities into natural surface waters and terrestrial environments via treated WW and land application of biosolids. Reentry of drugs from the natural environment into the built water environment occurs during uptake of source water for the water supply. Removal of drugs during DW treatment is incomplete, causing the occurrence of ng/L to μg/L concentrations of certain pharmaceuticals in finished DW. Population growth and a shortening of the natural water cycle likely will lead to an increase in trace levels of drugs in DW unless pharmaceutical pollution will be managed more proactively in the future.

Risk reduction opportunities and continuing research needs exist in the areas of:

- Long-term effects of low-level pharmaceutical contamination on human health;
- Short- and long-term effects of pharmaceutical contamination on non-target organisms;
- Comprehensive evaluations of treatment works (DWTPs and WWTPs) to identify infrastructure that improves the removal efficiencies of pharmaceuticals, and minimizes associated costs;
- Enhanced monitoring of pharmaceuticals in the built water environment to facilitate accurate evaluation of removal efficiencies of treatment works and to identify all possible avenues by which pharmaceuticals enter the natural water environment (including sewer overflows, leachates from solid waste, and biosolids runoff);
- More targeted monitoring of pharmaceuticals that are toxic to indicator organisms, are produced in high-volume, and possess persistent physico-chemical properties (*i.e.*, long environmental half-life);
- Design of pharmaceuticals that are susceptible to transformation (or degradation) by treatment works and/or natural processes (e.g., photolysis);
- Identification and elimination of high-strength wastewaters from the pharmaceutical industry;
- Partnerships between academic institutions, pharmaceutical industries, and government to promote public education for proper use and disposal of pharmaceuticals.

Given the importance of preserving the quality of natural waters and the security of the drinking water supply, it may be desirable to provide additional funding mechanisms to support these ongoing and suggested activities (e.g., through leveling of a modest tax on pharmaceutical sales).

Acknowledgments

This study was supported in part by the National Institute of Environmental Health Sciences (NIEHS) through research grants 1R01ES015445 and 1R01ES020889, and by the Johns Hopkins University Center for a Livable Future. The content is solely the responsibility of the authors and does not necessarily represent the official views of the NIEHS or the National Institutes of Health (NIH).

Conflicts of Interest

The authors declare no conflict of interest.

References

1. Ruhoy, I.S.; Daughton, C.G. Beyond the medicine cabinet: An analysis of where and why medications accumulate. *Environ. Int.* **2008**, *34*, 1157–1169.
2. Daughton, C.G.; Ternes, T.A. Pharmaceuticals and personal care products in the environment: Agents of subtle change? *Environ. Health Perspect.* **1999**, *107*, 907–938.
3. Dong, Z.; Senn, D.B.; Moran, R.E.; Shine, J.P. Prioritizing environmental risk of prescription pharmaceuticals. *Regul. Toxicol. Pharmacol.* **2013**, *65*, 60–67.
4. Kummerer, K. The presence of pharmaceuticals in the environment due to human use—Present knowledge and future challenges. *J. Environ. Manag.* **2009**, *90*, 2354–2366.
5. Kummerer, K. Pharmaceuticals in the Environment. In *Annual Review of Environment and Resources*; Gadgil, A., Liverman, D.M., Eds.; Annual Reviews: Palo Alto, CA, USA, 2010; pp. 57–75.
6. Boxall, A.B.A.; Rudd, M.A.; Brooks, B.W.; Caldwell, D.J.; Choi, K.; Hickmann, S.; Innes, E.; Ostapyk, K.; Staveley, J.P.; Verslycke, T.; *et al.* Pharmaceuticals and Personal Care Products in the Environment: What Are the Big Questions? *Environ. Health Perspect.* **2012**, *120*, 1221–1229.
7. Monteiro, S.C.; Boxall, A.B.A. Occurrence and fate of human pharmceuticals in the environment. In *Reviews of Environmental Contamination and Toxicology*; Whitacre, D.M., Ed.; Springer: Summerfield, NC, USA, 2010; pp. 53–154.
8. Karnjanapiboonwong, A.; Suski, J.G.; Shah, A.A.; Cai, Q.S.; Morse, A.N.; Anderson, T.A. Occurrence of PPCPs at a Wastewater Treatment Plant and in Soil and Groundwater at a Land Application Site. *Water Air Soil Pollut.* **2011**, *216*, 257–273.
9. Loganathan, B.; Phillips, M.; Mowery, H.; Jones-Lepp, T.L. Contamination profiles and mass loadings of macrolide antibiotics and illicit drugs from a small urban wastewater treatment plant. *Chemosphere* **2009**, *75*, 70–77.
10. Wilson, B.; Chen, R.F.; Cantwell, M.; Gontz, A.; Zhu, J.; Olsen, C.R. The partitioning of triclosan between aqueous and particulate bound phases in the Hudson River Estuary. *Mar. Pollut. Bull.* **2009**, *59*, 207–212.
11. Katz, B.G.; Griffin, D.W.; Davis, J.H. Groundwater quality impacts from the land application of treated municipal wastewater in a large karstic spring basin: Chemical and microbiological indicators. *Sci. Total Environ.* **2009**, *407*, 2872–2886.
12. Yu, C.P.; Chu, K.H. Occurrence of pharmaceuticals and personal care products along the West Prong Little Pigeon River in east Tennessee, USA. *Chemosphere* **2009**, *75*, 1281–1286.
13. Guo, Y.C.; Krasner, S.W. Occurrence of primidone, carbamazepine, caffeine, and precursors for n-nitrosodimethylamine in drinking water sources impacted by wastewater. *J. Am. Water Resour. Assoc.* **2009**, *45*, 58–67.
14. Benotti, M.J.; Trenholm, R.A.; Vanderford, B.J.; Holady, J.C.; Stanford, B.D.; Snyder, S.A. Pharmaceuticals and endocrine disrupting compounds in US drinking water. *Environ. Sci. Technol.* **2009**, *43*, 597–603.
15. Daughton, C.G.; Ruhoy, I.S. The afterlife of drugs and the role of PharmEcovigilance. *Drug Saf.* **2008**, *31*, 1069–1082.

16. Daughton, C.G. Pharmaceutical ingredients in drinking water: Overview of occurrence and significance of human health considerations. In *Contaminants of Emerging Concern in the Environment: Ecological and Human Health Considerations*; Rolf, U.H., Ed.; American Chemical Society: Washington, DC, USA, 2010; pp. 9–68.

17. Snyder, A.S. Occurrence of pharmaceuticals in U.S. drinking water. In *Contaminants of Emerging Concern in the Environment: Ecological and Human Health Considerations*; Rolf, U.H., Ed.; American Chemical Society: Washington, DC, USA, 2010; pp. 69–80.

18. Aydin, E.; Talinli, I. Analysis, occurrence and fate of commonly used pharmaceuticals and hormones in the Buyukcekmece Watershed, Turkey. *Chemosphere* **2013**, *90*, 2004–2012.

19. Brodin, T.; Fick, J.; Jonsson, M.; Klaminder, J. Dilute concentrations of a psychiatric drug alter behavior of fish from natural populations. *Science* **2013**, *339*, 814–815.

20. Fair, P.A.; Lee, H.B.; Adams, J.; Darling, C.; Pacepavicius, G.; Alaee, M.; Bossart, G.D.; Henry, N.; Muir, D. Occurrence of triclosan in plasma of wild Atlantic bottlenose dolphins (Tursiops truncatus) and in their environment. *Environ. Pollut.* **2009**, *157*, 2248–2254.

21. Owen, S.F.; Huggett, D.B.; Hutchinson, T.H.; Hetheridge, M.J.; Kinter, L.B.; Ericson, J.F.; Sumpter, J.P. Uptake of propranolol, a cardiovascular pharmaceutical, from water into fish plasma and its effects on growth and organ biometry. *Aquati. Toxicol.* **2009**, *93*, 217–224.

22. Christenson, T. Fish on morphine: Protecting Wisconsin's natural resources through a comprehensive plan for proper disposal of pharmaceuticals. *Wiscon. Law Rev.* **2008**, 141–179.

23. Kwon, J.W.; Armbrust, K.L. Laboratory persistence and fate of fluoxetine in aquatic environments. *Environ. Toxicol. Chem.* **2006**, *25*, 2561–2568.

24. Mearns, A.J.; Reish, D.J.; Oshida, P.S.; Buchman, M.; Ginn, T.; Donnelly, R. Effects of pollution on marine organisms. *Water Environ. Res.* **2009**, *81*, 2070–2125.

25. Brozinski, J.-M.; Lahti, M.; Meierjohann, A.; Oikari, A.; Kronberg, L. The anti-inflammatory drugs diclofenac, naproxen and ibuprofen are found in the bile of wild fish caught downstream of a wastewater treatment plant. *Environ. Sci. Technol.* **2013**, *47*, 342–348.

26. Cecchini, M.; LoPresti, V. Drug residues store in the body following cessation of use: Impacts on neuroendocrine balance and behavior—Use of the Hubbard sauna regimen to remove toxins and restore health. *Med. Hypotheses* **2007**, *68*, 868–879.

27. Rudel, R.A.; Attfield, K.R.; Schifano, J.N.; Brody, J.G. Chemicals causing mammary gland tumors in animals signal new directions for epidemiology, chemicals testing, and risk assessment for breast cancer prevention. *Cancer* **2007**, *109*, 2635–2666.

28. Cunningham, V.L.; Binks, S.P.; Olson, M.J. Human health risk assessment from the presence of human pharmaceuticals in the aquatic environment. *Regul. Toxicol. Pharmacolog.* **2009**, *53*, 39–45.

29. Fent, K.; Weston, A.A.; Caminada, D. Ecotoxicology of human pharmaceuticals. *Aquat. Toxicol.* **2006**, *76*, 122–159.

30. Richardson, M.L.; Bowron, J.M. The fate of pharmaceutical chemicals in the aquatic environment. *J. Pharm. Pharmacolog.* **1985**, *37*, 1–12.

31. Daughton, C.G.; Brooks, B.W. Active pharmaceutical ingredients and aquatic organisms. In *Environmental Contaminants in Biota: Interpreting Tissue Concentrations*, 2nd ed.; Meador, W.B.J., Ed.; Taylor and Francis: Boca Raton, FL, USA, 2011; pp. 281–340.

32. Le-Minh, N.; Khan, S.J.; Drewes, J.E.; Stuetz, R.M. Fate of antibiotics during municipal water recycling treatment processes. *Water Res.* **2010**, *44*, 4295–4323.

33. Ternes, T.A.; Joss, A.; Siegrist, H. Scrutinizing pharmaceuticals and personal care products in wastewater treatment. *Environ. Sci. Technol.* **2004**, *38*, 392A–399A.

34. Petrovic, M.; de Alda, M.J.L.; Diaz-Cruz, S.; Postigo, C.; Radjenovic, J.; Gros, M.; Barcelo, D. Fate and removal of pharmaceuticals and illicit drugs in conventional and membrane bioreactor wastewater treatment plants and by riverbank filtration. *Philos. Trans. R. Soc. A* **2009**, *367*, 3979–4003.

35. Jones, O.A.H.; Voulvoulis, N.; Lester, J.N. Human pharmaceuticals in wastewater treatment processes. *Crit. Rev. Environ. Sci. Technol.* **2005**, *35*, 401–427.

36. Onesios, K.M.; Yu, J.T.; Bouwer, E.J. Biodegradation and removal of pharmaceuticals and personal care products in treatment systems: A review. *Biodegradation* **2009**, *20*, 441–466.

37. Snyder, S.A. Occurrence, treatment, and toxicological relevance of EDCs and pharmaceuticals in water. *Ozone Sci. Eng.* **2008**, *30*, 65–69.

38. Ikehata, K.; Naghashkar, N.J.; Ei-Din, M.G. Degradation of aqueous pharmaceuticals by ozonation and advanced oxidation processes: A review. *Ozone Sci. Eng.* **2006**, *28*, 353–414.

39. Yargeau, V.; Leclair, C. Impact of operating conditions on decomposition of antibiotics during ozonation: A review. *Ozone Sci. Eng.* **2008**, *30*, 175–188.

40. Werner, J.J.; McNeill, K.; Arnold, W.A. Photolysis of chlortetracycline on a clay surface. *J. Agric. Food Chem.* **2009**, *57*, 6932–6937.

41. Santoke, H.; Song, W.H.; Cooper, W.J.; Greaves, J.; Miller, G.E. Free-radical-induced oxidative and reductive degradation of fluoroquinolone pharmaceuticals: kinetic studies and degradation mechanism. *J. Phys. Chem. A* **2009**, *113*, 7846–7851.

42. Hu, L.; Martin, H.M.; Arcs-Bulted, O.; Sugihara, M.N.; Keatlng, K.A.; Strathmann, T.J. Oxidation of carbamazepine by Mn(VII) and Fe(VI): Reaction kinetics and mechanism. *Environ. Sci. Technol.* **2009**, *43*, 509–515.

43. Leech, D.M.; Snyder, M.T.; Wetzel, R.G. Natural organic matter and sunlight accelerate the degradation of 17 beta-estradiol in water. *Sci. Total Environ.* **2009**, *407*, 2087–2092.

44. Benotti, M.J.; Brownawell, B.J. Microbial degradation of pharmaceuticals in estuarine and coastal seawater. *Environ. Pollut.* **2009**, *157*, 994–1002.

45. Yu, T.H.; Lin, A.Y.C.; Lateef, S.K.; Lin, C.F.; Yang, P.Y. Removal of antibiotics and non-steroidal anti-inflammatory drugs by extended sludge age biological process. *Chemosphere* **2009**, *77*, 175–181.

46. Wu, C.X.; Spongberg, A.L.; Witter, J.D. Sorption and biodegradation of selected antibiotics in biosolids. *J. Environ. Sci. Health Tox. Hazard. Subst. Environ. Eng.* **2009**, *44*, 454–461.

47. Wu, C.X.; Spongberg, A.L.; Witter, J.D. Adsorption and degradation of triclosan and triclocarban in solis and biosolids-amended soils. *J. Agric. Food Chem.* **2009**, *57*, 4900–4905.

48. Chang, P.H.; Li, Z.H.; Yu, T.L.; Munkhbayer, S.; Kuo, T.H.; Hung, Y.C.; Jean, J.S.; Lin, K.H. Sorptive removal of tetracycline from water by palygorskite. *J. Hazard. Mater.* **2009**, *165*, 148–155.

49. Wilcox, J.D.; Bahr, J.M.; Hedman, C.J.; Hemming, J.D.C.; Barman, M.A.E.; Bradbury, K.R. Removal of organic wastewater contaminants in septic systems using advanced treatment technologies. *J. Environ. Qual.* **2009**, *38*, 149–156.

50. Phillips, P.J.; Smith, S.G.; Kolpin, D.W.; Zaugg, S.D.; Buxton, H.T.; Furlong, E.T.; Esposito, K.; Stinson, B. Pharmaceutical formulation facilities as sources of opioids and other pharmaceuticals to wastewater treatment plant effluents. *Environ. Sci. Technol.* **2010**, *44*, 4910–4916.

51. Nagarnaik, P.M.; Batt, A.L.; Boulanger, B. Healthcare facility effluents as point sources of select pharmaceuticals to municipal wastewater. *Water Environ. Res.* **2012**, *84*, 339–345.

52. Daughton, C.G.; Ruhoy, I.S. Environmental footprint of pharmaceuticals: The significance of factors beyond direct excretion to sewers. *Environ. Toxicol. Chem.* **2009**, *28*, 2495–2521.

53. Ruhoy, I.S.; Daughton, C.G. Types and quantities of leftover drugs entering the environment via disposal to sewage—Revealed by coroner records. *Sci. Total Environ.* **2007**, *388*, 137–148.

54. Glassmeyer, S.T.; Hinchey, E.K.; Boehme, S.E.; Daughton, C.G.; Ruhoy, I.S.; Conerly, O.; Daniels, R.L.; Lauer, L.; McCarthy, M.; Nettesheim, T.G.; *et al.* Disposal practices for unwanted residential medications in the United States. *Environ. Int.* **2009**, *35*, 566–572.

55. Seehusen, D.A.; Edwards, J. Patient practices and beliefs concerning disposal of medications. *J. Am. Board Fam. Med.* **2006**, *19*, 542–547.

56. Kotchen, M.; Kallaos, J.; Wheeler, K.; Wong, C.; Zahller, M. Pharmaceuticals in wastewater: Behavior, preferences, and willingness to pay for a disposal program. *J. Environ. Manag.* **2009**, *90*, 1476–1482.

57. USEPA. Summary of the Clean Water Act. Available online: http://www2.epa.gov/laws-regulations/summary-clean-water-act (accessed on 1 August 2013).

58. Heidler, J.; Halden, R.U. Meta-analysis of mass balances examining chemical fate during wastewater treatment. *Environ. Sci. Technol.* **2008**, *42*, 6324–6332.

59. McClellan, K.; Halden, R.U. Pharmaceuticals and personal care products in archived US biosolids from the 2001 EPA national sewage sludge survey. *Water Res.* **2010**, *44*, 658–668.

60. Fick, J.; Soderstrom, H.; Lindberg, R.H.; Phan, C.; Tysklind, M.; Larsson, D.G.J. Contamination of surface, ground, and drinking water from pharmaceutical production. *Environ. Toxicol. Chem.* **2009**, *28*, 2522–2527.

61. Larsson, D.G.J.; de Pedro, C.; Paxeus, N. Effluent from drug manufactures contains extremely high levels of pharmaceuticals. *J. Hazard. Mater.* **2007**, *148*, 751–755.

62. Edwards, M.; Topp, E.; Metcalfe, C.D.; Li, H.; Gottschall, N.; Bolton, P.; Curnoe, W.; Payne, M.; Beck, A.; Kleywegt, S.; *et al.* Pharmaceutical and personal care products in tile drainage following surface spreading and injection of dewatered municipal biosolids to an agricultural field. *Sci. Total Environ.* **2009**, *407*, 4220–4230.

63. Lapen, D.R.; Topp, E.; Metcalfe, C.D.; Li, H.; Edwards, M.; Gottschall, N.; Bolton, P.; Curnoe, W.; Payne, M.; Beck, A. Pharmaceutical and personal care products in tile drainage following land application of municipal biosolids. *Sci. Total Environ.* **2008**, *399*, 50–65.

64. Lin, A.Y.C.; Yu, T.H.; Lin, C.F. Pharmaceutical contamination in residential, industrial, and agricultural waste streams: Risk to aqueous environments in Taiwan. *Chemosphere* **2008**, *74*, 131–141.

65. Phillips, P.; Chalmers, A. Wastewater effluent, combined sewer overflows, and other sources of organic compounds to Lake Champlain. *J. Am. Water Resour. Assoc.* **2009**, *45*, 45–57.

66. Shala, L.; Foster, G.D. Surface water concentrations and loading budgets of pharmaceuticals and other domestic-use chemicals in an urban watershed (Washington, DC, USA). *Arch. Environ. Contam. Toxicol.* **2010**, *58*, 551–561.

67. Halden, R.U.; Paull, D.H. Co-occurrence of triclocarban and triclosan in US water resources. *Environ. Sci. Technol.* **2005**, *39*, 1420–1426.

68. Fram, M.S.; Belitz, K. Occurrence and concentrations of pharmaceutical compounds in groundwater used for public drinking-water supply in California. *Sci. Total Environ.* **2011**, *409*, 3409–3417.

69. Deo, R.P.; Halden, R.U. Effect of sample filtration on the quality of monitoring data reported for organic compounds during wastewater treatment. *J. Environ. Monit.* **2010**, *12*, 478–483.

70. Sunkara, M.; Wells, M.J.M. Phase II pharmaceutical metabolites acetaminophen glucuronide and acetaminophen sulfate in wastewater. *Environ. Chem.* **2009**, *7*, 111–122.

71. Jones, O.A.; Lester, J.N.; Voulvoulis, N. Pharmaceuticals: A threat to drinking water? *Trends Biotechnol.* **2005**, *23*, 163–167.

72. Yu, Y.; Wu, L.; Chang, A.C. Seasonal variation of endocrine disrupting compounds, pharmaceuticals and personal care products in wastewater treatment plants. *Sci. Total Environ.* **2013**, *442*, 310–316.

73. Wilde, M.L.; Kummerer, K.; Martins, A.F. Multivariate optimization of analytical methodology and a first attempt to an environmental risk assessment of beta-blockers in hospital wastewater. *J. Braz. Chem. Soc.* **2012**, *23*, 1732–1740.

74. Ferrari, B.; Paxeus, N.; Lo Giudice, R.; Pollio, A.; Garric, J. Ecotoxicological impact of pharmaceuticals found in treated wastewaters: Study of carbamazepine, clofibric acid, and diclofenac. *Ecotoxicol. Environ. Saf.* **2003**, *55*, 359–370.

75. Vazquez-Roig, P.; Andreu, V.; Onghena, M.; Blasco, C.; Pico, Y. Assessment of the occurrence and distribution of pharmaceuticals in a Mediterranean wetland (L'Albufera, Valencia, Spain) by LC-MS/MS. *Anal. Bioanal. Chem.* **2011**, *400*, 1287–1301.

76. Carlsson, C.; Johansson, A.K.; Alvan, G.; Bergman, K.; Kuhler, T. Are pharmaceuticals potent environmental pollutants? Part II: Environmental risk assessments of selected pharmaceutical excipients. *Sci. Total Environ.* **2006**, *364*, 88–95.

77. Agerstrand, M.; Ruden, C. Evaluation of the accuracy and consistency of the Swedish environmental classification and information system for pharmaceuticals. *Sci. Total Environ.* **2010**, *408*, 2327–2339.

78. Zheng, Q.; Zhang, R.J.; Wang, Y.H.; Pan, X.H.; Tang, J.H.; Zhang, G. Occurrence and distribution of antibiotics in the Beibu Gulf, China: Impacts of river discharge and aquaculture activities. *Mar. Environ. Res.* **2012**, *78*, 26–33.

79. Miller, K.J.; Meek, J. *Helena Valley Ground Water: Pharmaceuticals, Personal Care Products, Endocrine Disruptors (PPCPs) and Microbial Indicators of Faecal Contamination*; Montana Department of Environmental Quality: Helena, MT, USA, 2006.

80. Barnes, K.K.; Kolpin, D.W.; Furlong, E.T.; Zaugg, S.D.; Meyer, M.T.; Barber, L.B. A national reconnaissance of pharmaceuticals and other organic wastewater contaminants in the United States—I) Groundwater. *Sci. Total Environ.* **2008**, *402*, 192–200.

81. Bartelt-Hunt, S.; Snow, D.D.; Damon-Powell, T.; Miesbach, D. Occurrence of steroid hormones and antibiotics in shallow groundwater impacted by livestock waste control facilities. *J. Contam. Hydrol.* **2011**, *123*, 94–103.

82. Fang, Y.; Karnjanapiboonwong, A.; Chase, D.A.; Wang, J.F.; Morse, A.N.; Anderson, T.A. Occurrence, fate, and persistence of gemfibrozil in water and soil. *Environ. Toxicol. Chem.* **2012**, *31*, 550–555.

83. Watanabe, N.; Bergamaschi, B.A.; Loftin, K.A.; Meyer, M.T.; Harter, T. Use and environmental occurrence of antibiotics in freestall dairy farms with manured forage fields. *Environ. Sci. Technol.* **2010**, *44*, 6591–6600.

84. Mackie, R.I.; Koike, S.; Krapac, I.; Chee-Sanford, J.; Maxwell, S.; Aminov, R.I. Tetracycline residues and tetracycline resistance genes in groundwater impacted by swine production facilities. *Anim. Biotechnol.* **2006**, *17*, 157–176.

85. Zhao, S.; Zhang, P.F.; Crusius, J.; Kroeger, K.D.; Bratton, J.F. Use of pharmaceuticals and pesticides to constrain nutrient sources in coastal groundwater of northwestern Long Island, New York, USA. *J. Environ. Monit.* **2011**, *13*, 1337–1343.

86. Wang, C.A.; Shi, H.L.; Adams, C.D.; Gamagedara, S.; Stayton, I.; Timmons, T.; Ma, Y.F. Investigation of pharmaceuticals in Missouri natural and drinking water using high performance liquid chromatography-tandem mass spectrometry. *Water Res.* **2011**, *45*, 1818–1828.

87. Stackelberg, P.E.; Furlong, E.T.; Meyer, M.T.; Zaugg, S.D.; Henderson, A.K.; Reissman, D.B. Persistence of pharmaceutical compounds and other organic wastewater contaminants in a conventional drinking-watertreatment plant. *Sci. Total Environ.* **2004**, *329*, 99–113.

88. Stackelberg, P.E.; Gibs, J.; Furlong, E.T.; Meyer, M.T.; Zaugg, S.D.; Lippincott, R.L. Efficiency of conventional drinking-water-treatment processes in removal of pharmaceuticals and other organic compounds. *Sci. Total Environ.* **2007**, *377*, 255–272.

89. Snyder, S.A.; Adham, S.; Redding, A.M.; Cannon, F.S.; DeCarolis, J.; Oppenheimer, J.; Wert, E.C.; Yoon, Y. Role of membranes and activated carbon in the removal of endocrine disruptors and pharmaceuticals. *Desalination* **2007**, *202*, 156–181.

90. Kim, S.D.; Cho, J.; Kim, I.S.; Vanderford, B.J.; Snyder, S.A. Occurrence and removal of pharmaceuticals and endocrine disruptors in South Korean surface, drinking, and waste waters. *Water Res.* **2007**, *41*, 1013–1021.

91. Loraine, G.A.; Pettigrove, M.E. Seasonal variations in concentrations of pharmaceuticals and personal care products in drinking water and reclaimed wastewater in Southern California. *Environ. Sci. Technol.* **2006**, *40*, 687–695.

92. Trenholm, R.A.; Vanderford, B.J.; Snyder, S.A. On-line solid phase extraction LC-MS/MS analysis of pharmaceutical indicators in water: A green alternative to conventional methods. *Talanta* **2009**, *79*, 1425–1432.

93. PBT Profiler software. Available online: http://www.pbtprofiler.net (accessed on 1 August 2013).

94. Hernando, M.D.; Mezcua, M.; Fernandez-Alba, A.R.; Barcelo, D. Environmental risk assessment of pharmaceutical residues in wastewater effluents, surface waters and sediments. *Talanta* **2006**, *69*, 334–342.

95. Carlsson, C.; Johansson, A.K.; Alvan, G.; Bergman, K.; Kuhler, T. Are pharmaceuticals potent environmental pollutants? Part I: Environmental risk assessments of selected active pharmaceutical ingredients. *Sci. Total Environ.* **2006**, *364*, 67–87.

96. Hernando, M.D.; Gomez, M.J.; Aguera, A.; Fernandez-Alba, A.R. LC-MS analysis of basic pharmaceuticals (beta-blockers and anti-ulcer agents) in wastewater and surface water. *Trac Trends Anal. Chem.* **2007**, *26*, 581–594.

97. Clarke, B.O.; Smith, S.R. Review of 'emerging' organic contaminants in biosolids and assessment of international research priorities for the agricultural use of biosolids. *Environ. Int.* **2011**, *37*, 226–247.

98. Clarke, B.O.; Porter, N.A. Persistent organic pollutants in sewage sludge: Levels, sources, and trends. In *Contaminants of Emerging Concern in the Environment: Ecological and Human Health Considerations*; Rolf, U.H., Ed.; American Chemical Society: Washington, DC, USA, 2010; pp. 137–171.

99. Harrison, E.Z.; Oakes, S.R.; Hysell, M.; Hay, A. Organic chemicals in sewage sludges. *Sci. Total Environ.* **2006**, *367*, 481–497.

100. Rogers, H.R. Sources, behaviour and fate of organic contaminants during sewage treatment and in sewage sludges. *Sci. Total Environ.* **1996**, *185*, 3–26.

101. New England Bicycle Racing Association. *A National Biosolids Regulation, Quality, End Use & Disposal Survey*; Final Report; New England Bicycle Racing Association: Tamworth, NH, UK, 2007.

102. Xu, J.; Chen, W.P.; Wu, L.S.; Green, R.; Chang, A.C. Leachability of some emerging contaminants in reclaimed municipal wastewater-irrigated turf grass fields. *Environ. Toxicol. Chem.* **2009**, *28*, 1842–1850.

103. Xia, K.; Hundal, L.S.; Kumar, K.; Armbrust, K.; Cox, A.E.; Granato, T.C. Triclocarban, triclosan, polybrominated diphenyl ethers, and 4-nonylphenol in biosolids and in soil receiving 33-year biosolids application. *Environ. Toxicol. Chem.* **2010**, *29*, 597–605.

104. Smith, S.R. Organic contaminants in sewage sludge (biosolids) and their significance for agricultural recycling. *Philos. Trans. R. Soc. A* **2009**, *367*, 4005–4041.

105. Wu, C.X.; Spongberg, A.L.; Witter, J.D.; Fang, M.; Ames, A.; Czajkowski, K.P. Detection of Pharmaceuticals and Personal Care Products in Agricultural Soils Receiving Biosolids Application. *Clean Soil Air Water* **2010**, *38*, 230–237.

106. Dolliver, H.; Kumar, K.; Gupta, S. Sulfamethazine uptake by plants from manure-amended soil. *J. Environ. Qual.* **2007**, *36*, 1224–1230.

107. Kumar, K.; Gupta, S.C.; Baidoo, S.K.; Chander, Y.; Rosen, C.J. Antibiotic uptake by plants from soil fertilized with animal manure. *J. Environ. Qual.* **2005**, *34*, 2082–2085.

108. Holling, C.S.; Bailey, J.L.; Heuvel, B.V.; Kinney, C.A. Uptake of human pharmaceuticals and personal care products by cabbage (Brassica campestris) from fortified and biosolids-amended soils. *J. Environ. Monit.* **2012**, *14*, 3029–3036.

109. Klavarioti, M.; Mantzavinos, D.; Kassinos, D. Removal of residual pharmaceuticals from aqueous systems by advanced oxidation processes. *Environ. Int.* **2009**, *35*, 402–417.

110. Sharma, V.K. Oxidative transformations of environmental pharmaceuticals by Cl-2, ClO2, O-3, and Fe(VI): Kinetics assessment. *Chemosphere* **2008**, *73*, 1379–1386.

111. Ikehata, K.; Gamal El-Din, M.; Snyder, S.A. Ozonation and advanced oxidation treatment of emerging organic pollutants in water and wastewater. *Ozone Sci. Eng. J. Int. Ozone Assoc.* **2008**, *30*, 21–26.

112. Esplugas, S.; Bila, D.M.; Krause, L.G.T.; Dezotti, M. Ozonation and advanced oxidation technologies to remove endocrine disrupting chemicals (EDCs) and pharmaceuticals and personal care products (PPCPs) in water effluents. *J. Hazard. Mater.* **2007**, *149*, 631–642.

113. Ellis, W.C.; Tran, C.T.; Roy, R.; Rusten, M.; Fischer, A.; Ryabov, A.D.; Blumberg, B.; Collins, T.J. Designing green oxidation catalysts for purifying environmental waters. *J. Am. Chem. Soc.* **2010**, *132*, 9774–9781.

114. Daughton, C.G. Cradle-to-cradle stewardship of drugs for minimizing their environmental disposition while promoting human health. II. Drug disposal, waste reduction, and future directions. *Environ. Health Perspect.* **2003**, *111*, 775–785.

115. Kummerer, J.; Hempel, M. *Green and Sustainable Pharmacy*, 1st ed.; Springer-Verlag: Berlin Heidelberg, Germany, 2010.

Water **2013**, *5*, 292-311; doi:10.3390/w5010292

OPEN ACCESS

ISSN 2073-4441
www.mdpi.com/journal/water

Review

Sustainable Agro-Food Industrial Wastewater Treatment Using High Rate Anaerobic Process

Rajinikanth Rajagopal [1,*], **Noori M. Cata Saady** [1], **Michel Torrijos** [2], **Joseph V. Thanikal** [3] **and Yung-Tse Hung** [4]

[1] Dairy and Swine Research and Development Centre, Agriculture and Agri-Food Canada, Sherbrooke, Quebec J1M0C8, Canada; E-Mail: noori.saady@agr.gc.ca

[2] INRA, UR50, Laboratory of Environmental Biotechnology, Avenue des Etangs, Narbonne, F-11100, France; E-Mail: michel.torrijos@supagro.inra.fr

[3] Caledonian Centre for Scientific Research, Department of Built and Natural Environment, Caledonian College of Engineering, P.O. Box 2322, CPO Seeb 111, Muscat, Sultanate of Oman; E-Mail: joseph@caledonian.edu.om

[4] Department of Civil and Environmental Engineering, Cleveland State University, Cleveland, OH 44115, USA; E-Mail: yungtsehung@yahoo.com

* Author to whom correspondence should be addressed; E-Mail: rajinikanth.rajagopal@agr.gc.ca and rrajinime@yahoo.co.in; Tel.: +1-819-780-7303; Fax: +1-819-564-5507.

Received: 5 January 2013; in revised form: 15 February 2013 / Accepted: 5 March 2013 / Published: 15 March 2013

Abstract: This review article compiles the various advances made since 2008 in sustainable high-rate anaerobic technologies with emphasis on their performance enhancement when treating agro-food industrial wastewater. The review explores the generation and characteristics of different agro-food industrial wastewaters; the need for and the performance of high rate anaerobic reactors, such as an upflow anaerobic fixed bed reactor, an upflow anaerobic sludge blanket (UASB) reactor, hybrid systems *etc.*; operational challenges, mass transfer considerations, energy production estimation, toxicity, modeling, technology assessment and recommendations for successful operation.

Keywords: anaerobic digestion; agro-food wastewater; biogas; high-rate systems; sustainable wastewater treatment

---

1. Introduction

Agro-industries are major contributors to worldwide industrial pollution. Effluents from many agro-food industries are a hazard to the environment and require appropriate and a comprehensive management approach. Worldwide, environmental regulatory authorities are setting strict criteria for discharge of wastewaters from industries. As regulations become stricter, there is now a need to treat and utilize these wastes quickly and efficiently. With the tremendous pace of development of sustainable biotechnology, substantial research has been devoted recently to cope with wastes of ever increasing complexity generated by agro-industries. Anaerobic digestion is an environmentally friendly green biotechnology to treat agro-food industrial effluents. In addition, the carbon emission and, therefore, the carbon footprint of water utilities is an important issue nowadays. In this perspective, it is essential to consider the prospects for the reduction of the carbon footprint from small and large wastewater treatment plants. The use of anaerobic treatment processes rather than aerobic would accomplish this purpose, because no aeration is required and the biogas generated can be used within the plant. Anaerobic digestion is unique, as it reduces waste and produces energy in the form of methane. Not only does this technology have a positive net energy production, but the biogas produced can also replace fossil fuel; therefore, it has a direct positive effect on greenhouse gas reduction. Thus, the carbon-negative anaerobic digestion process is considered as a sustainable wastewater treatment technology, which also provides the best affordable (low-cost) process for public health and environmental protection, as well as resource recovery. The attractiveness of biogas technology for large scale applications has been limited, essentially, because of the slow rate and process instability of anaerobic digestion. The slow rate means large digester volumes (and consequently, greater costs and space requirements) and process instability means the lack of assurance for a steady energy supply. These two major disadvantages of conventional anaerobic processes have been overcome by high rate anaerobic reactors, which employ cell immobilization techniques, such as granules and biofilms. Thus, various reactor designs that employ various ways of retaining biomass within the reactor have been developed over the past two decades. The purpose of this article is to summarize the current status of the research on high rate anaerobic treatment of agro-food industrial wastewater and to provide strategies to overcome some of the operational problems.

2. Agro-Food Industrial Wastewaters

About 65%–70% of the organic pollutants released in the water bodies in India are from food and agro-product industries, such as distilleries, sugar factories, dairies, fruit canning, meat processing and pulp and paper mills [1]. Similarly, the pulp and paper industry is one of the most significant industries in Sweden, as well as many other countries around the world, and the products constitute important industrial trade in terms of value of production [2]. Wine production is one of the leading agro-food industries in Mediterranean countries, and it has also attained importance in other parts of the world, such as Australia, Chile, the United States, South Africa and China, with increasing influence on the economy of these countries [3]. The wine industry generates huge volumes of wastewater that are mainly originated from several washing operations, e.g., during crushing and pressing of the grapes, cleaning the fermentation tanks, containers, other equipment and surfaces. In addition to this, olive oil industries have gained fundamental economic importance for many Mediterranean countries [4]. Malaysia presently accounts for 39% of world palm oil production and 44% of world exports [5]. Due to its surplus production, a huge amount of polluted wastewater, commonly referred to as palm oil mill effluent (POME) is generated. Fia *et al.* [6] reported that in coffee producing regions, such as Brazil, Vietnam and Colombia, the final effluent produced from this process has become a large environmental problem, creating the need for low cost technologies for the treatment of wastewater. Nieto *et al.* [7] determined the potential for methane production from six agro-food wastes (beverage waste, milled apple waste, milk waste, yogurt waste, fats and oils from dairy wastewater treatment and cattle manure). The wastewater generation varies from country to country. For instance, world wine production in 2011 was estimated to be about 270 million hectoliters, with the European Union producing 152 million hL (France with 50.2 million hL; Italy, with 40.3 million hL; Spain, with 35.4 million hL), whereas it was estimated to be 18.7, 15.5, 11.9, 10.6, 9.3 and 2.3 million hL for countries, such as the USA, Argentina, Australia, Chile, South Africa and New Zealand, respectively [8]. The annual worldwide production of olive oil is estimated to be about 1750 million metric tons, with Spain, Italy, Greece, Tunisia and Portugal being the major producers, and about 30 million cubic meters of oil mill wastes (OMW) are generated annually in the Mediterranean area during the seasonal extraction of olive oil [4,9].

The composition and concentration of different agro-food wastewaters vary from low (wash water from sugar mill or dairy effluents) to high strength substrates (cheese, winery and olive mill wastewaters), particularly in terms of organic matter, acids, proteins, aromatic compounds, available nutrients, *etc.* [1,3,10]. The main parameters of the agro-food industrial wastewater, such as total solids (TS), total nitrogen (TN), total phosphorus (TP) and biochemical and chemical oxygen demand (BOD and COD), respectively are given in Table 1.

Table 1. Characteristics of typical agro-food industrial wastewater.

Industry	TS (mg L^{-1})	TP (mg L^{-1})	TN (mg L^{-1})	BOD (mg L^{-1})	COD (mg L^{-1})	Reference
Food processing[a]	-	3	50	600–4,000	1,000–8,000	[11]
Palm oil mill	40	-	750	25	50	[12]
Sugar-beet processing	6100	2.7	10	-	6,600	[13]
Dairy	1,100–1,600	-	-	800–1,000	1,400–2,500	[14]
Corn milling	650	125	174	3,000	4,850	[15]
Potato chips	5,000	100	250	5,000	6,000	[16]
Baker's yeast	600	3	275	-	6,100	[17]
Winery	150–200	40–60	310–410	-	18,000–21,000	[1,3]
Dairy	250–2,750	-	10–90	650–6,250	400–15,200	[18]
Cheese dairy	1,600–3,900	60–100	400–700	-	23,000–4,0000	[1]
Olive mill	75,500	-	460	-	130,100	[19]
Cassava starch	830	90	525	6,300	10,500	[20]

Notes: [a] contains flour, soybean, tomato, pepper and salt. TS: total solids; TN: total nitrogen; TP: total phosphorus; BOD: biochemical oxygen demand; COD: chemical oxygen demand.

3. High Rate Anaerobic Reactors

High-rate anaerobic digesters receive increasing interest, due to their high loading capacity and low sludge production [21]. The commonly used high-rate anaerobic digesters include: anaerobic filters, upflow anaerobic sludge blanket (UASB) reactors, anaerobic baffled, fluidized beds, expanded granular sludge beds (EGSB), sequencing batch reactors and anaerobic hybrid/hybrid upflow anaerobic sludge blanket reactors [1,21].

Rajagopal *et al.* [22] developed high-rate upflow anaerobic filters (UAFs) packed with low-density polyethylene media for the treatment of wastewater discharged from various agro-food industries with different composition and COD concentrations *viz.* synthetically prepared low strength ($\sim$1.9 g COD L^{-1}), fruit canning ($\sim$10 g COD L^{-1}), winery ($\sim$20 g COD L^{-1}) and cheese-dairy ($\sim$30 g COD L^{-1}) wastewaters. High organic loading rates (OLRs) (12–27 g COD L^{-1} d^{-1}) were reported in this study. Low-density polyethylene support (29 mm high; 30–35mm diameter; density: 0.93 kg m^{-3}; specific area: 320 m^2 m^{-3}) was able to retain between 0.7 and 1.6 g dried solids per support. This study concluded that the low-density polyethylene support is a good colonization matrix to increase the quantity of biomass in the reactor compared to conventional treatment systems.

In a similar study, Ganesh *et al.* [3] investigated the performance of upflow anaerobic fixed-bed reactors filled with low density supports of varying size and specific surface area for the treatment of winery wastewater. They found that efficiency of reactors increased with decrease in size and increase in specific surface area of the supports. A maximum OLR of 42 g L^{-1} d^{-1} with 80% COD removal efficiency was attained, when supports with the highest specific surface area were used. However, for long-term operation, clogging might occur in the reactors packed with small size supports.

Esparza *et al.* [23] reported that a pilot-scale upflow anaerobic sludge blanket (UASB) reactor treating cereal-processing wastewater at 17 °C with OLR 4–8 kg COD m^{-3} d^{-1} and a hydraulic retention time (HRT) of 5.2 h removed 82%–92% of the COD. Shastry *et al.* [24] investigated the feasibility of a UASB reactor system as a pretreatment for hydrogenated vegetable oil wastewater. COD removal efficiency of 99%–80% at OLR varying in the range 1.3–10 g COD L^{-1} d^{-1} was obtained with a specific methane yield of 0.30–0.35 m^3 CH_4 kg^{-1} COD.

Won and Lau [25] operated an anaerobic sequencing batch reactor (ASBR) to investigate the effect of pH, HRT and OLR on biohydrogen production at 28 °C. For a carbon-rich substrate, a maximum hydrogen production rate and yield of 3 L H_2 L^{-1} reactor d^{-1} and 2.2 mol H_2 mol^{-1} hexose, respectively, were achieved at pH 4.5, HRT 30 h and OLR 11.0 g L^{-1} d^{-1}. A mesophilic anaerobic sequencing batch biofilm reactor (ASBBR) treating lipid-rich wastewater with OLR as much as 12.1 g COD L^{-1} d^{-1} was achieved with 90% COD removal efficiency [26].

Shanmugam and Akunna [27] investigated the performance of a granular bed baffled reactor (GRABBR) for the treatment of low-strength wastewaters at increasing OLRs (up to 60 g COD L^{-1} d^{-1}). They showed experimentally that GRABBR encouraged different stages of anaerobic digestion in separate vessels longitudinally across the reactor, and it had greater process stability at relatively short HRTs (1 h) with 86% methane in biogas. Bialek *et al.* [28] assessed the performance of two kinds of reactors (inverted fluidized bed (IFB) and expanded granular sludge bed (EGSB) reactors) treating simulated dairy wastewater. At 37 °C, they obtained more than 80% of COD and protein removals with an OLR of 167 mg COD L^{-1} h^{-1} and a HRT of 24 h.

Rajagopal *et al.* [29] proposed a modified UAF reactor design, called "hybrid upflow anaerobic sludge-filter bed (UASFB) reactor", to overcome problems faced by fixed bed reactors, such as clogging, short circuiting and biomass washout. This configuration contains a sludge bed in the lower part of the reactor and a filter bed in the upper part. For the treatment of wine distillery vinasse (21.7 g COD L^{-1}) they achieved a high OLR (18 g COD L^{-1} d^{-1}) at a short HRT (26 h), while maintaining high COD removal efficiencies of about 85%. However, an aerobic post-treatment is required to make the effluent fit for final disposal, especially in terms of nitrogen [3,14,22]. Table 2 summarizes the performance of anaerobic digesters for the treatment of various agro-food wastewaters in terms of design and applied operational conditions, process efficiency and energy characteristics.

Other Treatment Strategies to Enhance the Reactor Performance

Biogas production can be augmented by using ample lignocellulose materials *viz.* agricultural and forest residues [30,31]. However, the complex lignocellulose structure limits the accessibility of sugars in cellulose and hemicellulose. Consequently, a pretreatment is essential and several potential pretreatment techniques have been developed for lignocellulose material [32].

Nkemka and Murto [31] evaluated the biogas production in batch and UASB reactors from pilot-scale acid catalyzed steam-pretreated and enzymatic-hydrolysed wheat straw. They showed that the pre-treatment increased the methane yield [0.28 m^3 kg^{-1} volatile solids (VS) fed] by 57% compared to untreated straw. The treatment of straw hydrolysate with nutrient supplementation in a UASB reactor resulted in a high methane production rate (2.70 m^3 m^{-3} d^{-1}) at an OLR of 10.4 g COD L^{-1} d^{-1} and with 94% COD reduction.

Badshah *et al.* [2] proved that the methanol condensate can be efficiently converted into biogas in a UASB reactor instead of methanol, with much of the smell of the feed eliminated. Bosco and Chiampo [33] reported polyhydroxyalkanoates (PHAs) (biodegradable plastic) production from milk whey and dairy wastewater activated sludge. They defined the suitable C/N ratio, the pre-treatments required to lower the protein content and the effect of pH correction.

Garcia *et al.* [34] investigated the effect of polyacrylamide (PAM) for the biomass retention in an UASB reactor treating liquid fraction of dairy manure at several organic loading rates. They have concluded that PAM addition enhanced sludge retention and reactor performance (with a total COD removal of 83% compared to 77% removal efficiencies for UASB without polymer addition).

In order to remove ammonia nitrogen, Yu *et al.* [35] attempted to enrich anammox bacteria in sequencing batch biofilm reactors (SBBRs) with different inoculations. The maximum total nitrogen loading rate of SBBR gradually reached 1.62 kg N m^{-3} d^{-1}, with a removal efficiency higher than 88%. Few researchers [36,37] reported that anammox bacteria have slow growth rate kinetics, and hence, they are vulnerable to external conditions, such as low temperature, high dissolved oxygen and some inhibitors. Although this technique faces some complications in full-scale applications, such as long start-up and instability, anammox is still an innovative technological development in reducing ammonia from wastewater [35,38].

Table 2. Performance of anaerobic digesters for the treatment of various agro-food wastewaters.

Substrate	Reactor type[a]	TCOD (g L^{-1})	pH	HRT (h)	OLR (g COD L^{-1} d^{-1})	TCOD removal (%)	SCOD removal (%)	VFA/ alkalinity	Comments	Reference
Fruit canning	UAF	9–11.6	6–6.5	12	19	68	81	0.35	Specific sludge loading rate (SSLR): 0.56 g COD g^{-1} VSS d^{-1}	[1]
Cheese dairy	UAF	23–40	6–6.5	40	17	71	82	0.6	SSLR: 0.63 g COD g^{-1} VSS d^{-1}	
Winery wastewater	FBR	18–21	8–11	15–23	22–42	65–70	80	0.8	SSLR: 0.93–1.2 g COD g^{-1} VSS d^{-1}	[3]
Dairy manure	UASB	16.5–20.4	7.4	36–48	9–12.7	76.5–83.4	-	-	- Methane yield: 0.296–0.312 L CH$_4$ g^{-1} COD; - Higher values corresponded to UASB with polymer addition	[34]
Dairy wastewater	AF/BAF	1.8–2.4	6.6–8.4	18.2–38.6	0.66–0.72	98–99 (Overall)	-	-	- Recirculation ratio ranged 100%–300%; - Specific growth rate of the integrated biomass: 0.621–1.208 d^{-1}	[39]
Distillery vinasse	UASFB	3.1–21.7	6–6.5	25–27	17.9–18.2	80–82	84–87	<0.4	SSLR: 0.43–0.47 g COD g^{-1} VS d^{-1}	[29]
Food processing	JBILAFB	0.96–7.9	3.4–11.2	24	1.6–5.6	80	-	0.2–0.5	- Methane production: 0.414 m^3 L^{-1} reactor volume day^{-1}; - Specific energy input for the anaerobic reactor: 0.12 kWh m^{-3}	[11]
Cassava starch wastewater	UMAR	10.5	4.5–4.92	6	10.2–40	77.5–92	-	-	- The optimum HRT was 6.0 h at influent COD of 4000 mg L^{-1}; - Specific methanogenic activity: 0.31 and 0.73 g COD$_{CH4}$ g^{-1} VSS d^{-1} for the first and second feed	[20]
Olive mill effluent	UMAR	5–48	~7	240–120	5–48	81–87	-	-	Maximum biogas production: 1.4 m^3 m^{-3} d^{-1}	[19]

Notes: [a] UAF: upflow anaerobic filter; FBR: anaerobic fixed bed reactor; UASB: upflow anaerobic sludge blanket reactor; AF/BAF: integrated anaerobic/aerobic filter system; UASFB: hybrid upflow anaerobic sludge-filter bed reactor; JBILAFB: full-scale jet biogas internal loop anaerobic fluidized bed; UMAR: upflow multistage anaerobic reactor; TCOD: total chemical oxygen demand; HRT: hydraulic retention time; OLR: organic loading rates; SCOD: soluble chemical oxygen demand; VFA: volatile fatty acids; VSS: volatile suspended solids; VS: volatile solids.

4. Specific Design and Operational Considerations

Within the high rate anaerobic treatment technologies, the immobilization of the microorganisms in the fixed bed reactors and the formation of the granules in UASB have been recognized as the two most frequently used anaerobic techniques to reduce the HRT during the anaerobic digestion of wastewater having low organic matter concentration [1,21]. The upflow fixed bed and UASB reactors are discussed in the subsequent sections.

4.1. Upflow Fixed Bed Reactors

The upflow fixed bed reactors or upflow anaerobic filter (UAF) is one of the earlier designs, and its characteristics are well defined [3,22,40]. The UAF is a relatively simple technology; and in engineering terms, it is not as complex as fluidized bed reactors, and in biological terms, it does not require the formation of a granular sludge, a prerequisite for the UASB reactor, which is usually very difficult to maintain. Also, fixed-film processes are inherently stable and resistant to organic and hydraulic shock loading conditions [6]. Since there is no provision made for intentional wastage of excess biomass from the filters similar to UASB reactors, clogging occurs with continued operation [1].

To accommodate the accumulation of non-attached biomass without plugging of the bed, the early designs of low voidage, rock-packed reactors have largely been replaced by systems that incorporate synthetic packing materials [6,41]. The microorganisms' immobilization on surfaces of synthetic carrier material is an increasingly used strategy to enhance biological treatment. The preferable features of the immobilized cells in comparison to the non-immobilized counterparts includes lower sensitivity to toxic loads, greater catalytic stability, longer microbial residence time, more tolerance to oligotrophic conditions and lower biomass washout risk [42,43].

Nikolaeva *et al.* [40] used waste tire rubber and zeolite as microorganism immobilization supports; Fia *et al.* [6] used blast furnace cinders, polyurethane foam and crushed stone as the supporting materials with porosities of 53%, 95% and 48%, respectively. Other materials that have been utilized for the adhesion of biomass included brick ballasts [44], Raschig rings (1.2 cm diameter $\times$ 1.2 cm) [45], low and high-density polyethylene media [3,22,35,41], polypropylene pall rings [46], non-woven disks [47], porous polyurethane foam [48] and modified porphyritic andesite (WRS) as ammonium adsorbent and bed material [49]. Rajagopal [1] suggested the unclogging procedures with fluidization by gas re-circulation or using liquid, which can be applied whenever clogging occurs. Such a problem can also be overcome by using UAF filled (80% or lower by volume) with low-density floating media [22]. Recently, biofilm reactors have also attracted more attention, especially for treatment of wastewaters containing bio-recalcitrant, inhibitory and toxic compounds [50,51]. However, there is a necessity of (i) post-treatment to reach the discharge standards for organic matter, nutrients (e.g., NH_4^+, PO_4^{3-}, S^{2-}) and pathogens; and (ii) purification of biogas [22,52].

4.2. UASB Reactors

The chief characteristics of a UASB reactor that makes it the established high-rate anaerobic digester worldwide (particularly in tropical countries) is the availability of granular or flocculent sludge, allowing it to achieve high organic matter (COD) removal efficiencies without the need of a support material [21,34,52,53]; the natural turbulence caused by the rising gas bubbles enhances the reactor content mixing and provides efficient wastewater and biomass contact. Therefore, mechanical mixing is not required, thus significantly reducing the energy demand and its associated cost.

Nevertheless, there are still unresolved issues in the anaerobic treatment technology. One of the major drawbacks of these reactors has been the requirement of a long solids retention time, which is not associated with the increasing volume of sludge produced from industrial and human activities [21]. Garcia *et al.* [34] listed various factors affecting granulation and, then, efficiency in UASB reactors *viz.* composition and concentration of organic matter in wastewater to be treated, operating temperature, pH, high ammonia nitrogen concentrations, presence of polyvalent cations, hydrodynamic conditions, inoculated seed and the production of exo-cellular polymeric substances by anaerobic bacteria. Rajagopal *et al.* [29] mentioned that with poor sedimentation characteristics in high loaded anaerobic reactors with suspended solids, the active biomass can be washed out from the reactor with the effluent, causing digester instability. Other drawbacks have been the long start-up period, flotation and disintegration of granular sludge, sludge bulking, deterioration of performance at low temperatures, high sulfate concentration, impure biogas and insufficient removal of organic matters, pathogens and nutrients in the final effluent, thereby failing to comply with the local standards for discharge or reuse [39,52,54,55].

4.3. Integrated Approach: Modified Configurations and Combined Systems

The researchers have been trying to find out new technologies to enhance the performance of anaerobic digesters, especially on the effluent quality, start up and biogas purification, in order to develop a global sustainable wastewater treatment technology. Rajagopal *et al.* [29] proposed the hybrid upflow anaerobic sludge-filter bed (UASFB) reactors, in which the filter bed is located above the sludge bed zone, to reduce the risk of biomass washout from the reactor, especially at high loading rates. This configuration also improved the hydrodynamics characteristics by minimizing the clogging and short circuiting problems inside the reactors.

Lim and Fox [39] developed an anaerobic/aerobic filter (AF/BAF) system for the treatment of dairy wastewater, primarily to remove organic matter and nitrogen simultaneously. The influent was blended with recirculated effluent (100%–300%) to allow for pre-denitrification in the AF, followed by nitrification in the BAF. The average COD removal efficiency was 79.8%–86.8% in the AF, and the average total nitrogen removal efficiency was 50.5%–80.8% in the AF/BAF system. They have concluded that linear velocity was a critical parameter to determine sloughing of biomass in the AF.

Huang *et al.* [56] studied the mesophilic two-phase anaerobic system for the treatment of maize ethanol wastewater, particularly in terms of start-up, cultivation and the morphology of mature granular sludge in an improved methanogenic UASB reactor. By gradually increasing volumetric loading rate and regulating internal circulation, the UASB reactor developed bigger size granules (more than 1.3 mm), and hence, a quick start-up was achieved. They observed that the microfloras of mature methanogenic granular sludge were mainly *Brevibacterium* and filamentous bacteria.

A three-phase separator configuration increased the COD removal by 20%, biogas yield by 29% and decreased biomass wash out by 25% in an UASB treating fruit canning wastewater [57]. An intermittent feeding strategy at an OLR of 4.8 g COD L^{-1} d^{-1} of olive mill wastewater applied to an inverted anaerobic sludge blanket (IASB) reactor improved long chain fatty acids mineralization, prevented biomass washout and yielded 1.4 m^3 CH_4 m^{-3} d^{-1} [19]. Diez *et al.* [58] reported that an anaerobic membrane bioreactor (AnMBR) removed 97% of the COD of oil and greasy wastewater (COD of 22 g L^{-1}) at an OLR of 5.1 kg COD m^{-3} d^{-1}. Kim *et al.* [59] implemented a high-rate two-phase system (OLR = 6.5 g COD L^{-1} d^{-1}) of an anaerobic sequencing batch reactor (ASBR) and an upflow anaerobic sludge blanket (UASB) reactor in series to treat synthetic dairy wastewater treatment. The overall lipid and COD removals were about 80%.

A three-stage configuration of a pre-acidification tank and sequential upflow anaerobic sludge bed reactors (UASBRs) at OLRs of 2.8–7 g COD $L^{-1}d^{-1}$ has been used to treat raw cheese whey with an effluent recycling achieved COD removal of 50%–92% and fat removal of 63%–89% [60]. An upflow anaerobic packed bed (UAPB) reactor filled with seashell treated cheese whey at OLRs of 1.6 to 9.9 g COD L^{-1} d^{-1}, and HRTs of 6–24 h at 25 °C removed 95% of COD [61].

In order to upgrade the quality of anaerobically treated effluent to a level recommended for irrigation, Yasar and Tabinda [62] integrated the UASB reactor with UV and AOPs (advanced oxidation processes) (Ozone, H_2O_2/UV, Fenton and photo-Fenton) primarily for complete color and COD removal and disinfection of pathogens.

5. Mass Transfer Considerations

Rajagopal [1] described that substrate mass transfer into biomass occurs most rapidly when the bio-particle surface-to-volume ratio is high. According to that research, the suspended growth anaerobic processes generate relatively small bio-particles with optimal surface-to-volume ratios. Fixed film or attached growth reactors are often considered to be susceptible to bio-film surface area limitations. In this respect, Chen *et al.* [63] concluded that the bed expansion ratio of a super-high rate anaerobic bioreactor correlated positively with superficial gas and liquid velocities, while maximum bed sludge content and maximum bed contact time between sludge and liquid correlated negatively. Ganesh *et al.* [3] addressed the effect of media design on treatment performance and stated that physical parameters, like the type of media, its size and shape, affect the performance of waste treatment. Chen *et al.* [64] modeled the dynamic behavior and concentration distribution of granular sludge in a super-high-rate spiral anaerobic bioreactor (SSAB) and found that these two parameters depend on the ecological environment of microbial communities and substrate degradation efficiency along the bed's height. The sludge transport efficiency of up-moving biogas in SSAB is less than that in a UASB reactor. Yang *et al.* [65] analyzed the mass transfer in tubular membrane anaerobic bioreactors operated under gas-lift two-phase flow using fluid dynamic modeling. They found that the results with water were in contrast to those with sludge. The sludge filterability strongly influences the transmembrane pressure, and there is a difference between the mass transfer capacity at the noses and the tails of the gas bubbles. Feng *et al.* [66] reported a rapid-mass transfer in a fluidized bed reactor using brick particles as carrier materials; at increasing OLRs from 7.37 to 18.52 kg COD m^{-3} d^{-1} and HRT of 8 h, COD removals between 65% and 75% was obtained. To minimize non-idealities in reactor hydraulics, most anaerobic reactor designs utilize proprietary systems for enhancing process mixing. Anaerobic contact systems rely on mechanical or gas recirculation systems that must be properly sized for the specific application [1].

6. Energy Production Estimation

The production of a useful and valuable product during agro-food wastewater treatment, such as hydrogen and/or methane, could help to lower treatment costs [1,67]. Using a single-chamber microbial electrolysis cell with a graphite-fiber brush anode, Wagner *et al.* [67] generated hydrogen gas at the rate of 0.9–1.0 m^3 $m^{-3}day^{-1}$ using a full-strength or diluted swine wastewater. Under the best conditions, the specific hydrogen production rate of 270 mL H_2 g^{-1} MLVSS d^{-1} (or 3310 mL H_2 L^{-1} d^{-1}) and a hydrogen yield of 172 mL H_2 g^{-1} COD removed were obtained for the treatment of alcohol distillery wastewater containing high potassium and sulfate in an anaerobic sequencing batch reactor [68].

Nieto *et al.* [7] obtained methane yields ranging from 202 to 549 mL $CH_4 \cdot gVS_{fed}^{-1}$ at standard temperature and pressure (STP) from six wastes (beverage waste, milled apple waste, milk waste, yogurt waste, fats and oils (F&O) from dairy wastewater treatment, F&O and cattle manure). They reported that methane content in biogas ranged from 58% to 76%. Rajagopal *et al.* [69] developed a process combining anaerobic digestion and anoxic/oxic treatment to treat swine wastewater and obtained 5.9 Nm^3 of CH_4 m^{-3} of slurry added. In another study, Labatut *et al.* [70] observed that co-digestion of dairy manure with easily-degradable substrates increases the specific methane yields when compared to manure-only. For instance, co-digestion of dairy manure with used oil (ratio: 75:25 on volatile solids basis) produced more methane yield (361 Nm^3 $ton^{-1}VS_{added}$) than dairy manure with cheese whey (ratio: 75:25 on VS basis; methane yield: 252 Nm^3 $ton^{-1}VS_{added}$). Ogejo and Li [71] obtained a biogas yield ranging from 0.072 to 0.8 m^3 g^{-1} VS (methane content ranging from 56% to 70%) by co-digesting flushed dairy manure and turkey processing wastewater. The biogas produced was enough to run a 50 kW generator to produce electricity for about 5.5 and 9 h for the 1:1 and 1:2 feed mixtures. In addition to the electricity to be produced, other possible revenues, such as carbon credits, renewable energy credits, green tags for electricity, putting a value to the environmental benefits of anaerobic digestion or subsidies from grants or other incentives programs to make the system economically viable should be considered. On average, 18.5–40 kg of VS added in to the anaerobic digestion system can produce a biogas yield of 10 ± 5 m^3, when 65% VS removal is achieved [72]. This indicates a daily electricity generation of 12.5–33.6 kWh from biogas, on the assumption that the generator efficiency is 35%–50%. In addition, a daily heat energy profit of 17.8–46.5 kWh from biogas can be estimated. One cubic meter of biogas obtained while co-digesting dairy manure and animal fat is equivalent to 20 MJ of heat energy [73]. When used as fuel for a co-generator, 1 m^3 of biogas can produce 1.7 kWh of electricity and 7.7 MJ of heat. In addition to co-digestion, Esposito *et al.* [74] suggested several pre-treatment techniques that can be applied to increase further the biogas production, such as mechanical comminution, solid–liquid separation, bacterial hydrolysis and alkaline addition at high temperature, ensilage, alkaline, ultrasonic and thermal pre-treatments.

7. Toxicity

Sodium toxicity is a common problem causing inhibition of anaerobic digestion. Digesters treating highly concentrated wastes, such as food and concentrated animal manure, are likely to suffer from partial or complete inhibition of methane-producing consortia, including methanogens [75,76]. Zhao *et al.* [10] confirmed that organofluorine compounds, such as 4-fluorophenol (p-FP), 4-fluorobenzoic acid (p-FB) and 4-fluoroaniline (p-FA), have a potential toxicity on methanogenesis and biodegradability. Procházka *et al.* [77] described that high ammonia nitrogen concentration (especially the unionized form) (4.0 g L^{-1}) would inhibit methane production, while low ammonia nitrogen concentration (0.5 g L^{-1}) could cause low methane yield, loss of biomass (as VSS) and loss of the acetoclastic methanogenic activity. Chen *et al.* [78] indicated that certain ions, such as Na^+, Ca^{2+} and Mg^{2+}, were found to be antagonistic to ammonia inhibition, a phenomenon in which the toxicity of one ion is decreased by the presence of other ion(s). At high concentrations, potassium, light metals ions (Na, K, Mg, Ca and Al) and other salts can also interrupt cell function [78].

Toxicity Control Strategies

Vyrides *et al.* [79] indicated that glycine-betaine (GB), an organic compound, can cause antagonism against sodium toxicity. However, using GB to decrease sodium toxicity in commercial scale anaerobic digesters would be too costly [75]. Suwannoppadol *et al.* [75] described that when grass clippings were added at the onset of anaerobic digestion of acetate containing a sodium concentration of 7.8 g $Na^+ L^{-1}$, a total methane production of about 8 L $CH_4 L^{-1}$ was obtained, whereas no methane was produced in the absence of grass leaves. Another way of tackling the sodium salts problem is by allowing the anaerobic sludge to acclimate to high sodium concentrations [80], but this technique requires time for the methanogens to adapt to the saline conditions, which in turn, results in a prolonged period before the anaerobic reactor can achieve its full-loading capacity. Zhao *et al.* [10] concluded that adsorption was the main removal mechanism for the three F-substituent aromatics, such as 4-fluorophenol (p-FP), 4-fluorobenzoic acid (p-FB) and 4-fluoroaniline (p-FA). To overcome ammonia toxicity, many strategies have been suggested: chemical precipitation, pH and temperature control [78]; use of carbon fiber textiles [81]; acclimation of methanogenic consortia to high ammonia levels [82]; and ammonia stripping and adjustment of the C:N ratio of feedstock [83]. Uludag-Demirer *et al.* [84] and Wang *et al.* [49] described the physical, chemical and biological methods, such as addition of ammonium-selective adsorbent, ammonium removal by forming struvite precipitation or a biological anoxic/oxic (A/O) process. Among these methods, ammonium removal by adding ammonium-selective adsorbents could be the most attractive and practical, because of its easier operation and economic impact, and the ammonium-saturated adsorbents can be further used as nitrogen fertilizer [49].

8. Modeling Advances

Various kinetic equations reported for anaerobic processes [3,85] generally relied on Monod's equation. Monod's equation is based on the growth rate and the substrate utilization rate during biodegradation. Wong *et al.* [86] determined the biological kinetic constants for a laboratory anaerobic bench scale reactors (ABSR) treating palm oil mill effluent. The investigation showed that the growth yield (Y_G), specific biomass decay (b), maximum specific biomass growth rate (μ_{max}), saturation constant (K_s) and critical retention time (Θ_c) were in the range of 0.990 g VSS g^{-1} $COD_{removed}$ d^{-1}, 0.024 d^{-1}, 0.524 d^{-1}, 203.433 g COD L^{-1} and 1.908 d, respectively.

Fuzzato *et al.* [26] used simplified first-order kinetics for modeling a mesophilic anaerobic sequencing batch biofilm reactor (ASBBR) treating lipid-rich wastewater. Nikolaeva *et al.* [40] also observed that a first-order kinetic model described the experimental results obtained for the upflow fixed bed digesters treating dairy manure well. In addition, they also concluded that the first-order model was adequate for assessing the effect of HRT on the removal efficiency and methane production. Ganesh *et al.* [3] used a modified Stover-Kincannon kinetic model to predict the performance of anaerobic fixed bed reactors treating winery wastewater. In a similar study, Rajagopal *et al.* [87] applied bio-kinetic models, such as the modified Stover-Kincannon and second-order kinetic models for the upflow anaerobic filters treating high strength fruit canning and cheese dairy wastewaters. Fia *et al.* [6] also used the modified Stover-Kincannon and second-order kinetic models for the experimental data obtained from the upflow anaerobic fixed bed reactors treating coffee bean processing wastewater. Abdurahman *et al.* [85] employed kinetic equations from Monod, Contois and Chen and Hashimoto to describe the kinetics of palm oil mill effluent (POME) treatment in a membrane anaerobic system (MAS) at organic loading rates ranging from 1 to 11 kg COD m^{-3} d^{-1}. Kaewsuk *et al.* [88] conducted a pilot scale experiment to investigate the performance of the membrane sequencing batch reactor (MSBR) treating dairy wastewater and checked the suitability of the kinetics for an engineering design. The kinetic coefficients K_s, k, k_d, Y and mm were found to be 174-mg COD L^{-1}, 7.42 d^{-1}, 0.1383 d^{-1}, 0.2281 d^{-1} and 1.69 d^{-1}, respectively.

Recently, there has been a move by the International Water Association's (IWA) Task Group for Mathematical Modeling of Anaerobic Digestion Processes to develop a common model called Anaerobic Digestion Model No. 1 (ADM1) that can be used by researchers and practitioners [89,90]. Lee *et al.* [91] examined the application of the ADM1 for mathematical modeling of anaerobic process using a lab-scale temperature-phased anaerobic digestion (TPAD) process. Sensitivity analysis showed that $k_{m.process}$ (maximum specific uptake rate) and $K_{S.process}$ (half saturation value) had high sensitivities to model components. They have concluded that simulation with estimated parameters showed good agreement with experimental results in the case of methane production, uptake of acetate, soluble (SCOD) and total chemical oxygen demand (TCOD). The structure and properties of a microbial community may be influenced by process operation and, in their turn, also determine the reactor functioning. In order to adequately describe these phenomena, Ramirez *et al.* [89] emphasized that mathematical models need to consider the underlying microbial diversity. In order to demonstrate this contribution, they have extended the ADM1 to describe microbial diversity between organisms of the same functional group. Boubaker and Ridha [92] used the ADM1 model to simulate the mesophilic anaerobic co-digestion of olive mill wastewater (OMW) with olive mill solid waste (OMSW). The results indicated that the ADM1 model could simulate with good accuracy: gas flows, methane and carbon-dioxide contents, pH and total volatile fatty acids (TVFA) concentrations of effluents for various feed concentrations digested at different HRTs and especially at HRTs of 36 and 24 days. Furthermore, effluent alkalinity and ammonium nitrogen were also successfully predicted.

9. Technology Assessments

Onsite industrial wastewater anaerobic treatment requires systems with a significant capital cost and incur increasing expenses for successful long-term operation, control and maintenance. Farhan [93] evaluated a high-rate digestion system for brewery wastewater technically and economically. The technical evaluating criteria consist of, among commonly used engineering design criteria, such as hydraulic and organic loading rates, wastewater characteristics and layout and space requirements, factors that reflect the dynamics of technology development. The abilities of the anaerobic high rate bioreactors to meet the regulatory requirements reliably with flexibility for future upgrading are some of the technical evaluating factors [93]. The energy savings and renewable energy credits are among the economical assessment criteria [93].

Gebrezgabher *et al.* [94] analyzed the economic performance of anaerobic digestion of a biogas plant using the net present value (NPV) and internal rate of return (IRR). They conclude that the uncertainty of the increasingly tightened regulations regarding the effluent of anaerobic treatment, the quality and value of the digestate and the high investment and operating costs limit the on-farm applications of anaerobic digestion of agro wastes.

10. Summary and Conclusions

A critical analysis of the literature reveals that there is a strong possibility and need to enhance the performance of high rate anaerobic biogas reactors. This technique has many advantages over other conventional methods. However, the challenges associated with the digester operation at lower HRT and higher OLR need to be addressed, which include biomass washout, clogging, short-circuiting, process inhibitions, poor final effluent and biogas quality. Different materials (polyethylene, polypropylene pall rings, polyurethane foam, carbon felt and waste tire rubber) had been tried as packing material in the anaerobic fixed bed reactors, depending upon their availability and other specifications, such as material properties, cost, *etc.* These packing materials would help to reduce hydraulic retention time, which in turn lessens the required volume of the reactor, and ultimately, the cost could be reduced. The practical aspects of using pure microbial film as magnifying microbial layers should be looked into. In the case of UASB reactors, the following important aspects should be looked into: (i) enhancing the start-up and granulation in UASB reactors; (ii) coupling with post-treatment unit to overcome the temperature constraint; and (iii) improving the removal efficiencies of the organic matter, nutrients and pathogens in the final effluent.

When talking about toxicity, there are many soluble organic and inorganic materials that can be either stimulatory or inhibitory. A good example of this is the effect of ammonia nitrogen on the anaerobic digestion process. When potential inhibitory materials are slowly increased within the environment, many biological organisms can rearrange their metabolic resources, thus overcoming the metabolic block produced by the normally inhibitory material. Under shock load conditions, sufficient time is not available for this rearrangement to take place. Finally, there is the possibility of antagonism and synergism effects of using different organic wastes as co-substrates. Antagonism is defined as a reduction of the toxic effect of one substance by the presence of another. Synergism is defined as an increase in the toxic effect of one substance by the presence of another. This is an important consideration when designing for potential cation toxicity. Additional research efforts are essential to get more insight about the stable performance of the digesters against various process inhibitors, such as ammonia, sodium, sulfur, *etc.* While lessening the economic losses, vigilant substrate management and early detection of inhibitions are critical.

References

1. Rajagopal, R. Treatment of Agro-Food Industrial Wastewaters Using UAF and Hybrid UASB-UAF Reactors. Ph.D. Thesis, Indian Institute of Technology Roorkee, Roorkee, India, 2008.

2. Badshah, M.; Parawira, W.; Mattiasson, B. Anaerobic treatment of methanol condensate from pulp mill compared with anaerobic treatment of methanol using mesophilic UASB reactors. *Bioresour. Technol.* **2012**, *125*, 318–327.

3. Ganesh, R.; Rajagopal, R.; Torrijos, M.; Thanikal, J.M.; Ramanujam, R. Anaerobic treatment of winery wastewater in fixed bed reactors. *Bioprocess Biosyst. Eng.* **2010**, *33*, 619–628.

4. Meksi, N.; Haddar, W.; Hammami, S.; Mhenni, M.F. Olive mill wastewater: A potential source of natural dyes for textile dyeing. *Ind. Crops Prod.* **2012**, *40*, 103–109.

5. MPOC (Malaysian Palm Oil Council). Available online: http://www.mpoc.org.my (accessed on 5 March 2013).

6. Fia, R.L.; Matos, A.T.; Borges, A.C.; Fia, R.; Cecon, P.R. Treatment of wastewater from coffee bean processing in anaerobic fixed bed reactors with different support materials: Performance and kinetic modeling. *J. Environ. Manag.* **2012**, *108*, 14–21.

7. Nieto, P.P.; Hidalgo, D.; Irusta, R.; Kraut, D. Biochemical methane potential (BMP) of agro-food wastes from the Cider Region (Spain). *Water Sci. Technol.* **2012**, *66*, 1842–1848.

8. Veronafiere (Ente Autonomo per le Fiere di Verona). Vinitaly analysis. Worldwide Wine: The Sector Scenario, Production, Consumption and Trade on a World Scale under the Magnifying Glass at Vinitaly. Available online: http://www.vinitaly.com/pdf/cartellaStampa/5gbCsVinitaly12_SituazioneMondoItalia_23marzo.pdf (accessed on 5 March 2013).

9. Un, U.T.; Altay, U.; Koparal, A.S.; Ogutveren, U.B. Complete treatment of olive mill wastewaters by electrooxidation. *Chem. Eng. J.* **2008**, *139*, 445–452.

10. Zhao, Z.-Q.; Xu, L.-L.; Li, W.-B.; Wang, M.-Z.; Shen, X.-L.; Mae, G.-S.; Shena, D.-S. Toxicity of three F-substituent aromatics in anaerobic systems. *J. Chem. Technol. Biotechnol.* **2012**, *87*, 1489–1496.

11. Wei, C.; Zhang, T.; Feng, C.; Wu, H.; Deng, Z.; Wu, C.; Lu, B. Treatment of food processing wastewater in a full-scale jet biogas internal loop anaerobic fluidized bed reactor. *Biodegradation* **2011**, *22*, 347–357.

12. Rupani, P.F.; Singh, R.P.; Ibrahim, M.H.; Esa, N. Review of current palm oil mill effluent (POME) treatment methods: Vermicomposting as a sustainable practice. *World Appl. Sci. J.* **2010**, *10*, 1190–1201.

13. Alkaya, E.; Demirer, G.N. Anaerobic acidification of sugar-beet processing wastes: Effect of operational parameters. *Biomass Bioenergy* **2011**, *35*, 32–39.

14. Gotmare, M.; Dhoble, R.M.; Pittule, A.P. Biomethanation of dairy waste water through UASB at mesophilic temperature range. *Int. J. Adv. Eng. Sci. Technol.* **2011**, *8*, 1–9.

15. Ersahin, M.E.; Ozgun, H.; Dereli, R.K.; Ozturk, I. Anaerobic treatment of industrial effluents: An overview of applications. In *Waste Water—Treatment and Reutilization*; Einschlag, F.S.G., Ed.; InTech: New York, NY, USA, 2011; Available online: http://www.intechopen.com/books/waste-water-treatment-and-reutilization/anaerobic-treatment-of-industrial-effluents-an-overview-of-applications (accessed on 5 March 2013).

16. Senturk, E.; Ince, M.; Engin, O.G. Treatment efficiency and VFA composition of a thermophilic anaerobic contact reactor treating food industry wastewater. *J. Hazard. Mater.* **2010**, *176*, 843–848.

17. Ersahin, M.E.; Dereli, R.K.; Ozgun, H.; Donmez, B.G.; Koyuncu, I.; Altinbas, M.; Ozturk, I. Source based characterization and pollution profile of a baker's yeast industry. *Clean-Soil Air Water* **2011**, *39*, 543–548.

18. Passeggi, M.; Lopez, I.; Borzacconi, L. Integrated anaerobic treatment of dairy industrial wastewater and sludge. *Water Sci. Technol.* **2009**, *59*, 501–506.

19. Gonçalves, M.R.; Costa, J.C.; Marques, I.P.; Alves, M.M. Strategies for lipids and phenolics degradation in the anaerobic treatment of olive mill wastewater. *Water Res.* **2012**, *46*, 1684–1692.

20. Sun, L.; Wan, S.; Yu, Z.; Wang, Y.; Wang, S. Anaerobic biological treatment of high strength cassava starch wastewater in a new type up-flow multistage anaerobic reactor. *Bioresour. Technol.* **2012**, *104*, 280–288.

21. Chong, S.; Sen, T.K.; Kayaalp, A.; Ang, H.M. The performance enhancements of upflow anaerobic sludge blanket (UASB) reactors for domestic sludge treatment—A State-of-the-art review. *Water Res.* **2012**, *46*, 3434–3470.

22. Rajagopal, R.; Ganesh, R.; Escudie, R.; Mehrotra, I.; Kumar, P.; Thanikal, J.V.; Torrijos, M. High rate anaerobic filters with floating supports for the treatment of effluents from small-scale agro-food industries. *Desalin. Water Treat.* **2009**, *4*, 183–190.

23. Esparza, S.M.; Solís, M.C.; Herná, J.J. Anaerobic treatment of a medium strength industrial wastewater at low-temperature and short hydraulic retention time: A pilot-scale experience. *Water Sci. Technol.* **2011**, *64*, 1629–1635.

24. Shastry, S.; Nandy, T.; Wate, S.R.; Kaul, S.N. Hydrogenated vegetable oil industry wastewater treatment using UASB reactor system with recourse to energy recovery. *Water Air Soil Pollut.* **2010**, *208*, 323–333.

25. Won, S.G.; Lau, A.K. Effects of key operational parameters on biohydrogen production via anaerobic fermentation in a sequencing batch reactor. *Bioresour. Technol.* **2011**, *102*, 6876–6883.

26. Fuzzato, M.C.; Tallarico Adorno, M.A.; de Pinho, S.C.; Ribeiro, R.; Tommaso, G. Simplified mathematical model for an anaerobic sequencing batch biofilm reactor treating lipid-rich wastewater subject to rising organic loading rates. *Environ. Eng. Sci.* **2009**, *26*, 1197–1206.

27. Shanmugam, A.S.; Akunna, J.C. Comparing the performance of UASB and GRABBR treating low strength wastewaters. *Water Sci. Technol.* **2008**, *58*, 225–232.

28. Bialek, K.; Kumar, A.; Mahony, T.; Lens, P.N.L.; Flaherty, V.O. Microbial community structure and dynamics in anaerobic fluidized-bed and granular sludge-bed reactors: Influence of operational temperature and reactor configuration. *Microb. Biotechnol.* **2012**, *5*, 738–775.

29. Rajagopal, R.; Mehrotra, I.; Kumar, P.; Torrijos, M. Evaluation of a hybrid upflow anaerobic sludge-filter bed reactor: Effect of the proportion of packing medium on performance. *Water Sci. Technol.* **2010**, *61*, 1441–1450.

30. Gnansounou, E. Production and use of lignocellulosic bioethanol in Europe: Current situation and perspectives. *Bioresour. Technol.* **2010**, *101*, 4842–4850.

31. Nkemka, V.N.; Murto, M. Biogas production from wheat straw in batch and UASB reactors: The roles of pretreatment and seaweed hydrolysate as a co-substrate. *Bioresour. Technol.* **2013**, *128*, 164–172.

32. Alvira, P.; Tomás-Pejó, E.; Ballesteros, M.; Negro, M.J. Pretreatment technologies for an efficient bioethanol production process based on enzymatic hydrolysis: A review. *Bioresour. Technol.* **2010**, *101*, 4851–4861.

33. Bosco, F.; Chiampo, F. Production of polyhydroxyalcanoates (PHAs) using milk whey and dairy wastewater activated sludge Production of bioplastics using dairy residues. *J. Biosci. Bioeng.* **2010**, *109*, 418–421.

34. Garcia, H.; Rico, C.; Garcia, P.A.; Rico, J.L. Flocculants effect in biomass retention in a UASB reactor treating dairy manure. *Bioresour. Technol.* **2008**, *99*, 6028–6036.

35. Yu, Y.-C.; Gao, D.-W.; Tao, Y. Anammox start-up in sequencing batch biofilm reactors using different inoculating sludge. *Appl. Microbiol. Biotechnol.* **2012**, doi:10.1007/s00253-012-4427-z.

36. Zhang, T.; Yan, Q.M.; Ye, L. Autotrophic biological nitrogen removal from saline wastewater under low DO. *J. Chem. Technol. Biot.* **2010**, *85*, 1340–1345.

37. Hendrickx, T.L.G.; Wang, Y.; Kampman, C.; Zeeman, G.; Temmink, H.; Buisman, C.J.N. Autotrophic nitrogen removal from low strength waste water at low temperature. *Water Res.* **2012**, *46*, 2187–2193.

38. Dosta, J.; Fernandez, I.; Vazquez-Padin, J.R.; Mosquera-Corral, A.; Campos, J.L.; Mata-Alvarez, J.; Mendez, R. Short- and long-term effects of temperature on the anammox process. *J. Hazard. Mater.* **2008**, *154*, 688–693.

39. Lim, S.J.; Fox, P. A kinetic analysis and experimental validation of an integrated system of anaerobic filter and biological aerated filter. *Bioresour. Technol.* **2011**, *102*, 10371–10376.

40. Nikolaeva, S.; Sanchez, E.; Borja, R.; Raposo, F.; Colmenarejo, M.F.; Montalvo, S.; Jiménez-Rodríguez, A.M. Kinetics of anaerobic degradation of screened dairy manure by upflow fixed bed digesters: Effect of natural zeolite addition. *J. Environ. Sci. Health Part A Toxic/Hazard. Substan. Environ. Eng.* **2009**, *44*, 146–154.

41. Koupaie, E.H.; Moghaddam, M.R.A.; Hashemi, S.H. Evaluation of integrated anaerobic/aerobic fixed-bed sequencing batch biofilm reactor for decolorization and biodegradation of azo dye Acid Red 18: Comparison of using two types of packing media. *Bioresour. Technol.* **2013**, *127*, 415–421.

42. Gómez-De Jesús, A.; Romano-Baez, F.J.; Leyva-Amezcua, L.; Juárez-Ramírez, C.; Ruiz-Ordaz, N.; Galíndez-Mayer, J. Biodegradation of 2,4,6-richlorophenol in a packed-bed biofilm reactor equipped with an internal net draft tube riser for aeration and liquid circulation. *J. Hazard. Mater.* **2009**, *161*, 1140–1149.

43. González, A.J.; Gallego, A.; Gemini, V.L.; Papalia, M.; Radice, M.; Gutkind, G.; Planes, E.; Korol, S.E. Degradation and detoxification of the herbicide 2,4-dichlorophenoxyacetic acid (2,4-D) by an indigenous Delftia sp. strain in batch and continuous systems. *Int. Biodeter. Biodegr.* **2012**, *66*, 8–13.

44. Satyawali, Y.; Pant, D.; Singh, A.; Srivastava, R.K. Treatment of rayon grade pulp drain effluent by upflow anaerobic fixed packed bed reactor (UAFPBR). *J. Environ. Biol.* **2009**, *30*, 667–672.

45. Ahn, J.-H. Nitrogen requirement for the mesophilic and thermophilic upflow anaerobic filters of a simulated paper mill wastewater. *Korean J. Chem. Eng.* **2008**, *25*, 1022–1025.

46. Deshannavar, U.B.; Basavaraj, R.K.; Naik, N.M. High rate digestion of dairy industry effluent by upflow anaerobic fixed-bed reactor. *J. Chem. Pharma. Res.* **2012**, *4*, 2895–2899.

47. Gao, F.; Zhang, H.; Yang, F.; Qiang, H.; Zhang, G. The contrast study of anammox-denitrifying system in two non-woven fixed-bed bioreactors (NFBR) treating different low C/N ratio sewage. *Bioresour. Technol.* **2012**, *114*, 54–61.

48. Ji, G.; Wu, Y.; Wang, C. Analysis of microbial characterization in an upflow anaerobic sludge bed/biological aerated filter system for treating microcrystalline cellulose wastewater. *Bioresour. Technol.* **2012**, *120*, 60–69.

49. Wang, Q.; Yang, Y.; Li, D.; Feng, C.; Zhang, Z. Treatment of ammonium-rich swine waste in modified porphyritic andesite fixed-bed anaerobic bioreactor. *Bioresour. Technol.* **2012**, *111*, 70–75.

50. Bajaj, M.; Gallert, C.; Winter, J. Biodegradation of high phenol containing synthetic wastewater by an aerobic fixed bed reactor. *Bioresour. Technol.* **2008**, *99*, 8376–8381.

51. Farhadian, M.; Duchez, D.; Vachelard, C.D.; Larroche, C. Monoaromatics removal from polluted water through bioreactors—A review. *Water Res.* **2008**, *42*, 1325–1341.

52. Mahmoud, N. High strength sewage treatment in a UASB reactor and an integrated UASB-digester system. *Bioresour. Technol.* **2008**, *99*, 7531–7538.

53. Elangovan, C.; Sekar, A.S.S. Application of Upflow anaerobic sludge blanket (UASB) reactor process for the treatment of dairy wastewater—A review. *Nat. Environ. Pollut. Technol.* **2012**, *11*, 409–414.

54. Lew, B.; Lustig, I.; Beliavski, M.; Tarre, S.; Green, M. An integrated UASB-sludge digester system for raw domestic wastewater treatment in temperate climates. *Bioresour. Technol.* **2011**, *102*, 4921–4924.

55. Li, J.; Hu, B.; Zheng, P.; Qaisar, M.; Mei, L. Filamentous granular sludge bulking in a laboratory scale UASB reactor. *Bioresour. Technol.* **2008**, *99*, 3431–3438.

56. Huang, J.P.; Liu, L.; Shao, Y.M.; Song, H.J.; Wu, L.C.; Xiao, L. Study on cultivation and morphology of granular sludge in improved methanogenic UASB. *Appl. Mechan. Mater.* **2012**, *209–211*, 1152–1157.

57. Wongnoi, R.; Songkasiri, W.; Phalakornkule, C. Influence of a three-phase separator configuration on the performance of an upflow anaerobic sludge bed reactor treating wastewater from a fruit-canning factory. *Water Environ. Res.* **2007**, *79*, 199–207.

58. Diez, V.; Ramos, C.; Cabezas, J.L. Treating wastewater with high oil and grease content using an anaerobic membrane bioreactor (AnMBR). Filtration and cleaning assays. *Water Sci. Technol.* **2012**, *65*, 1847–1853.

59. Kim, S.-H.; Shin, H.-S. Enhanced lipid degradation in an upflow anaerobic sludge blanket reactor by integration with an acidogenic reactor. *Water Environ. Res.* **2010**, *82*, 267–272.

60. Erdirencelebi, D. Treatment of high-fat-containing dairy wastewater in a sequential UASBR system: Influence of recycle. *J. Chem. Technol. Biotechnol.* **2011**, *86*, 525–533.

61. Najafpour, G.D.; Komeili, M.; Tajallipour, M.; Asadi, M. Bioconversion of cheese whey to methane in an upflow anaerobic packed bed bioreactor. *Chem. Biochem. Eng. Q.* **2010**, *24*, 111–117.

62. Yasar, A.; Tabinda, A.B. Anaerobic treatment of industrial wastewater by UASB reactor integrated with chemical oxidation processes: An overview. *Pol. J. Environ. Stud.* **2010**, *19*, 1051–1061.

63. Chen, X.-G.; Zheng, P.; Cai, J.; Qaisar, M. Bed expansion behavior and sensitivity analysis for super-high-rate anaerobic bioreactor. *J. Zhejiang Univ. Sci. B.* 2010, *11*, 79–86.

64. Chen, X.G.; Zheng, P.; Qaisar, M.; Tang, C.J. Dynamic behavior and concentration distribution of granular sludge in a super-high-rate spiral anaerobic bioreactor. *Bioresour. Technol.* **2012**, *111*, 134–140.

65. Yang, J.; Vedantam, S.; Spanjers, H.; Nopens, I.; van Lier, J.B. Analysis of mass transfer characteristics in a tubular membrane using CFD modeling. *Water Res.* **2012**, *46*, 4705–4712.

66. Feng, Y.; Lu, B.; Jiang, Y.; Chen, Y.; Shen, S. Anaerobic degradation of purified terephthalic acid wastewater using a novel, rapid mass-transfer circulating fluidized bed. *Water Sci. Technol.* **2012**, *65*, 1988–1993.

67. Wagner, R.C.; Regan, J.M.; Oh, S.E.; Zuo, Y.; Logan, B.E. Hydrogen and methane production from swine wastewater using microbial electrolysis cells. *Water Res.* **2009**, *43*, 1480–1488.

68. Searmsirimongkol, P.; Rangsunvigit, P.; Leethochawalit, M.; Chavadej, S. Hydrogen production from alcohol distillery wastewater containing high potassium and sulfate using an anaerobic sequencing batch reactor. *Int. J. Hydrog. Energ.* **2011**, *36*, 12810–12821.

69. Rajagopal, R.; Rousseau, P.; Bernet, N.; Béline, F. Combined anaerobic and activated sludge anoxic/oxic for piggery wastewater treatment. *Bioresour. Technol.* **2011**, *102*, 2185–2192.

70. Labatut, R.A.; Angenent, L.T.; Scott, N.R. Biochemical methane potential and biodegradability of complex organic substrates. *Bioresour. Technol.* **2011**, *102*, 2255–2264.

71. Ogejo, J.A.; Li, L. Enhancing biomethane production from flush dairy manure with turkey processing wastewater. *Appl. Energ.* **2010**, *87*, 3171–3177.

72. Rajagopal, R.; Lim, J.W.; Mao, Y.; Chen, C.L.; Wang, J.Y. Anaerobic co-digestion of source segregated brown water (feces-without-urine) and food waste: For Singapore context. *Sci. Total Environ.* **2013**, *443*, 877–886.

73. Alberta Agriculture and Rural Development. *Economic Feasibility of Anaerobic Digesters*; Department of Agriculture and Rural Development: Edmonton, AB, Canada, 2008; Available online: http://www1.agric.gov.ab.ca/$department/deptdocs.nsf/all/agdex12280 (accessed on 5 March 2013).

74. Esposito, G.; Frunzo, L.; Giordano, A.; Liotta, F.; Panico, A.; Pirozzi, F. Anaerobic co-digestion of organic wastes. *Rev. Environ. Sci. Biotechnol.* **2012**, *11*, 325–341.

75. Suwannoppadol, S.; Ho, G.; Cord-Ruwisch, R. Overcoming sodium toxicity by utilizing grass leaves as co-substrate during the start-up of batch thermophilic anaerobic digestion. *Bioresour. Technol.* **2012**, *125*, 188–192.

76. Hierholtzer, A.; Akunna, J.C. Modelling sodium inhibition on the anaerobic digestion process. *Water Sci. Technol.* **2012**, *66*, 1565–1573.

77. Procházka, J.; Dolejš, P.; Máca, J.; Dohányos, M. Stability and inhibition of anaerobic processes caused by insufficiency or excess of ammonia nitrogen. *Appl. Microbiol. Biotechnol.* **2012**, *93*, 439–447.

78. Chen, Y.; Cheng, J.J.; Creamer, K.S. Inhibition of anaerobic digestion process: A review. *Bioresour. Technol.* **2008**, *99*, 4044–4064.

79. Vyrides, I.; Santos, H.; Mingote, A.; Ray, M.J.; Stuckey, D.C. Are compatible solutes compatible with biological treatment of saline wastewater? Batch and continuous studies using submerged anaerobic membrane bioreactors (SAMBRs). *Environ. Sci. Technol.* **2010**, *44*, 7437–7442.

80. Vyrides, I.; Stuckey, D.C. Adaptation of anaerobic biomass to saline conditions: Role of compatible solutes and extracellular polysaccharides. *Enzym. Microb. Technol.* **2009**, *44*, 46–51.

81. Sasaki, K.; Morita, M.; Hirano, S.; Ohmura, N.; Igarashi, Y. Decreasing ammonia inhibition in thermophilic methanogenic bioreactors using carbon fiber textiles. *Appl. Microbiol. Biotechnol.* **2011**, *90*, 1555–1561.

82. Abouelenien, F.; Nakashimada, Y.; Nishio, N. Dry mesophilic fermentation of chicken manure for production of methane by repeated batch culture. *J. Biosci. Bioeng.* **2009**, 107, 293–295.

83. Resch, C.; Wörl, A.; Waltenberger, R.; Braun, R.; Kirchmayr, R. Enhancement options for the utilisation of nitrogen rich animal by-products in anaerobic digestion. *Bioresour. Technol.* **2011**, *102*, 2503–2510.

84. Uludag-Demirer, S.; Demirer, G.N.; Frear, C.; Chen, S., Anaerobic digestion of dairy manure with enhanced ammonia removal. *J. Environ. Manag.* **2008**, *86*, 193–200.

85. Abdurahman, N.H.; Rosli, Y.M.; Azhari, N.H. Development of a membrane anaerobic system (MAS) for palm oil mill effluent (POME) treatment. *Desalination* **2011**, *266*, 208–212.

86. Wong, Y.S.; Kadir, M.O.A.B.; Teng, T.T. Biological kinetics evaluation of anaerobic stabilization pond treatment of palm oil mill effluent. *Bioresour. Technol.* **2009**, *100*, 4969–4975.

87. Rajagopal, R.; Torrijos, M.; Kumar, P.; Mehrotra, I. Substrate removal kinetics in high-rate upflow anaerobic filters packed with low-density polyethylene media treating high-strength agro-food wastewaters. *J. Environ. Manag.* **2013**, *116*, 101–106.

88. Kaewsuk, J.; Thorasampan, W.; Thanuttamavong, M.; Seo, J.T. Kinetic development and evaluation of membrane sequencing batch reactor (MSBR) with mixed cultures photosynthetic bacteria for dairy wastewater treatment. *J. Environ. Manag.* **2010**, *91*, 1161–1168.

89. Ramirez, I.; Volcke, E.I.P.; Rajagopal, R.; Steyer, J.P. Application of ADM1 towards modelling biodiversity in anaerobic digestion. *Water Res.* **2009**, *43*, 2787–2800.

90. Girault, R.; Bridoux, G.; Nauleau, F.; Poullain, C.; Buffet, J.; Steyer, J.-P.; Sadowski, A.G.; Béline, F. A waste characterisation procedure for ADM1 implementation based on degradation kinetics. *Water Res.* **2012**, *46*, 4099–4110.

91. Lee, M.-Y.; Suh, C.-W.; Ahn, Y.-T.; Shin, H.-S. Variation of ADM1 by using temperature-phased anaerobic digestion (TPAD) operation. *Bioresour. Technol.* **2009**, *100*, 2816–2822.

92. Boubaker, F.; Ridha, B.C. Modelling of the mesophilic anaerobic co-digestion of olive mill wastewater with olive mill solid waste using anaerobic digestion model No. 1 (ADM1). *Bioresour. Technol.* **2008**, *99*, 6565–6577.

93. Farhan, M.H. High Rate Anaerobic Digester Systems for Brewery Wastewater Treatment and Electricity Generation: Engineering Design Factors and Cost Benefit Analysis. In Proceedings of The World Brewing Congress, Oregon Convention Centre, Portland, OR, USA, 28 July–1 August 2012; Available online: http://www.worldbrewingcongress.org/2012/Abstracts/ AbstractsDetail.cfm?AbstractID=318 (accessed on 5 March 2013).

94. Gebrezgabher, S.A.; Meuwissen, M.P.M.; Prins, B.A.M.; Lansink, A.G.J.M.O. Economic analysis of anaerobic digestion—A case of green power biogas plant in The Netherlands. *NJAS Wagening. J. Life Sci.* **2010**, *57*, 109–115.

MDPI AG
Grosspeteranlage 5
4052 Basel
Switzerland
Tel.: +41 61 683 77 34

Water Editorial Office
E-mail: water@mdpi.com
www.mdpi.com/journal/water